ELEMENTS OF REFRIGERATION
AND
AIR CONDITIONING

ELEMENTS OF REFRIGERATION
AND
AIR CONDITIONING

Prasanta Kumar Satapathy

(MTech and PhD, IIT Kharagpur)

Professor, Mechanical Engineering

Odisha University of Technology and Research

(Formerly College of Engineering and Technology, Bhubaneswar)

I.K. International Pvt. Ltd.

NEW DELHI

Published by
I.K. International Pvt. Ltd.
4435-36/7, Ansari Road, Daryaganj
New Delhi-110 002 (India)
E-mail: info@ikinternational.com
Website: www.ikbooks.com

ISBN: 978-93-90620-98-2

Published by Krishan Makhijani for I.K. International Pvt. Ltd., 4435-36/7, Ansari Road, Daryaganj, New Delhi-110 002 and Printed by Rekha Printers Pvt. Ltd., Okhla Industrial Area, Phase II, New Delhi-110 020.

PREFACE

Elements of Refrigeration and Air Conditioning is specifically intended to provide the fundamentals of refrigeration and air conditioning derived from the first principle of thermodynamics, heat and mass transfer and fluid mechanics. In other words, this subject is an application part of the above principles. Keeping in view to its wide industrial and domestic applications, it has been emphasized on physical understanding of the fundamental concept of conventional and non-conventional refrigeration processes in a simple, yet concise manner. A genuine attempt has been made to meet the need of the students, practising engineers as well.

This book contains thirteen chapters. Chapter 1 recapitulates the basic principles and laws of thermodynamics, units, different measuring devices and early methods of refrigeration in a simplified manner. Chapter 2 presents different refrigerants and refrigerant mixtures, their properties and their impact on environment. Aircraft refrigeration system is explained in Chapter 3. Chapters 4 and 5 give a detailed picture of vapour compression refrigeration system and design of its components, respectively. Starting from development of vapour compression cycle to production of dry ice is explained with simplified line diagrams in a precise manner so that students can have an easy grasp over these important topics. How to read the refrigerant tables is also included because this is a great draw back with the students. Vapour absorption refrigeration system is discussed in chapter 6. Use of refrigerant mixture in this system raises many doubts in students' mind in understanding the system performance. This has been taken care of with illustrative examples. Non-conventional refrigeration systems like thermoelectric, ejector, magnetic, etc. have been elaborated in chapter 7. Chapter 8 to 12 air conditioning, starting from psychrometry to conditioned air distribution. They include properties of moist air and generation of psychrometric chart, psychrometric calculations, cooling load calculation, Human comfort and other industrial applications, air distribution, components of an air conditioning system. Each explanation is followed by line diagrams which give an easy access to the topics. Chapter 13 discusses different aspects of power saving in refrigeration and air conditioning by adopting advanced techniques and new refrigerants. This leads to the sustainability of refrigeration sectors.

This text is based upon the lecture notes prepared for UG and PG students since last 30 years. Its continuous improvement is done from students' feed back each year.

This book is specifically meant for diploma and undergraduate students. This book will be of great value as a handbook for GATE appearing and postgraduate students. Chapters have been designed with a strict adherence to the syllabi of mechanical, agricultural sciences in India and abroad. More than 100 practical problems and a number of review and multiple-choice questions in SI unit at the end of each chapter, also 100 MCQs at the end of the book have been provided for better understanding of the concepts.

I am very much obliged to my beloved teacher Late Prof. Ramesh Chandra Arora of IIT Kharagpur whose invaluable teaching is the foundation of this text. I am indebted to Prof. M. Ramgopal of IIT Kharagpur for his constant guidance in developing the concept on the subject during my PhD. I express my heartfelt thanks to Prof. Manoj Ranjan Nayak, President, SoA University for his inspiration to get me to the track of writing the book. I am also grateful to Vice Chancellor, Prof. B.B. Biswal and all my colleagues of OUTR and professors of other institutes for their encouragement and support in preparing this text.

I express my deep gratitude to my parents for their blessings and for the support extended by my wife Mrs Gita Satapathy and son Sri Prem Prasad Satapathy.

Prasanta Kumar Satapathy

CONTENTS

1

Introduction

1.1 WHAT IS REFRIGERATION

A major application area of principles of thermodynamics and heat transfer is 'refrigeration'. Creation of lower temperature in a system than that of the surrounding and to maintain that low temperature for a desirable time period is called refrigeration. It requires the heat to be extracted from a low temperature zone (source/system) and reject it to a high temperature zone (sink/surrounding).

A thermodynamic system performs on the principles of laws of thermodynamics and laws of heat transfer. Most of its performances follow the very fundamental law of nature, i.e., the principle of conservation of energy. Refrigeration is accomplished through a thermodynamic system, which performs on the basis of laws of thermodynamics and heat transfer. Hence, it is very important to recapitulate the fundamentals of these laws before discussing the refrigeration process.

1.2 TERMS FREQUENTLY USED IN THERMODYNAMICS (Refer Fig. 1.1)

System: It is defined as a quantity of matter or a region in space chosen for study.

Surrounding: The mass or the region outside the system is the surroundings.

Boundary: The real or imaginary surface that separates the system and the surroundings is called boundary. This may be fixed or moving. Truly speaking, boundary has zero thickness. It contains no mass or energy.

Universe: System, surroundings and boundary, together are called universe.

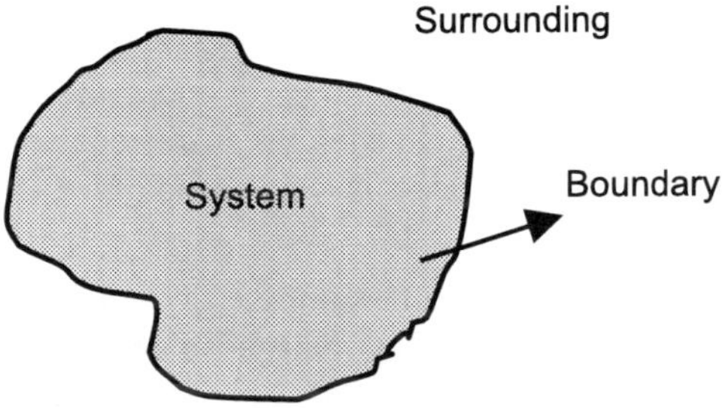

Figure 1.1: Concept of system

1.2.1 Thermodynamic Systems

These are of three types: (1) closed system, (2) open system, and (3) isolated system.

Closed system (also called control mass): In this system no mass can cross the boundary, i.e., no mass can enter or leave the boundary of a closed system. Only energy (in any form, i.e., heat, work, etc.) can cross the boundary in the form of heat or work, e.g., a pressure cooker (before gas escapes through the valve). These systems may have fixed boundary or moving boundary as shown in Fig. 1.2 (a)The boundary consists of a cylinder wall on three sides and piston on one side. Piston is moving sidewise. So the boundary is moving boundary. Figure 1.2 (b) is a closed container. So it is fixed boundary system.

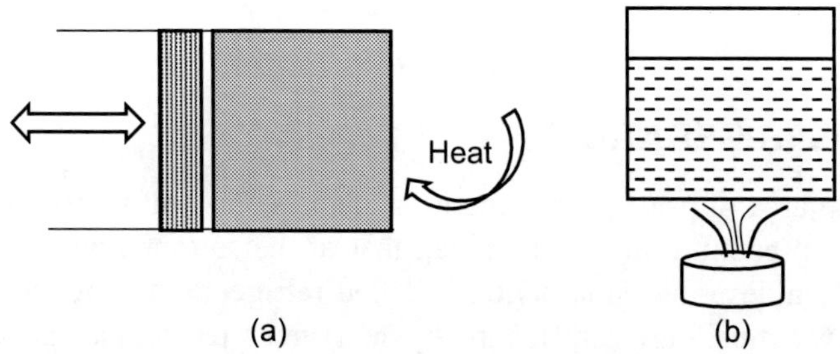

Figure 1.2: Closed system (a) moving, (b) fixed boundary

Open system (also called flow system): (Refer Fig. 1.3). In this system, both mass and energy cross the boundary. Hence, it becomes difficult to calculate the thermodynamic state of the fluid as the process occurs. To ease the calculation, we fix our attention in a particular portion of the whole open system enclosed by an imaginary line and calculate the mass and energy those cross that imaginary line. The portion, selected for the mass and energy transfer calculation, is called the control volume, e.g., flow through an air compressor, a turbine, a nozzle, etc.

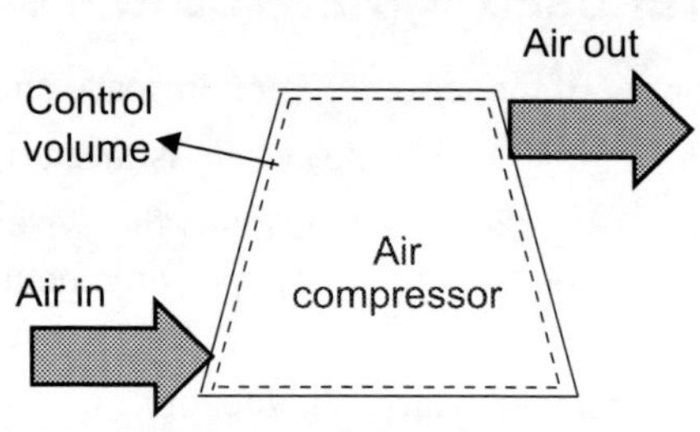

Figure 1.3: Open system with control volume

In the above figure, the mass of air enters the compressor and comes out of it being compressed (mass of air transferred from inlet to outlet and energy transferred to the mass of air to compress it) to higher pressure.This compression process is analysed taking the control volume—marked by the dotted line.

Isolated system: In certain systems neither mass nor energy is allowed to cross the boundary. Such system is called an isolated system. A thermally insulated sealed chamber is an isolated system.

Thermodynamic equilibrium: A system is said to be in thermodynamic equilibrium when it is in: (i) mechanical equilibrium (the resultant force and moment on the system is zero),

(ii) chemical equilibrium (no net chemical reaction inside the system or across the boundary) and, (iii) thermal equilibrium (temperature is equal to that of the surroundings so that there is no net heat transfer between the system and the surroundings).

1.3 USEFUL THERMODYNAMIC PROPERTIES OF A FLUID

A fluid, specifically gas and vapour, has many physical, chemical and thermodynamic properties. Out of these six properties have practical importance for the study of refrigeration process. Those are pressure, temperature, specific volume, internal energy, specific enthalpy, and specific entropy (specific term is used when the property is calculated per unit mass). The first three properties are basic properties and are directly measurable. Other three properties are calculated and termed derived properties.

Internal energy (u) (kJ/kg) of a fluid is the energy stored in the fluid by virtue of its temperature and is expressed as $C_V T$; where C_V is the specific heat at constant volume and T is its temperature.

Specific enthalpy (h) (kJ/kg): Specific enthalpy is the total energy contained in a fluid per unit mass. This is the summation of internal energy and the flow energy. Flow energy means it is the energy stored in a fluid by virtue of which it has a tendency to flow.

This is expressed as

$$h = u + pv \quad [1.1]$$

where u is the internal energy and pv is the flow work stored in the fluid or the ability of the fluid to do the work. For example, when the water at 100°C is heated it becomes vapour at same temperature and pressure. The total heat supplied to the water is stored as internal energy and the work done on the water to expand it from liquid specific volume (v_f) to vapour–specific volume, v_g.

$\Delta u = 2087.6$ kJ/kg and $p\Delta v = 101.325 \times (1.6729 - 0.001044) = 169.401$ kJ/kg

So $\Delta u + p\Delta v = 2257.001$ which is the enthalpy of vaporisation of water at 100°C.

Specific entropy (s) (kJ/kg-K): Like specific enthalpy, specific entropy is also a derived property. This is expressed as total energy supplied to the fluid per degree of absolute temperature to bring it from a reference state to the present state. $\Delta s = \Delta Q / T$; where ΔQ is the total heat supplied to the fluid at an average temperature of T in absolute scale (K). Entropy is an extensive property of the system. In differential form it can be expressed as $ds = dQ_{\text{Rev}}/T$; i.e., this equality is valid only for reversible process (even for internal reversible process). If it is extended for cyclic process, the following inferences are drawn.

$$\text{Cyclic}\oint dQ/T = 0 \text{ for reversible cycle} \quad [1.2a]$$

$$\text{Cyclic}\oint dQ/T < 0 \text{ for irreversible cycle} \quad [1.2b]$$

$$\text{Cyclic}\oint dQ/T > 0 \text{ is not possible} \quad [1.2c]$$

From entropy principle it can be stated that entropy of an isolated system never decreases.

When a system undergoes any process, there is energy/heat interaction within the system (friction or chemical reaction) or with the external (like heat or work transfer). This causes a change in entropy (ds) of the system and is expressed as $ds \geq dQ/T$; i.e., equal for reversible process and greater for irreversible process.

$$(s_{\text{final}} - s_{\text{initial}}) - dQ/T = s_{\text{gen}} \qquad [1.3]$$

s_{gen} is the entropy generation which is equal to zero for reversible process and >0 for irreversible process.

For different property calculations, Tds relationship is used.

$$dQ_{\text{Rev}} = Tds \qquad [1.4]$$

$$dQ = dU + pdV$$

This gives $$Tds = dU + pdV \qquad [1.5]$$

Again $$h = U + pV; \text{ So } dh = dU + pdV + Vdp$$

$$dh = Tds + Vdp \qquad [1.6]$$

So for an isentropic flow process, where $ds = 0$, $dh = Vdp$

1.4 LAWS THAT GOVERN THE REFRIGERATION PROCESSES

As it is discussed, refrigeration process is based upon the principles of thermodynamics and heat transfer, those laws are to be recapitulated.

1.4.1 Laws of Thermodynamics

Zeroth law: When body A is in thermal equilibrium with body B and body B is in thermal equilibrium with body C then body C is also in thermal equilibrium with body A. Measurement of temperature of an object by thermometer is a good example of the law. Let us assume, object is body A; bulb of the thermometer is body B; and the thermometric fluid (mercury or alcohol etc.) is body C. Now the bulb is in contact with the object. So the bulb and the object attain the thermal equilibrium, i.e., $t_A = t_B$. The bulb is in contact with thermometric fluid and both of them attain the thermal equilibrium, i.e., $t_B = t_C$. Hence, $t_A = t_C$, i.e., body A is in thermal equilibrium with body C. This ensures that the response of the thermometric fluid is in accordance with the temperature of A and that is the true temperature of the object though it is not in direct contact with the thermometric fluid.

First law: It is based upon the conservation of energy principle. It states energy can neither be created nor be destroyed. It can only be converted from one form to another. We can consider an example of an I.C. engine. Fuel has certain amount of chemical energy in it. When it burns it releases this chemical (internal) energy into heat energy. In the engine heat energy is converted to mechanical work by pushing out the piston (i.e., from *tdc* to *bdc*) and ultimately by moving the vehicle. Thus, the first law says that total energy generated due to the combustion of fuel (E) = Work output at the wheel of the vehicle + Energy lost due to friction in the engine components + Energy lost in the form of heat to coolant and radiation loss to environment, wheel friction, etc. Similarly, for a refrigeration system, energy goes into the system through compressor and evaporator. Energy goes out of the system through condenser. There is no loss or gain of energy during throttling process as this is an isenthalpic process. However, for a closed system (Fig. 1.4) undergoing a cyclic process (A-D-B-C-A) first law is expressed as: $\oint dW = J\oint dQ$; where J is a constant and known as the mechanical equivalent of heat. Its value in SI unit is

1.0. But for a change of state (Path *A* to *B* through *D* or Path *A* to *B* through *C*), first law is expressed as

$$dQ = dU + dW \qquad [1.7]$$

where dQ is the change in energy across the boundary, dU is the change in internal energy, and dW is the work done by/on the system.

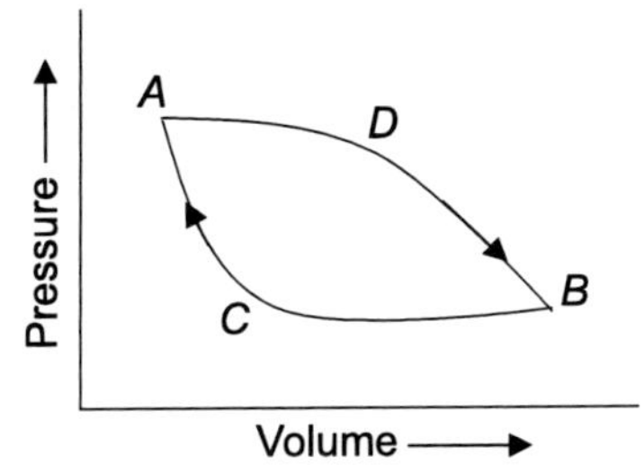

Figure 1.4: Thermodynamic cycle consists of two paths

Second Law: This is directional in nature and states the quality of energy. It states that heat can flow only from high temperature to low temperature spontaneously, but it flows from low temperature to high temperature with the help of external energy (Clausius statement). Here comes the concept of high grade energy and low grade energy. All of the high grade energy (work) can be converted to low grade energy (heat). But all the low grade energy (heat) cannot be converted to high grade energy (work) because during conversion some amount of energy (heat) is rejected to environment/sink (Kelvin-Planck statement) (Discussed in detail in Section 1.7.)

Third Law: It is impossible to attain absolute zero of temperature by following any finite number of idealized procedures. Many authors call it the derivation of 2nd law.

Let us consider a household refrigerator. If whole of the refrigerator is taken for analysing the energy consumption, cooling effect, heat rejection, etc., this can be considered a closed system and energy interactions occur on the basis of the first law of thermodynamics. If individual component is taken, then it is considered an open system and both first and second laws of thermodynamics are applied to get the quantitative as well as qualitative state of that component. By second law analysis the losses at different components can be estimated separately so that improvisation of the system can be made by reducing the losses at that component. This is otherwise called exergy analysis of the system.

1.4.2 Laws of Heat Transfer

Heat conduction and Fourier's Law: Heat is conducted through solids or fluids at rest obeying the Fourier's law which is expressed as the amount of heat conducted is directly proportional to temperature gradient ($\Delta T/\Delta x$) and the area (A) perpendicular to the heat flow direction. The proportionality constant (k) is the property of the substance and is known as thermal conductivity.

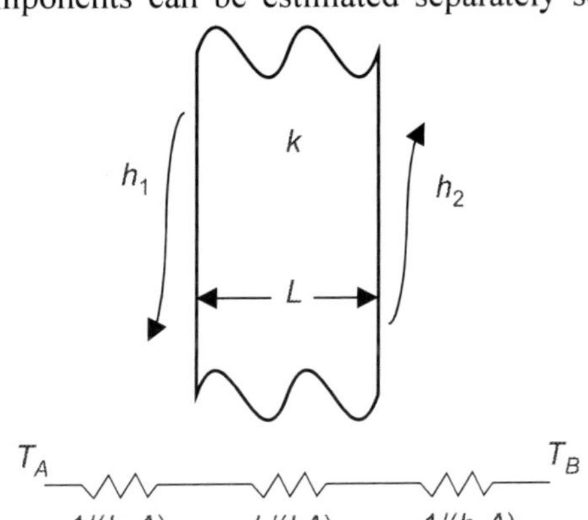

Figure 1.5: Resistance concept to heat flow

Mathematically it is expressed as

$$Q = -kA(\Delta T/\Delta x) \qquad [1.8]$$

(For quantitative calculation of the amount of heat conducted, '– ve' sign is not mandatory.)

For a solid block, as shown in Fig. 1.5, amount of heat transfer can be calculated as follows.

The solid block has a width L, and thermal conductivity, k. The two sides of the block is exposed to convective current having convective heat transfer coefficients h_1 and h_2. Temperature of the fluids on both the sides are T_A and T_B, respectively. Calculate the amount of heat transfer across the block. ($T_A > T_B$).

Resistance offered due to convective current h_1 is $1/(h_1A)$; Resistance offered due to block material is $L/(kA)$; Resistance offered by convective current h_2 is $1/(h_2A)$. The thermal potential difference is the difference in two extreme temperatures, i.e., $T_A - T_B$; so the heat flow Q is expressed as $\Delta T/\Sigma R_{th}$.

$$Q = \frac{T_A - T_B}{\left(\frac{1}{h_1 A}\right) + \left(\frac{L}{kA}\right) + \left(\frac{1}{h_2 A}\right)} \qquad [1.9]$$

Or heat transfer Q can be expressed as $Q = UA\Delta T$; where U is the overall heat transfer coefficient, and U is expressed as

$$\frac{1}{U} = \frac{1}{h_1} + \frac{L}{k} + \frac{1}{h_2} \qquad [1.10]$$

Convective heat transfer and Newton's Law of Cooling: Amount of heat lost/gained by a body due to convection is directly proportional to temperature difference (ΔT) between the body and the environment and the exposed surface area of the body to the environment (A). The proportionality constant (h) is known as the convective heat transfer coefficient which depends upon fluid thermo-physical properties, velocity, mass flow rate, etc. Mathematically, it is expressed as

$$Q = hA(\Delta T) \qquad [1.11]$$

Convective heat transfer coefficient is calculated from various correlations available for different condition and types of flow. For incompressible fluid flow through tubes, in turbulent condition, the most useful correlation is:

$\mathrm{Nu} = 0.023\ \mathrm{Re}^{0.8}\mathrm{Pr}^{n}$ where, $n = 0.4$ for heating and 0.3 for cooling. [1.12]

where Nusselt number Nu is expressed as hD_h/k, (D_h is the hydraulic diameter and k is the thermal conductivity of the conducting fluid).

D_h, hydraulic diameter is expressed as $4A_C/P_W$, where A_C is the cross-sectional area through which flow occurs and P_W is the wetted perimeter.

Thus, when there is heat flow due to combined conduction and convection, overall heat transfer coefficient (h) for different types of flow (laminar or turbulent) from available relations and correlations.

Radiation and Stefan-Boltzmann Law: This law governs the amount of heat loss from a black body by radiation. The hemispherical heat flow from a black body by radiation is directly proportional to surface area (A) and fourth power of the temperature in absolute scale (T^4). Mathematically, this is expressed as

$$Q = \sigma A T^4 \qquad [1.13]$$

The proportionality constant (σ) is called Stefan-Boltzmann constant and its value is 5.667×10^{-8} W/m^{2}K^4. For real bodies the total amount of radiations is expressed as $Q = \varepsilon\sigma A T^4$; where ε is the emissivity of the surface.

1.4.3 Laws of Mass Transfer

Fick's Law: This law states how much mass of one species (A) is diffused through another stationary mass (medium) which depends upon its concentration gradient. It is expressed as

$$\frac{m_A}{A} = -D\left(\frac{\partial C_A}{\partial_x}\right)$$

where, A – normal area; D – Diffusion coefficient; $\frac{\partial C_A}{\partial_x}$ – Concentration gradient

Unit of diffusion coefficient is m^2/s.

For convective mass transfer, mass transfer coefficient (h_M) is given by

$$\frac{m_A}{A} = h_M(C_W - C_\infty) = \frac{k_W}{\rho}(\omega_W - \omega_\infty)$$

where, $C_W - C_\infty$ is the concentration difference between the species A at wall and that at free stream. ($\omega_W - \omega_\infty$) is the specific humidity difference. This helps in determining the evaporation rate. Unit of h_M is m/s.

1.4.4 Universal Gas Law

This law is applicable to ideal gases with the following most important assumptions:

1. Intermolecular attraction among the gas molecules is negligible.
2. Space occupied by the gas molecules is negligible.

No gas on the earth is ideal. Thus, this becomes valid for a real gas only when the gas is at low pressure and high temperature because at low pressure and high temperature the above two assumptions are valid for a real gas.

$$PV = mRT \qquad [1.14a]$$

where P is pressure (kPa), V is volume (m^3), m is mass (kg), R is the characteristic gas constant (kJ/kg-K) and T is temperature in absolute scale.

or

$$PV = n \infty T \qquad [1.14b]$$

∞ is the universal gas constant and its value is 8.314 kJ/kgmole -K; and n is the number of moles.

And $\infty = R \times M$ where M is the molecular weight of the gas.

For example, air has a molecular mass of around 29.00.

So characteristic gas constant R for air is 8.314/29.00 = 0.287 kJ/kg-K

Some other laws like Kirchhoff's law of thermal radiation (emissivity of a body is equal to absorptivity for same temperature and wavelength), Stokes' law (governs the fluid flow) are of importance.

1.5 SOME USEFUL THERMODYNAMIC PROPERTY RELATIONSHIPS

Thermodynamic relationships amongst different properties of a system are discussed.

1.5.1 For Closed Systems

As discussed in Section 1.4.1, first law of thermodynamics for a cyclic process is expressed mathematically (in SI unit) as

Cyclic $\int Q$ = Cyclic $\int W$

But for a process $\partial Q = dU + \partial W$

∂Q is the heat interaction between the system and the surroundings, dU is the change in internal energy, i.e., $C_V dT$, ∂W is the external work done by/on the system.

For our range of study, in a closed system $\partial W = pdV$ (All other types of external, electrical, shaft works, etc., are not considered.)

So $\quad \partial Q = dU + pdV$

Thus, in closed or non-flow systems:

For constant volume process, $V = C$, (i.e., exponent index, n is ∞, in $pV^n = C$)

Work done = 0 and Heat transfer, $\partial Q = dU = C_V dT$

So $\quad Q = C_V(T_2 - T_1) \quad$ [1.15]

For Constant pressure process, $p = C$,(i.e., exponent index, n is 0, in $pV^n = C$)

$$\partial Q = C_P dT, \text{ So } C_P dT = C_V dT + pdV$$

$$Q = C_P (T_2 - T_1) \quad [1.16]$$

Work done $\quad = \int pdV = p(V_2 - V_1) \quad$ [1.17]

For isothermal (Constant temperature) process, $pV - C$ (i.e., exponent index, n is 1 in $pV^n = C$)

$$dU = C_V(T_2 - T_1) = 0, \text{ So } dQ = pdV$$

So $\quad Q = \int pdV = \int CdV/V = \text{C}\ln(V_2/V_1) = p_1V_1 \ln(V_2/V_1) = mRT_1\ln(p_1/p_2) \quad$ [1.18]

Work done (dW) is also equal to $mRT_1\ln(p_1/p_2)$ [1.19]

For polytropic process, $pV^n = \text{C}$

$$Q = C_V (T_2 - T_1) + \int pdV$$

$$\int pdV = \int C/V^n dV = pV^n \int (1/V^n) dV = (p_1V_1 - p_2V_2)/(n-1) = mR(T_1 - T_2)$$

$$Q = C_V(T_2 - T_1) + mR(T_1 - T_2)/(n-1), \text{ putting } R = C_P - C_V$$

$$Q = \{m(nC_V - C_P)/(n-1)\}(T_2 - T_1) \quad [1.20]$$

For reversible adiabatic or isentropic process exponent index n is replaced by γ,
$(pV^\gamma) = C$ (i.e., exponent index n is γ in $pV^n = C$)
In adiabatic process heat interaction is zero (i.e., system is perfectly insulated), $Q = 0$
So $\quad dU = \int pdV$, or $C_V(T_2 - T_1) = (p_1V_1 - p_2V_2)/(\gamma - 1) = mR(T_1 - T_2)/(\gamma - 1)$ [1.21]

1.5.2 For Open or Flow Systems

Conservation of energy equation gives

$$h_1 + V_1^2/2 + gZ_1 + Q = h_2 + V_2^2/2 + gZ_2 + W \qquad [1.22]$$

If the kinetic energy and potential energy are neglected, the equation will be

$$h_1 + Q = h_2 + W, \text{ So } dQ = dh + dW = dh + pdV$$

Tds relationship gives $Tds = dQ = dU + pdV$
Enthalpy is expressed as $h = U + pV$, So $dh = dU + pdV + Vdp$;
So $\quad Tds = dh - Vdp$ [1.23]
For an isentropic or reversible adiabatic process, $(pV^\gamma) = C$; $ds = 0$; $Tds = 0$, Q also equal to 0.
So external work done = dh and $dh = Vdp$. (This is for isentropic compression in flow process.)
So $h_2 - h_1 = \int Vdp = \gamma\,(p_1V_1 - p_2V_2)/(\gamma - 1) = \gamma\, mR(T_1 - T_2)/(\gamma - 1)$
For polytropic process, γ is replaced by n
and $\quad h_2 - h_1 = n\,(p_1V_1 - p_2V_2)/(n - 1) = nR(T_1 - T_2)/(n - 1)$ [1.24]
For constant pressure process, $Q = h_2 - h_1 = C_P\,(T_2 - T_1)$
But $\int Vdp = 0$

1.6 STATES OF A FLUID

The working fluids used for the refrigeration systems change phases during the process. Let us discuss how does it happen and some important definitions associated with it.

A fluid passes through three states: solid, liquid and vapour. For the application in refrigeration system, normally two states are dealt with, i.e., liquid and vapour. Hence, these two states will be discussed here.

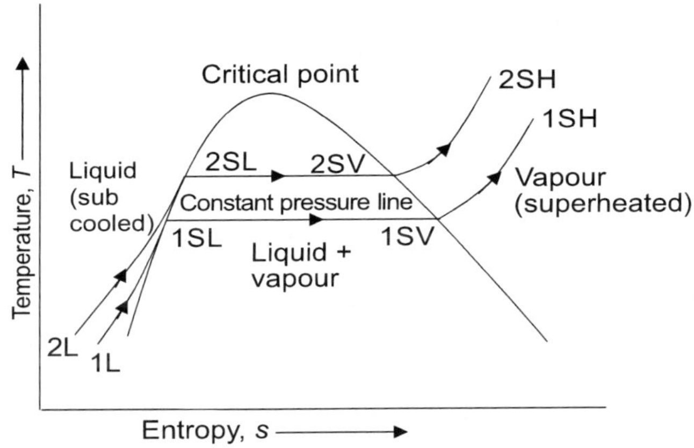

Figure 1.6: Phase diagram of a fluid on *T-s* plane

Refer Fig. 1.6. The diagram on *T-s* plane is considered because in the refrigeration cycle analysis, diagrams on *T-s* plane or *p-h* plane are suitable for cycle analysis from energy point of view and most of our system analysis are based upon energy conservation.

Let us take a fluid at its initial liquid state 1L at pressure p_1 and temperature t_1. The fluid is heated at constant pressure. Its temperature increases and reaches a state point 1SL, where boiling starts instead of increase in temperature. This state is called the liquid at saturated state or saturated liquid. The supplied to it is called sensible heat. As the heat supply continues, the liquid state will change into vapour state till it reaches a state 1SV. At this point, all the liquid is converted to vapour by absorbing the heat supplied, known as latent heat of vaporisation and the state is called saturated vapour. As the heat supply still continues, temperature of vapour increases and the state point it reaches, may it be 1SH, is called superheated vapour. This is called sensible heating of vapour. Similarly, the fluid at state 2L, if heated at constant pressure (p_2), will reach a state of superheated vapour at 2SH through saturated liquid at 2SL and saturated vapour at 2SV. Now all the saturated liquid states for different pressures on left hand side are joined and is called 'saturated liquid line'. Similarly, all the saturated vapour states for different pressures on right-hand side are joined and is called 'saturated vapour line'. Then both these lines are extended to meet at a point called 'critical point'.

Critical point is defined by 'critical temperature and critical pressure'. Critical temperature of a vapour is the maximum temperature beyond which the vapour cannot be condensed by application of any amount of pressure. Critical pressure of a vapour is the minimum pressure below which the vapour cannot exist in liquid state at its critical temperature. Critical point is different for different gases/vapours. Example, for water, critical temperature is 374.15°C, critical pressure is 221.2 bar and corresponding critical volume is 0.00317 m^3/kg.

1.7 HEAT ENGINE, HEAT PUMP AND REFRIGERATION SYSTEMS

Heat engine (HE) (Fig. 1.7(a)) draws heat (Q_1) from a high temperature heat source (T_1), does some amount of work (W_{OUT}) and reject the rest amount of heat (Q_2) to heat sink at lower temperature (T_2). Thus, it follows the Kelvin-Planck statement of second law of thermodynamics. The efficiency of the heat engine is expressed as (Desired effect/Work input), i.e., $W_{OUT}/(Q_1 - Q_2)$. So efficiency of a heat engine, even that of a Carnot engine can never be 100%, otherwise that

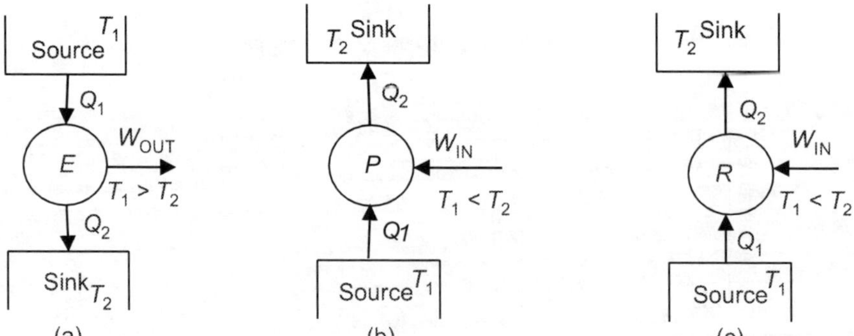

Figure 1.7: (a) Heat engine, (b) Heat pump, (c) Refrigerator

will be the violation of 2nd law of thermodynamics. Working principle of both **the heat pump and refrigerator** are basically same as shown in Fig. (1.7 (b) and (c) respectively). Both of them draw heat from a heat source at lower temperature (T_1) and deliver to a heat sink at higher temperature (T_2). The difference is in the desired effect. In heat pump system, heat rejection to the high temperature heat sink (Q_2) is the desired effect (e.g., heating a room in cold climatic condition) whereas heat extraction from low temperature heat source (Q_1) is the desired effect (e.g., cooling a room in hot climatic condition) for the refrigeration system.

Efficiency and COP: For any system, its performance is judged from its efficiency which is expressed as the ratio of desired effect obtained to the work input to the system. The term efficiency is used to evaluate the performance of a heat engine whereas the coefficient of performance (COP) is used for a refrigeration or a heat pump system.

Efficiency of the heat engine = Desired (work) effect/Work input

$$= W/Q_1 = (Q_1 - Q_2)/Q_1 \qquad [1.25]$$

Efficiency of heat engine is also expressed as $(T_1 - T_2)/T_1$

COP of a heat pump system = Desired (heating) effect / Work input to the system

$$= Q_2/W = Q_2/(Q_2 - Q_1) \qquad [1.26]$$

COP of heat pump is also expressed as $T_2/(T_1 - T_2)$

And COP of a refrigeration system = Desired (cooling) effect / Work Input to the system

$$= Q_1/W = Q_1/(Q_2 - Q_1) \qquad [1.27]$$

COP of refrigeration system is also expressed as $T_1/(T_2 - T_1)$.

Hence, COP of heat pump = COP of refrigeration system + 1.

This shows **COP of refrigeration system may be smaller than one** depending upon amount cooling effect and work input, but COP of heat pump is always greater than one.

Both the heat pump and refrigerator work on the principle of reverse Carnot cycle (RC cycle). It is worth mentioning here that the performance of any reversible cycle, may be Carnot or reverse Carnot cycles, is independent of the working fluid.

1.8 REVERSE CARNOT CYCLE

The working of Carnot cycle is best described in thermodynamics book. It consists of four reversible processes. Isothermal heat addition / expansion, isentropic expansion, isothermal heat rejection and isentropic compression. This cycle is internally as well as externally reversible, i.e., totally a reversible cycle. So the cycle can be reversed and in this case it becomes Carnot refrigeration cycle or reverse Carnot cycle. This is shown on *T-s* and *p-v* plane in Fig. 1.8. The diagrams are same as that of the Carnot cycle. Only the directions of processes are reversed. The area enclosed by the cyclic processes in both the *p-v* plane and *T-s* plane are same and represent the net work output of the system and also equal to the net heat transfer for the cycle. Because for any cyclic process change in internal energy is zero.

So, Cyclic ΣQ_{Net} = Cyclic $\sum W_{\text{Net}}$

COP of a reverse Carnot cycle for a refrigerator is $T_L / (T_H - T_L) = 1/ \{(T_H / T_L)\} - 1)$; where T_L is the temperature of source and T_H is the temperature of sink in absolute scale (kelvin). It can be shown that in any fraction the fractional value varies much with the variation of denominator than that of numerator. This shows, the COP will be affected more with the variation of T_L than the variation of T_H. COP goes on increasing as the difference in T_H and T_L decreases.

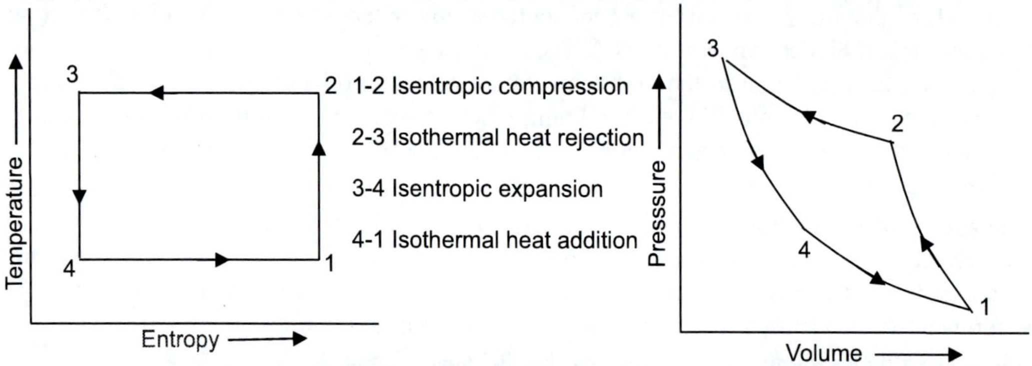

Figure 1.8: Reverse Carnot cycle on *T-s* plane and *P-v* plane

Besides COP, there is another term 'refrigerating efficiency' which compares the actual refrigerating cycle, described in forthcoming chapters, with RC cycle.

Refrigerating efficiency, η_{Ref} = Actual COP/COP$_{RC}$.

1.9 MEASUREMENTS AND UNITS

The properties of the refrigerant at different state points are to be measured and quantified. The primary or fundamental measurable quantities are mass (m), pressure (p), temperature (t), volume (v), etc. The most important quantities, to be measured, for refrigeration system are pressure, temperature and mass flow rate. To quantify any of these properties, knowledge of unit is also important. There are different units followed throughout the world. Universally, it is accepted to adopt the SI (Standard International) units though in some parts of the world MKS units and FPS units of measurement are still in practice.

1.9.1 Pressure Measurement

Pressure is expressed as force exerted normal to the surface per unit area of the surface. This is measured by different devices as discussed below.

Barometer is used to measure the atmospheric pressure. The working principle of the simplest one is given in Fig. 1.9.

A glass tube, whose one end is open and the other end is closed, is made vacuum. Then the open end of the tube is put inside a dish containing mercury. The mercury in the dish will flow into the tube and rise as long as the pressure exerted by the mercury in the tube becomes equal to the pressure exerted by the atmospheric air on the mercury in the dish. Now the height of the mercury in the glass tube is the atmospheric pressure, expressed in mm of mercury.

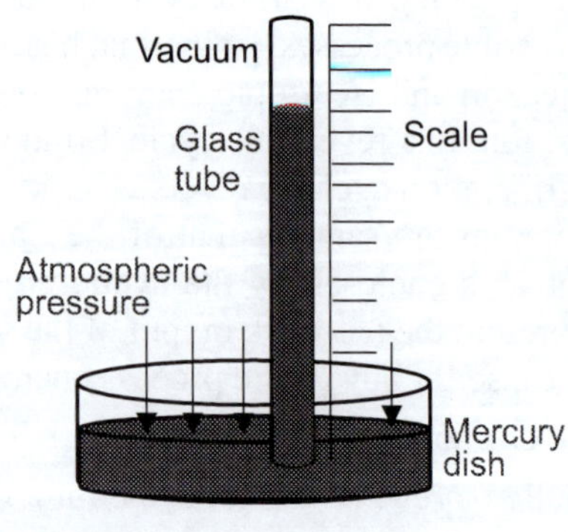

Figure 1.9

Manometers: Manometers are usually two legged 'U'-shaped tubes as shown in Fig. 1.10. One leg is opened to atmosphere and the other end is connected to the system whose pressure is to be determined. The tube contains some manometric fluid (may be water, alcohol, mercury, etc). The principle of this device is, at equilibrium position, the pressure exerted by the chamber (i.e., P) is equal to the atmospheric pressure + the pressure exerted by the differential height 'h'.

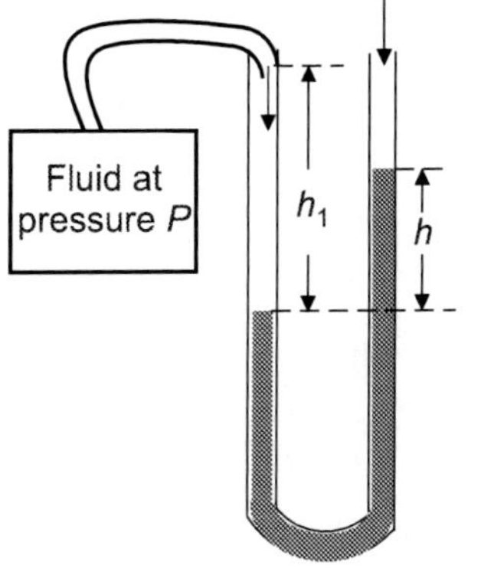

Figure 1.10

So $P + \rho_f g h_1 = P_a + \rho_m g h$; $P - P_a = g(h\rho_m - h_1\rho_f)$ [1.28]

ρ_f and ρ_m are the mass density of the fluid and the manometric fluid, respectively; g is the acceleration due to gravity. Thus, it measures the gauge pressure.

In a manometer if one end is closed and the other is connected to the system, then it measures the absolute pressure.

Example: An U-tube manometer contains a fluid whose specific gravity is 0.87. This tube is used to measure the pressure inside a closed cylinder which contains a certain gas at 40°C. The differential height (h) shown by the manometer is 18 cm. The value of h_1 is 25 cm. Calculate pressure in the cylinder. Atmospheric pressure is 101.325 kPa.

Ans: The density of the manometric fluid (ρ_m) = Sp. gravity × Density of water

$$= 0.87 \times 1000 \text{ kg/m}^3 = 870 \text{ kg/m}^3$$

The atmospheric pressure is 101.325 kPa.

Density of the gas (ρ_f) at 40°C is calculated from (mass of gas/ volume) = 7.143 kg/m^3

So $P - P_a = g(h\rho_m - h_1\rho_f) = 9.81 \times (0.18 \times 870 - 0.25 \times 7.143) =$ Pa = 1.519 kPa

So $P = 101.325 + 1.519 = 102.844$ kPa

Hence, the pressure of the gas inside the cylinder is 102.844 kPa.

From the above example, it is clear that if gas pressure is to be measured, then ρ_f can be neglected as $\rho_f << \rho_m$.

Bourden-tube pressure gauge: This measures the gauge pressure. It finds a very wide range of applications. This instrument has a 'C' shape configuration as shown in Fig. 1.11. When external pressure is applied to the inside of the tube, an elastic deformation occurs which is directly proportional to the applied pressure. This deformation is amplified by a rack-pinion arrangement and transforms it into angular rotation. This is displayed on a dial gauge.

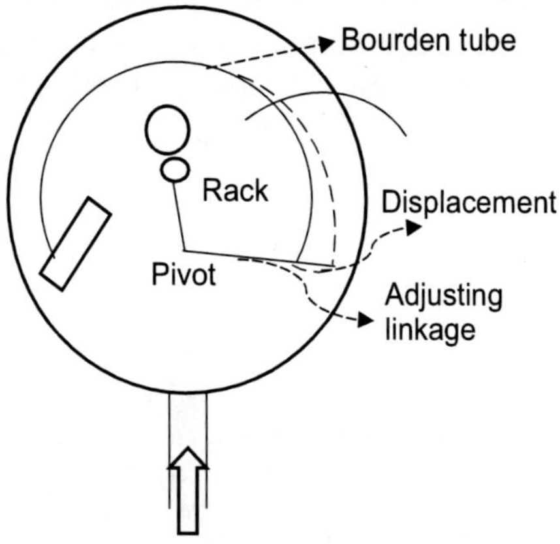

Figure 1.11

Unit in SI is 1 Pa = 1 N/m^2. Since it is very small, for practical purpose kPa, MPa are used. Though not SI unit, bar and atm are also used for pressure.

1 bar = 10^5 Pa = 100 kPa = 0.1 MPa

1 atm = 101.325 kPa = 1.01325 bar

Sometimes pressure is expressed in terms of mm of Hg (mercury) and this is also called 'torr'.

1 mm of Hg = 1 torr = 133 Pa = 1.3366 bar

1 atmospheric pressure is 760 mm of Hg or 1.01325 bar or 101.325 kPa.

1 pound per square inch (lb/in^2) = 6.894 kPa.

Pressure, when measured relative to perfect vacuum, is called absolute pressure and when it is measured relative to atmospheric pressure is called gauge pressure (Fig. 1.12).

So it can be stated: Absolute pressure = Gauge pressure + Atmospheric pressure.

When the pressure of a system is below atmospheric, the gauge pressure will be negative. So it is expressed in terms of vacuum, i.e., how much lower than he atmospheric pressure.

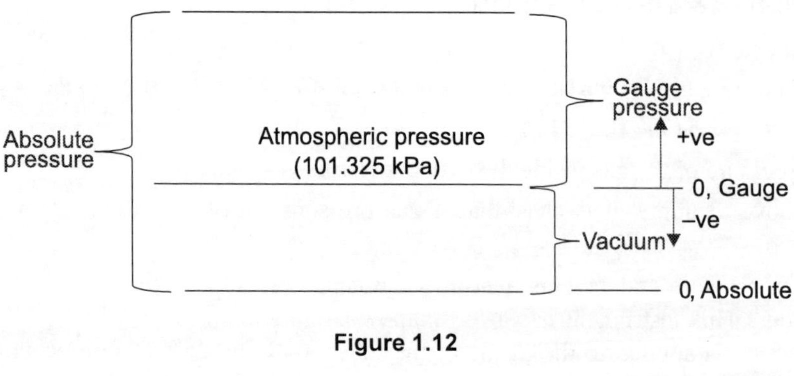

Figure 1.12

For example, 50 kPa vacuum means the pressure is 50 kPa lower than the atmospheric pressure. It is read from 0 gauge pressure down to 0 absolute pressure.

So 50 kPa vacuum is (101.325 – 50) = 51.325 kPa absolute.

In any case, if not mentioned, the pressure is the absolute pressure. The gauge pressure is mentioned like 20.0 kPa gauge. It means 121.325 kPa absolute.

1.9.2 Temperature Measurement

The most commonly used temperature measuring instrument is liquid in glass thermometer. A thermometric fluid, most commonly mercury, and sometimes alcohol, is put inside a glass tube. The thermal linear expansion of the fluid is the indicator of the temperature. The calibration is done by taking the ice at 0°C as reference point. The most common example is the clinical thermometer we use to measure our body temperature. The application of these types of thermometers are limited to very low temperature range (–20°C to 150°C). For higher temperatures, very long glass tube is required and this is not practicable.

Thermocouples: (Refer Fig. 1.13) These are very commonly used for industrial applications. It is based upon the principle of 'Seebeck effect' that says, 'when two dissimilar metals are joined together and their two ends are exposed to two different temperatures, there is generation of emf in the metallic conductors. This emf is directly proportional to the difference in temperature. Mathematically, it can be expressed as emf = $\alpha_{ab}\ \Delta T$, where α_{ab} is the Seebeck coefficient between the two materials 'a' and 'b' and Δt is the temperature difference between two ends. The amount of emf is the measure of the temperature difference. If one end is taken as the reference, the temperature of the other end, which is in contact with the body whose temperature is to be measured, can be known.

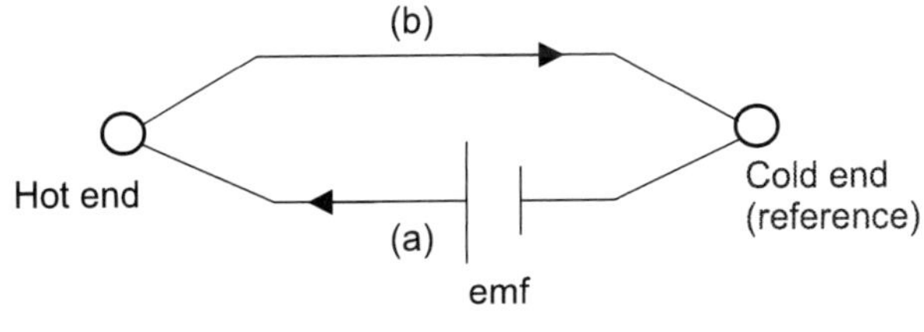

Figure 1.13

Listed below are some of the material combinations and their range of temperatures it can measure.

Material combination (Type of thermocouple)	Range of temperature (°C)	Range of emf generated (mV) (Reference junction, 0°C)
Copper – Constantan (T)	–150 to 400	–4.648 to 20.872
Chromel – Alumel (K)	–150 to 1200	–4.913 to 48.838
Iron – Constantan (J)	–150 to 1200	–6.500 to 69.553
Chromel – Constantan (E)	–150 to 1000	–7.279 to 76.373
Platinum – Platinum 10% + Rhodium (S)	–50 to 1750	–0.236 to 18.503

Still lower temperatures, (for cryogenic applications), like measurement of –200°C (73 K), can be done by measuring the size of the atomic cloud. Atomic cloud is illuminated with laser light. The atoms absorb the laser light (photons) and they cast a shadow.

Or by measuring the kinetic energy of the atom which increases with increase in temperature. This is done by suddenly switching off the magnetic trap. In the absence of the magnetic force, the atoms fly away and the atomic cloud expands. This expansion increases with time. This is a direct measure of the velocity of the atom and therefore the temperature.

Very high temperatures, i.e., of the range of 1000°C or more are measured by optical pyrometer which measures the temperature basing on the principle of Wein's displacement law. According to this law as the temperature increases the wavelength (of maximum amplitude) of the radiation emitted from the body decreases. By measuring the wavelength by wavemeter/ interferometer, temperature can be calculated from the law which states $\lambda_{max}T \approx 2900$ where, λ_{max} is the wavelength having maximum amplitude, T is temperature in K. Besides there is another thermometer known as resistance thermometer. The principle is that the electrical resistance of the conducting wire increases with increase in temperature. Thus, the resistance of the wire is an indicator of the temperature.

Temperature is normally measured in °C (degree centigrade), °F, or K (kelvin)

The relationship among these three temperature scales is as follows.

°C/5 = (°F – 32)/9; °C + 273.15 = K

Thus, 40°C = 104°F = 313.15 K

Volume, in MKS and SI , it is m^3 and in FPS it is ft^3.

1 ft^3 = 0.028217 m^3 or 1 m^3 = 35.3147 ft^3.

Mass, in SI and MKS it is kilogram (kg) and in FPS it is pound mass (lbm)

1 lbm = 0.454 kg

1.9.3 Flow Measurement

The amount of mass/volume flow is measured by many devices like orifice meter, venturi meter, rotameter, turbine meter etc. Out of these, venturi meter and rotameter are mostly used for flow measurement in refrigeration systems.

Venturi meter, very often used for mass flow measurement of liquid, works on the principle of Bernoulli (energy) equation (Fig. 1.14).

$P_1/\rho_1 + v_1^2/2 = P_2/\rho_2 + v_2^2/2$; assuming same elevation at 1 and 2, no loss of energy.

Mass flow rate, $m = \rho_1 A_1 v_1 = \rho_2 A_2 v_2$

For incompressible flow (liquids), $\rho_1 = \rho_2 = \rho_f$

Substituting this value $$P_1 - P_2 = \frac{\rho v_2^2}{2}\left[1 - \left(\frac{A_2}{A_1}\right)^2\right] \qquad [1.29]$$

$P_1 - P_2$ is determined from the differential manometer fixed across the Chapters 1 – 2. As per the principle of manometer $P_1 + \Delta h g \rho_f = P_2 + \Delta h g \rho_m$;

This gives

$$P_1 - P_2 = (\rho_m - \rho_f)\Delta hg \quad [1.30]$$

Hence, from above equation v_2 is calculated.

(If $\rho_f << \rho_m$, the above equation can be written as $P_1 - P_2 = \rho_m \Delta hg$)

Volumetric flow rate, $Q = A_2 \times v_2$ m^3/s and mass flow rate, $m = \rho Q$ kg/s

For all practical cases, actual discharge is less than theoretical discharge due to frictional loss. Taking this into consideration, a multiplying factor, C_d (coefficient of discharge) is incorporated.

C_d = Actual discharge/Theoretical discharge

Example: A venturi meter is used to measure the air flow rate through a duct. A water tube manometer is used to measure the pressure drop across the venturi. Tube and throat cross-sectional area are 0.5 m^2 and 0.36 m^2, respectively. Densities of air and water are 1.2 kg/m^3 and 1000 kg/m^3, repectively. Deflection in the manometer tube is 24 mm. Calculate (a) the pressure difference across the venturi, and (b) the air mass flow rate.

Ans: Refer Fig. 1.14. Δh is given 24 mm = 0.024 m.

So pressure drop/difference across 1 – 2 is

$$(\rho_m - \rho_f)g\Delta h = (1000 - 1.2) \times 9.81 \times 0.024 = 235.15 \text{ Pa}$$

Velocity of air at throat, $v_2 = \sqrt{\dfrac{2 \times 235.15}{1.2\left[1 - \left(0.36/0.5\right)^2\right]}} = 28.53$ m/s

Discharge, $m = \rho A_2 v_2 = 1.2 \times 0.36 \times 28.53 = 12.325$ kg/s

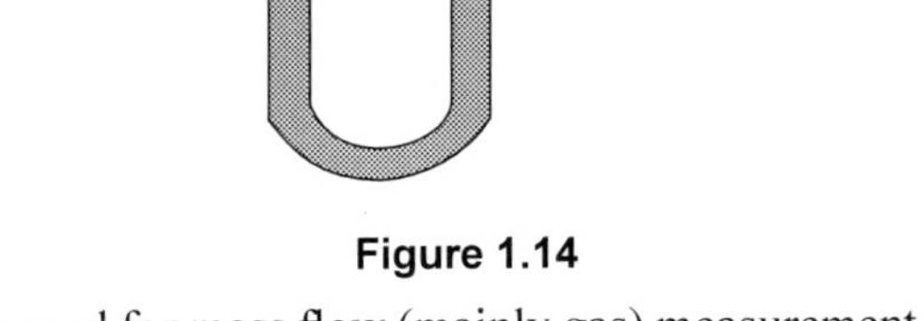

Figure 1.14

Rotameter is commonly used for mass flow (mainly gas) measurement. Its working principle is as follows. Refer Figure 1.15.

A metallic bob is placed inside a tapered tube. As the fluid flows through it, the bob rises up due to buoyancy and remains stationary in floating position at an equilibrium position. The bob remains in so under the forces (1) the buoyancy acting upward $\rho_f V_b g$; (2) drag force acting upward F_D; and (3) gravitational force, i.e., its weight acting downward $\rho_b V_b g$. Mathematically, it can be expressed as

$$\rho_f V_b g + F_D = \rho_b V_b g \quad [1.31]$$

V_b is the volume of the bob, ρ_f and ρ_b are the densities of fluid and the bob, respectively.

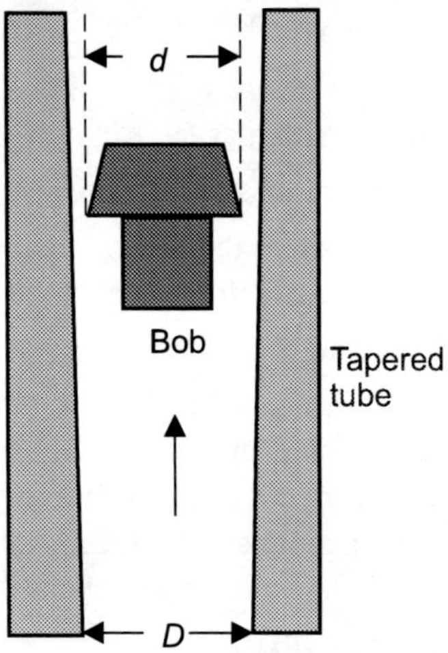

Figure 1.15

The drag force (F_D) is expressed as $F_D = C_d A_b \dfrac{\rho_f v_m^2}{2}$ [1.32]

where, C_d is the drag coefficient, A_b is the frontal area of the bob and v_m is the average velocity of flow through the annular space between the tube and the bob. C_d depends upon the Reynolds number and hence on the fluid viscosity.

From the above two equations, $v_m = \left[\dfrac{2gV_b}{C_d A_b}\left(\dfrac{\rho_b}{\rho_f} - 1\right)\right]^{1/2}$ [1.33]

$Q = Av_m$; where, A is the annular area [1.34]

i.e., $A = \dfrac{\pi}{4}\left[(D + ah)^2 - d^2\right]$. Here D is the entry diameter of tube, a is the taper angle in radian, h is the height the bob has moved up from the base, d is the maximum diameter of the bob.

However, if the bob can be made in such a way that C_d can be kept constant for a small range of velocity, temperature and for a particular fluid (v_m) can be kept constant. Ultimately, the discharge rate (Q) varies proportionally with 'h' only and the proportionality constant is said to be instrument constant. This should be supplied by the manufacturer.

1.9.4 Work and Energy

Work and energy are obtained by the dot product of force and displacement. In MKS unit work is expressed in kgfm and heat energy is expressed in kcal (kilocalorie). The conversion of heat and work is made by a factor 'mechanical equivalent of heat' which is equal to 427 kcal/kgfm.

So 1 kcal = 427 kgfm = 427 × 9.81 ≈ 4190 J (joule) = 4.19 kJ

In SI unit work is expressed in newton-meter.

1 N-m = 1 joule and energy is also expressed in joule. Most frequently kJ or MJ is used.

In FPS the energy unit is British thermal unit (Btu).

Hence, 1 kJ = 0.948 Btu = 0.239 kcal

Unit power: In SI power unit is watt (W)

1 W = 1 J/s =1 volt × 1 ampere

1 horsepower (hp) (imperial) = 746 W

1 hp (metric) = 736 W

1 kWh (a conventional unit used for electric power consumption) = 3600 kJ

The units of enthalpy and entropy in SI are expressed in kJ/kg and kJ/kg-K respectively. Specific heat is expressed in kJ/kg-K.

1.9.5 Refrigerating Effect, Refrigerating Capacity and Tonne of Refrigeration

The amount of refrigerating effect or cooling effect is expressed by the unit tonne of refrigeration (TR) and is defined as the amount of heat required to be extracted from one US tonne of water at 32°F to convert it to ice at same temperature in 24 hours.

Putting the numerical values in SI unit: 1 US tonne = 2000 lb = 2000 × 0.454 kg = 908 kg

Converting water to ice at same temperature means it is only the latent heat of fusion of ice at 32°F (0°C) = 334.65 kJ/kg is to be extracted.

24 hours = 24 × 3600 seconds

$$\text{So } 1 \text{ TR} = \frac{908 \times 334.65}{24 \times 3600} \approx 3.5167 \text{ kW}$$

Refrigerating effect is the cooling produced by the system per unit mass of refrigerant. Refrigerant effect multiplied by mass of refrigerant is the refrigeration capacity of the system.

1.9.6 Units of Viscosity

Viscosity is a term used to describe the frictional force arises due to cohesive force between the fluid molecules as they flow past over each other. This is expressed as $F = \mu\ (dU/dy)$; where, U is the velocity of the fluid and y is the distance along vertical direction and μ is the proportionality constant and termed 'dynamic viscosity'.

The unit of dynamic viscosity (μ) is centi poise, i.e., gm/(m-s) = 0.01 poise = 0.001Pa-s or 0.001 Ns/m^2.

Or 1 poise = 1 dyne-s/cm^2 = 1 gm/cm-s

At 20°C dynamic viscosity of water and air are 1 centipoise and 1.8 × 10^{-2} centipoise.

The unit of kinematic viscosity (ν) = μ/ρ = m^2/s = 10^4 stokes (cm^2/s).

1.10 NATURAL COOLING METHODS

Earlier days people adopted different methods of natural cooling process for comfort and preserving food items. Some of those methods are discussed below.

1.10.1 Natural Ice

Northern and southern hemispheres of earth and places of higher altitudes are the deposits of natural ice. Over centuries ice was harvested, stored with good insulation and transported to warmer places lacking natural ice.

1.10.2 Nocturnal Cooling

Earth is covered with different spheres (stratosphere, ozonosphere, etc) and there is heat exchange between these spheres and earth by radiation. A thin layer of water (5 mm thick) in a well-insulated (covered with thick layer, around 30 cm of hays) earthen tray is kept under the cloudless sky during night time. This water in the tray, which is at environmental temperature, exchanges heat with the stratosphere (11 km from earth surface) which is at –55°C by radiation. The water cools down to form ice by the morning. This is the ancient method of making ice. Such practice of making ice is hampered if there is a high convective heat transfer due to high velocity wind flow.

1.10.3 Evaporative Cooling

Evaporative cooling is the most observable natural cooling method. The minimum temperature to which air can be cooled is equal to its wet bulb temperature. The factors affecting evaporative cooling are: (i) environmental temperature and humidity, (ii) wind velocity, and (iii) surface tension of the fluid. Best examples are cooling of water in an earthen pot, cooling of our body when we perspire. Animals keep themselves cool by natural cooling methods. A dog brings out its tongue so that there is evaporation from the tongue's surface and the dog feels cool. The dark and light coloured strips of a zebra get heated unequally. Thus, it induces convective current between these strips and it cools the body. The minimum temperature that can be attained by evaporative cooling is equal to the wet bulb temperature (defined in Chapter 8) of the air.

1.10.4 Cooling by Salt Solution

Water gets cooled to some extent when salt is added to it. The salt absorbs the heat of solution from water and cools it. Theoretically, in a well insulated container, temperature up to –20°C with NaCl and temperature up to –50°C with $CaCl_2$ can be attained. But it is difficult to recover the salts, once being dissolved, for cyclic process.

These natural methods of cooling are of very short duration and intermittent in nature. These had very low capacity and temperature below –50°C (practically –20°C) was not possible. So attempt was made to produce artificial cooling.

1.11 ARTIFICIAL METHODS OF REFRIGERATION

With the development of science and due to the increase in demand for refrigeration, mankind has resorted to artificial cooling techniques.

The following artificial systems for refrigeration or for heat pump are now in practice.

1. Vapour compression refrigeration system (external energy is provided by the compressor run by electric power)
2. Vapour absorption refrigeration system (external energy is provided by the pump and the absorbing affinity)
3. Air/gas cycle refrigeration system (compressor)
4. Water refrigeration system (vacuum pump)
5. Thermoelectric refrigeration system (battery)
6. Magnetic refrigeration system and some others to name (energy generated due to sudden demagnetisation)

From the above discussion it is clear that external energy from any source has to be provided to drive the heat from lower temperature to higher temperature and the name of the refrigeration system has been given as per the source of external energy from which it is provided.

Of the types of refrigeration systems outlined above the most prevalent are vapour compression and vapour absorption refrigeration systems.

The first attempt was made in 1775 when Cullen built an apparatus to make ice artificially by vaporizing water at low pressure. The first development to produce cold mechanically took place when Perkin developed a hand operated compressor working on ether in 1834.

1.11.1 Other Methods of Producing Cold

1. When two dissimilar metals are joined together and an emf is applied across them, one end of the joint gets cooler and the other gets hotter. This is called Peltier effect. This principle is the basis of the development of 'thermoelectric refrigerators'.
2. In the year 1824, Faraday conducted an experiment. One beaker contains AgCl powder and another one contains NH_3–H_2O solution. Now the two beakers are connected. NH_3 from NH_3–H_2O solution bubbles out and gets absorbed by the AgCl powder as it has a high affinity for the later leaving the water cold. This principle leads to the development of vapour absorption refrigeration.
3. Throttling (bringing down the pressure of a liquid by passing it through a constricted passage adiabatically so that enthalpy remains constant) of high pressure gas is called as Joule-Thompson effect. The change in temperature and pressure during throttling is expressed as $\mu = (\partial T/\partial P)_h$. Here, μ is called Joule-Thompson coefficient. Throttling does not always lead to cooling. When $\mu > 0$, it produces cooling effect and when $\mu < 0$, it produces heating effect. So cold can be produced by keeping $\mu > 0$ and air liquefaction (temperature less than –150°C) is done by this method.
4. Temperature near to 0 K can be obtained by adiabatic demagnetisation of paramagnetic salts. This is called 'magnetic refrigeration'. When randomly oriented atoms of a paramagnetic salt are exposed to a strong magnetic field, they align themselves in the direction of the magnetic field. Now if the magnetic field is suddenly removed, the atoms reorient themselves to the original random positions at the expense of the internal energy. This makes the salt cool.

5. When high pressure air expands through a turbine, it does some work at the expense of the pressure energy. This decreases the air temperature and cooling effect is obtained. Air cycle refrigeration, specifically for aircraft refrigeration, works on this principle.
6. Molecules of a gas move randomly at different velocities. High velocity molecules have higher temperature and low velocity molecules have lower temperatures. Air, at high pressure, is fed to a vortex tube tangentially through nozzle. The air has a vortex motion inside the tube so that the high speed molecules (at higher temperature) moves outward and low speed molecules (lower temperature) moves inward. Inner core also loses momentum to the outer vortex. Thus, the vortex tube divides the flow into two streams, a low temperature and low kinetic energy stream and a high kinetic energy warm stream. By suitably positioning the valves hot and cold air stream can be collected separately. This phenomenon was first observed by George Ranque in 1928 and he invented this vortex tube. Some improvement in this tube was done by Rudolph Hilsch in the later part.

Commonly Used Abbreviations

HVAC: Heating, Ventilating and Air Conditioning

ASHRAE: American Society of Heating, Refrigerating and Air Conditioning Engineers

IIR: International Institute of Refrigeration

Example 1.1

In a solar energy operated refrigeration system water is cooled at the rate of 2.5 kg/s from 35°C to 15°C. If the incident solar energy is estimated to be 600 W/m^2 and the COP of the system is 0.7. Calculate the area of the solar collector needed.

Ans: For the water to be cooled the amount of heat required to be removed for the water is

$Q = m_W C_{P,W}\,\Delta T = 2.5 \times 4.19 \times (35 - 15) = 209.5$ kW

Refrigeration system COP = 0.7 = Refrigerating effect/Heat input = 209.5/Heat input

So required heat input = 209.5/0.7 = 299.3 kW

Hence the required area of solar collector is 299.3/600 = 0.5 m^2

Example 1.2

The method of producing ice by nocturnal cooling is described in Section 1.10.2. In such a case, the stratosphere and environment temperatures are –55°C and 20°C, respectively. The water is at 0°C. The thickness of hey is 0.3 m and its thermal conductivity is 0.2 W/m-K. The convective heat transfer coefficient for environmental condition is 6 W/m^2-K. Calculate the amount of heat loss from the water surface if the emissivity of that is 0.87 (Fig. 1.16).

Ans: Heat is transferred from the water surface

1. By radiation heat exchange between water surface and the stratosphere, when the tray sees the sky fully, the heat lost by the water is
   $Q_R/A = \varepsilon\,\sigma\,(T_W^4 - T_{St}^4)$; $= 0.87 \times 5.67 \times 10^{-8} \times (273^4 - 218^4) = 162.59$ W/m^2
2. Heat gained by convection from the water surface is $Q_{C_1}/A = 6 \times (20 - 0) = 120$ W/m^2
3. Again heat gained through the insulated surfaces, $Q_{C_2} = U \times \Delta T = 1/\{(1/h + t/k)\} \times 20$
   $= 1/\{(1/6 + 0.3/0.2)\} \times 20 = 1/1.667 \times 20 = 12$ W/m^2

Hence the net heat transfer (heat loss) from the water surface = (162.59 – 120 – 12) = 30.59 W/m^2.

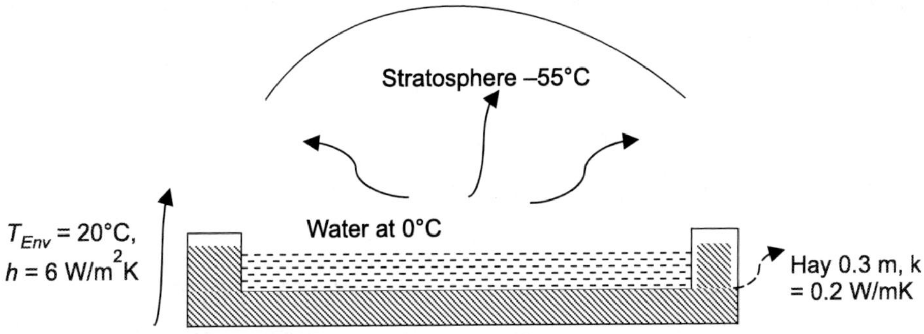

Figure 1.16

The latent heat of solidification for water is 335 kJ/kg.

So we can calculate what should be the area and what is the time requirement to produce 1 kg of ice.

Again from the above calculation it is clear that if the environmental temperature will be higher, net heat gain will be more than the heat lost due to radiation. So ice production will not be possible. That might be the reason why places of warm climatic condition were not able to produce ice by this method.

EXERCISE

1. Define one tonne of refrigeration. Show that 1 TR = 3.5167 kW.
2. Describe the early method of producing ice.
3. Calculate the heat flow through a slab, a composite slab and a composite cylinder as shown in the figures.

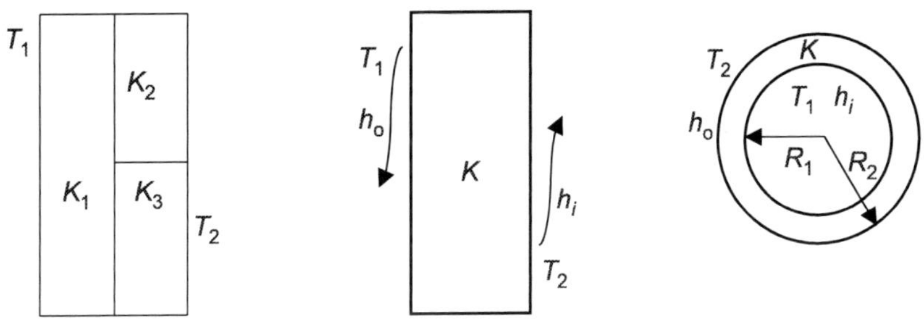

Numerical values for above figures: T_1 = 50°C, T_2 = 24°C, K = 12 W/m-°C, K_1, K_2, K_3 are 10, 15 and 12 W/m-°C, respectively, R_1 and R_2 are 11 and 12.5 mm respectively, h_0 and h_i are 18 W/m^2-°C and 8 W/m^2-°C.

4. Calculate the heat loss/m^2 from a furnace wall at 827°C to an environment at 27°C. Assume the emissivity of the furnace wall at that temperature is 0.75.

5. Water at 500 kPa and 150°C is passed through a pressure reducing valve and delivered to a separating tank at atmospheric pressure of 101.325 kPa. What is the state of the water entering the pressure reducing valve and what is the amount of vapour leaving the tank per kg of water entering the valve.
6. Is it possible to attain 100% efficiency in a Carnot engine? Justify your answer.
7. Write five means of producing cooling effect.
8. Describe the principle of fluid flow measurement by rotameter.
9. Describe the principle of fluid flow measurement by venturi meter.
10. Gas flow measurement is preferably made by rotameter or venturi meter.
11. What is the principle of measuring the temperature by thermocouple?
12. A Bourden gauge shows a pressure reading of 30 kPa. What is the pressure in absolute scale?
13. A vacuum gauge reads a pressure of 20 kPa. What is the pressure in absolute scale?

MULTIPLE–CHOICE QUESTIONS

1. In a closed system — is transferred across the boundary.

   (a) only mass (b) only energy
   (c) both mass and energy (d) neither mass nor energy

2. A system is said to be in thermodaynamic equilibrium when it is in — equilibrium.

   (a) thermal (b) mechanical
   (c) chemical (d) all of the three

3. Measuring the body temperature by a thermometer is an example of — law of thermodynamics.

   (a) zeroth (b) first
   (c) second (d) third

4. Heat flows spontaneously from high temperature to low temperature is — law of thermodynamics.

   (a) zeroth (b) first
   (c) second (d) third

5. Efficiency of a Carnot engine is

   (a) less than 100% (b) 100%
   (c) more than 100% (d) always 50%

6. Performance of any reversible cycle (Carnot or reverse Carnot) is — the working fluid.

   (a) dependent (b) independent
   (c) cannot say

7. COP of a refrigeration system can be

(a) >1 (b) <1

(c) never be <1 (d) both (a) and (b)

8. COP of a heat pump system is

(a) always >1 (b) always <1

(c) never be <1 (d) never be >1

9. A vacuum gauge reads 30 kPa. What is the approximate absolute pressure.

(a) 50 kPa (b) 70 kPa

(c) 100 kPa (d) 101 kPa

10. A dial pressure gauge reads 20 kPa. What is the approximate absolute pressure.

(a) 100 kPa (b) 80 kPa

(c) 140 kPa (d) 120 kPa

11. What is the temperature °F if a thermometer reads 100°C.

(a) 180 (b) 232

(c) 212 (d) 192

12. How much is 1 tonne of refrigeration in SI unit.

(a) 3.5 kJ (b) 1000 kg

(c) 10 kcal (d) 3.5 kW

13. COP of reverse Carnot cycle is more affected with variation of

(a) T_H (b) T_L

(c) $T_H - T_L$ (d) $T_H + T_L$

14. Measurement of temperature by thermocouple is based upon the principle of

(a) Joulean effect (b) Peltier effect

(c) Seebeck effect (d) Thompson effect

15. In nocturnal cooling heat loss occurs mainly due to

(a) conduction (b) convection

(c) radiation (d) evaporation

16. If sodium chloride is added to water, water temperature

(a) decreases (b) increases

(c) remains same (d) first increases, then decreases

17. Gas flow measurement is more preferably done by — in refrigeration system.

(a) venturi meter (b) orifice meter

(c) anemometer (d) rotameter.

18. For the given system, temperature, $T = 300$ K. When is the efficiency of engine (E) equal to the reciprocal of the COP of R?

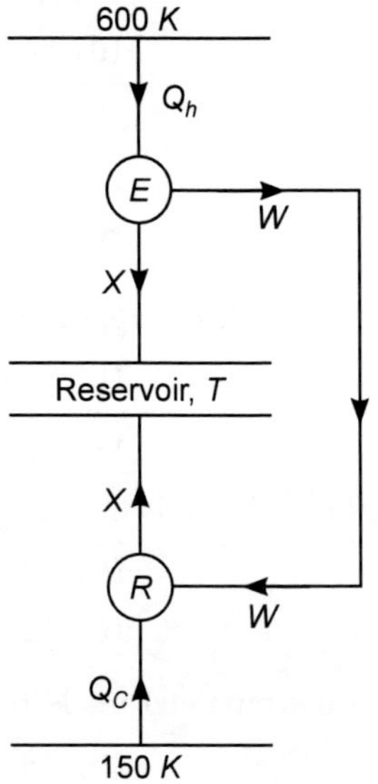

(a) When R acts as a heat pump

(b) When R acts as a refrigerator

(c) When R acts both as a heat pump and a refrigerator

(d) When R acts as neither a heat pump nor a refrigerator

19. A refrigerator of COP 5.0 is driven by a reversible heat engine with 30% efficiency. Find the net heat input to the engine for each MW of heat removed in the refrigerator?

(a) 666.67 kJ (b) 600 kJ

(c) 66.67 kJ (d) 6600 kJ

20. For the given system, E is a heat engine with efficiency of 0.4 and R is a refrigerator. Given that $Q_2 + Q_4 = 3Q_1$ the COP of the refrigerator is

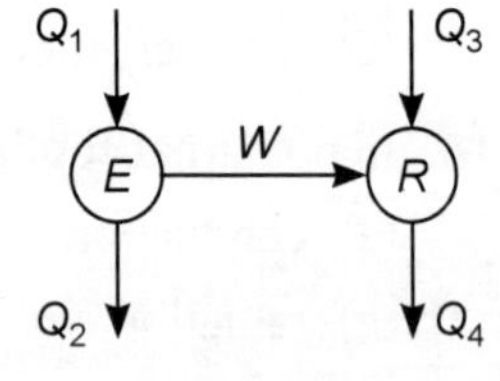

(a) 2.5

(b) 3.0

(c) 4.0

(d) 5.0

21. A reversible engine has thermal efficiency of 30%. When it is used as a refrigerating machine with all other conditions unchanged, the coefficient of performance will be

(a) 3.33 (b) 3.00

(c) 2.33 (d) 1.33

22. A reversed Carnot cycle working as a heat pump has a COP of 7. What is the ratio of minimum to maximum absolute temperatures?

(a) 7/8 (b) 1/6
(c) 6/7 (d) 1/7

ANSWERS

1. (b) 2. (d) 3. (a) 4. (c) 5. (a) 6. (b) 7. (d) 8. (a) 9. (b) 10. (d)
11. (c) 12. (d) 13. (b) 14. (c) 15. (c) 16. (a) 17. (d) 18. (a) 19. (a) 20. (d)
21. (c) 22. (c)

2

Refrigerants

2.1 INTRODUCTION

The thermodynamic (2nd law) analysis of a total reversible cycle shows that the performance and efficiency of a thermodynamic cycle is independent of the working substance. But this is not true for a real cycle. Because the working substance is the medium through which heat is transported from one part of the system to other and from the system to the environment or vice versa. There must be a temperature gradient between the working substance and the source/sink which increases the irreversibility and hence decreases the COP. Hence, judicious selection of a working substance for an actual refrigeration/heat pump system is important from performance as well as environment protection point of view. Working substance may be a pure fluid or a mixture of two or more fluids or electrons (as in thermoelectric refrigeration system). Thus a refrigerant can be defined as the working fluid, with desirable properties, that transports the heat from one part of the system to other in a refrigeration system. In this chapter we will discuss only about the refrigerant fluids.

2.2 HISTORY OF REFRIGERANTS

History gives us the experience of the past. The analysis of the shortcomings of the past research definitely gives an insight to its further development. Ether was the first refrigerant to be used in this system. Professor William Cullen of the University of Edinburgh in 1755 placed some water in thermal contact with ether under a receiver of a vacuum pump. The evaporation rate of ether increased due to the vacuum created or in a very low pressure and water could be frozen.

Jacob Perkins, in 1835 stated in his patent application "I have enabled volatile fluids for the purpose of producing the cooling or freezing of fluids, and at the same time constantly condensing such volatile fluids, and bringing them again into operation without waste."

The man responsible for making a practical vapour compression refrigeration system was James Harrison who made a patent in 1856 for a vapour compression system using ether, alcohol or ammonia.

2.2.1 Development of CFCs

Chloro Fluoro Carbons (CFCs) are vital to modern refrigeration and air conditioning practices. However, CFCs may not have been invented if the need of small and household refrigerator was not felt, mainly to replace the iceboxes. Engineers have been tinkering with this idea since 1880. It is a fact that the CFCs have already been phased out in many countries due to their adverse effects (ozone depletion and global warming) on the environment. But the knowledge of its gradual development to fit into the current refrigeration systems will help young scientists to develop substitutes for the CFCs.

In ice plants, warehouses and breweries leakage of refrigerants caused accidents and deaths. But once these problems enter the homes, it would be a different story. No one would like to be rousted out of home. Sulphur dioxide was a popular refrigerant, which even in small quantities, causes boots of chloughing and stinging of eyes and compels one out of doors. This was thought to be safe for this very reason since the leaks invariably became sensational. Methyl chloride (known as chlormethyl ether in early literature or by trade name ARTIC) was another popular refrigerant. It has sweet, ethereal odour. It has anaesthetic effect and delayed toxic effect. Isobutane (R601) was used in 1920s in household refrigerators and small refrigeration systems under the trade name of FREEZOL. Copeland Products Corporation of Detroit marketed isobutene up to 1930s. Then it was replaced by methyl chloride. Isopropane and propane were also used to some extent but they were highly flammable.

Dr. Midgley is particularly noted for his use of periodic table arrangement of elements to find solutions to industrial problems. His first success was the search of an antiknock agent for gasoline-tetra ethyl lead (TEL). The second major achievement was development of CFCs as refrigerants. It is ironical that both these exceptional industrial success stories of their time have turned out to be hazards for the atmosphere in present environmental condition.

Midgley observed that only a few elements were contained in the refrigerants in use and these elements were closely related in periodic table shown below.

Table 2.1: (Atomic numbers shown)

			H (1)	Noble elements
C (6)	N (7)	O (8)	F (9)	
		S (16)	Cl (17)	
			Br (35)	
			I (53)	

The elements of air, water, carbon dioxide, ammonia, sulphur dioxide, methyl chloride all fall in this part of the periodic table. Their objective was to find a new non-flammable refrigerant with NBP between –40°C and 0°C with good chemical stability and low toxicity.

The periodic table has elements with related properties arranged in vertical columns. The lighter elements at top are less toxic than the heavier elements at bottom not by themselves but in compounds. A look at the toxicity indicated that methyl iodide, methyl bromide and methyl chloride had threshold limit values (TLV) of 2 ppm, 5 ppm and 50 ppm respectively, that is, toxicity in organic compounds decreased for lighter halogens. The low level of toxicity for fluorinated compounds as hoped by Midgley was an original idea, since many inorganic fluorides are toxic; for example, hydrofluoric acid and fluorine itself is very reactive and toxic.

The heats of formation of compounds containing carbon and halogens were found to increase from iodine to fluorine. If the heat of formation is high, the resulting compound is generally stable.

Midgley then observed that the flammability decreases from left to right. Combination of carbon and hydrogen is most flammable. Nitrogen and hydrogen compounds like ammonia are less flammable. Next, hydrogen and oxygen compound (water) or hydrogen and fluorine compound (HF) are still less inflammable.

The normal boiling point of carbon tetrafluoride was reported to be –15°C at that time but later it was corrected to –128°C. Also fluorspar (CaF_2) has been known since fifteenth century to decrease the melting point of ores, ceramics and glass, etc. Fluorine is the thirteenth most abundant element available on earth's surface. The wrongly reported NBP of CF_4 was in the range sought and this led Midgley to seek organic fluorides as candidates.

It was thus established that fluorinated organic compounds would have all the desired properties for refrigerants.

The process of synthesizing these compounds was discovered by chemists under the guidance of Dr. F. Swarts of university of Ghent, Belgium in 1880. He found that the addition of small amount of pentavalent antimony to antimony trifluoride caused rapid fluorination. This is known as Swarts reaction for making organic fluoride products. Midgley synthesized R21 by Swarts reaction using antimony fluoride within three days of starting the work. A guinea pig was exposed to R21. The animal was completely unaffected, but it died when the test was repeated with another sample of R21. Midgley had procured five one-ounce bottles of antimony trifluoride, which was its total stock available in USA. An examination of the bottles revealed that all the bottles except the one used in the first experiment contained water. Water forms phosgene $COCl_2$ during the reaction of antimony trifluoride with CCl_4. Downing states that had the initial test been carried out using one of the bottles containing water; the discovery would have been delayed by years. Midgley developed many of the chlorofluorocarbons and perfected their manufacturing process. Finally in April 1930 Atlanta meeting of American Chemical Society made the history when these refrigerants were announced. All the parallel sessions of the meetings were adjourned and Midgley made the announcement, inhaled lungful of it and extinguished a lighted candle.

In addition to the original group of CFCs, namely, R21, R12, R11, R113, R114 and R22, the lower normal boiling point CFCs such as R13, R14 and R13b1 (Halon 1301) andazeotropic blends such as R500, 502 and R503 were also made during that period and extended the product line.

The simplest hydrocarbon, methane (CH_4), can be transformed to a total of 15 CFCs by appropriate substitution by various combinations of fluorine and chlorine atoms in place of hydrogen atoms. Similarly, ethane can be transformed to 28 different CFCs or 55 CFCs considering the isomers. The options increase considerably as one progresses on propane, butane, etc.

Apart from public and individual safety in use provided by these refrigerants, the refrigeration industry has benefited in several less obvious but significant ways. CFCs cover a wide range of normal boiling points and vapour pressures. For every specific application, a suitable CFC is available. This gives flexibility to the industry in adapting hardware and systems to specific applications common to all the CFCs. In fact, same compressors can give different capacities with different refrigerants avoiding the manufacture of different compressors for each cooling capacity. The thermodynamic and thermophysical properties of several CFCs proved to be especially well matched to the type of applications most sought after. Further, the inertness of CFCs to various materials gave a design flexibility, which is now taken for granted.

2.3 REFRIGERANTS CLASSIFICATION AND NOMENCLATURE

The basis of modern refrigeration is the ability of liquids to absorb enormous quantities of heat as they boil and evaporate. These liquids are called refrigerants. Refrigerants are working substances in any refrigeration system. These are also classified as primary and secondary refrigerants. Primary refrigerants produce refrigerating effect by absorbing heat from the surrounding at low temperature and pressure. Examples are ammonia, air, water, chlorofluorocarbons, hydrocarbons, etc. Secondary refrigerants are substance that is cooled by passing it over a cooling coil of a refrigeration system and transported it to the desired location for producing refrigerating effect. Examples are air, water, glycols, alcohols, etc.

The main group of compounds is halocarbons derived from hydrocarbons (mostly CH_4 or C_2H_6) by replacing hydrogen atoms with chlorine, fluorine and bromine. The term fully halogenated means all the hydrogen atoms are replaced with chlorine, fluorine or bromine atoms. However, following are the classifications of the organic refrigerants.

CFC (chlorofluorocarbon) are fully halogenated and contain chlorine atoms.

HCFC (hydrochlorofluorocarbon) are not fully halogenated. These contain a few hydrogen atoms and chlorine atoms.

HFC (hydrofluorocarbons) do not contain any chlorine atoms but have a few hydrogen atoms.

FC (fluorocarbons) do not contain any hydrogen or chlorine atoms.

HC (hydrocarbons) do not contain any halogen atoms.

Halons are fully halogenated CFC with bromine atoms.

Besides, there are inorganic compounds which are also as important as the organic compounds. Examples are ammonia, water, air, etc.

2.3.1 Nomenclature

Midgley et al., inventors of CFCs, developed a refrigerant number system to abbreviate their long names. This was later adopted and modified by ASHRAE. R, the abbreviation for refrigerants, replaces CFC, HCFC, HFC, HC, etc. Symbol R is still used with inorganic compounds, the seven hundred series: azeotropes, the 500 series and the newly named 400 series for non-azeotropic mixtures. Apart from this, new notations, namely, MP and HP have been introduced for medium pressure and high pressure respectively. For example, MP-717 is medium pressure inorganic refrigerant with molecular weight of 17, that is, ammonia. However, different numbers have been assigned to different refrigerants basing upon the following rules.

All the refrigerants are expressed by a common notation, R_{abc}

where, a is 'number of carbon atoms – 1'

b is 'number of hydrogen atoms + 1'

c is 'number of fluorine atoms'

Chlorine atoms satisfy the remaining valency of the carbon atom in saturated hydrocarbons (C_nH_{2n+2}) or unsaturated hydrocarbons (C_nH_{2n}). Hence, the number of chlorine atoms will be $2a - b - c + 5$ in saturated hydrocarbons and $2a - b - c + 3$ for unsaturated hydrocarbons. Most of the CFCs are either methane (CH_4) derivatives or ethane (C_2H_6) derivatives. For methane derivatives, where $C = 1$, a is 0 because C – 1 = 0. So the number of the corresponding refrigerant will be of two digits. Examples are R12, R22, etc. For R22, $a = 0$. So $C = 1$, $b = 2$, So H = 2 – 1 = 1, F = 2. Hence, Cl = 1. So R22 is $CHClF_2$. In reverse way if chemical formula is given then refrigerant nomenclature can be determined.

Similarly R134 is C – 1 = 1, So C = 2; H + 1 = 3, So H = 2; F = 4; Cl = 2 × 1 – 3 – 4 + 5 = 0

Thus the chemical formula will be $C_2H_2F_4$.

Isomers (having same chemical formula, but different molecular structure) are nominated with the same designation followed by *a*, *b*, etc. The commonly used refrigerant R134a is an isomer of R134.

Unsaturated hydrocarbons are distinguished from saturated ones by putting '1' before *a*. For example Dichloroethylene ($C_2H_2Cl_2$) or Dielene is designated as follows.

a is 2 – 1 = 1, *b* is 2 + 1 = 3, *c* is 0. Thus, designation, after putting 1 before *a*, will be R1130.

Brominated refrigerants are designated by putting an additional *Bn* at the end of *abc*, where *n* denotes the number of chlorine atoms replaced by bromine atoms. For example, R114B2 is $CBrF_2CBr_2F_2$ (Dibromo Tetrafluoro Ethane).

In case of common inorganic compounds, numbers have been given by adding their molecular weight to 700. Thus, NH_3, H_2O, CO_2 are designated as R717, R718, R744, respectively. For some hydrocarbons this numbering system is not valid. So an arbitrary number is given to those compounds. For example, n-butane is R600, methane is R50, propane is R290, etc.

Besides oxygen compounds have been assigned 610 series. R610 is ethyl ether ($C_2H_5OC_2H_5$), R611 is methyl formate ($HCOOCH_3$). Also nitrogen compounds are assigned 630 series. R630 is methylamine (CH_3NH_2), ethylamine ($C_2H_5NH_2$).

Azeotropic mixtures are nominated as R500 series. They are the mixtures of two refrigerants and due to mixing some of the undesirable properties of individual refrigerants can be eliminated. This behaves as a single refrigerant at a particular proportion of mixing. (discussed later in this chapter). Some of the azeotropes are given below.

R500, a mixture of 73.8% of R12 and 26.2% of R152a (by weight)

R502, a mixture of 48.8% of R22 and 51.2% of R115 (by weight)

R507, a mixture of 50% of R125 and 50% of 143a (by weight)

Zeotropic or non-azeotropic mixtures are nominated as R400 series. They are the mixtures of two or more refrigerants. The mixture has wider applications than the individual components due to improvement of the mixing properties over individual refrigerant properties. (Discussed later in this chapter.) Some of the non-azeotropes are given below:

R401A is a mixture of R22 (53%), R152a (13%), R124 (34%)

R404A is a mixture of R125 (44%), R143a (52%), R134a (4%)

R406A is a mixture of R22 (55%), R142b (41%), Isobutane (4%)

2.4 DESIRABLE PROPERTIES OF REFRIGERANTS

The desirable properties of refrigerants have been outlined by many researchers and scientists in early days. Now ozone friendliness and low global warming potential have to be added to this list in view of the crisis faced by mankind. As a rule, the selection of refrigerant is a compromise between conflicting desirable properties, for example, evaporator pressure should be high and at the same time condenser pressure should be low. Both of which are difficult to satisfy. Low viscosity and low surface tension are desirable but these make it difficult to provide drop-wise condensation that improves the heat transfer during condensation. Easy availability is also one of the criteria. In general, the desirable properties of a fluid to be used as refrigerant are discussed here.

Above atmospheric evaporator pressure: Refrigerant may leak from the refrigeration system if the evaporator gauge pressure is positive, that is, above atmospheric. Hence, the systems are made leak-tight. If the evaporator pressure is below atmospheric (negative gauge pressure) then atmospheric air and water vapour may leak into the system. If the water vapour is dissolved by refrigerant (like ammonia) then it does not pose a problem, otherwise it will remain as free water. This free water may freeze in the narrow orifice of the expansion valve where temperature may be less than 0°C. This will choke the flow in the refrigeration system. The air that leaks into the system occupies precious heat transfer area in the condenser. If the area decreases, then to reject the same heat, the condensation temperature must increase, which will increase the corresponding saturation pressure. Further, the presence of air will decrease the partial pressure of refrigerant hence; refrigerant will condense at higher total pressure. This decreases the volumetric efficiency. From above discussion it is clear that there is more risk in evaporator pressure being lower than the atmospheric pressure. Hence evaporator pressure should be maintained more than the atmospheric pressure so that atmospheric air and water vapour cannot enter the system and leak detection will be easier. A fluid having low normal boiling point will satisfy this condition.

Moderately low condensing pressure: This is required so that lightweight equipment and tubing can be used. Carbon dioxide has the highest condenser pressure among all the refrigerants. Power requirement is high if the condenser pressure is high. Refrigerant with low normal boiling points have higher condenser pressure and high vapour density, for example, NH_3 and R22 have pressures of 15.54 and 15.331 bar respectively at condenser temperature of 40°C, CO_2 has critical temperature and pressure of 31°C and 73.8 bar, respectively. Power requirement is also usually high if the condenser pressure is high. Reciprocating compressor is usually used if the condenser pressure is high, for example, R12, R22, R40, SO_2, CO_2, and NH_3. Refrigerants with low condenser pressure have high normal boiling points and low vapour densities. The volume flow rates are high for these refrigerants; hence centrifugal compressors are used for these. Examples are R11, R113 and R114. Rotary compressors are used for refrigerants with intermediate condensers pressures.

High critical point: As the condenser pressure approaches the critical pressure, the zone of condensation decreases and the heat rejection occurs mostly in the superheated vapour region. Since the zone for isothermal heat transfer decreases the irreversibility increases. This involves large temperature difference between refrigerant and the surroundings leading to large irreversibility. Power requirement also goes up since the area of superheat horn is large. The flash gas losses increase since the quality at the end of throttling process increases. The COP is small if the refrigeration cycle is operated near critical temperature. Hence, refrigerant should be such that the critical temperature is large compared to normal condenser temperature.

Specific heats: The specific heat of liquid (C_{PL}) refrigerant should be low. Saturated liquid line on T-s plane for low specific heat will be nearly vertical, i.e., having a small slope, as shown in Fig. 2.1. From the figure it is evident that the benefits of low C_{PL} are: (a) throttling loss will be decreased, (b) degree of subcooling is more for the same amount of heat rejection ($Q = m\ C_{PL}\ dT$). This increases the refrigerating effect.

Specific heat of vapour (C_{PV}) should be high. Saturated vapour line on T-s plane, as shown in Fig. 2.1 (a), will have a greater slope, i.e., nearly vertical. Higher the value of C_{PV}, flatter will be the constant pressure line. It is shown in Fig. 2.1 (b) that constant pressure line *a-b* has lower C_{PV} value than that of line *a-c*. From figure it is evident that the benefits of higher C_{PV} value are: (a) discharge temperature of the vapour after compression is not high which is evident from relation $Q = m\ C_{PV} dT$. This reduces the thermal stress over the compressor components and lubricating oil property is not affected. Requirement of condenser area decreases, (b) also work done is reduced as shown in Fig. 2.1 (b) as superheat horn area decreases. (superheat horn area *a-d-b* > *a-d-c*). Hence, refrigerant having saturation line (dome) of Figure 2.1 (b) is preferred over that shown in Fig. 2.1 (a).

$$C_p = T\left(\partial S/\partial T\right)_P \quad [2.1]$$

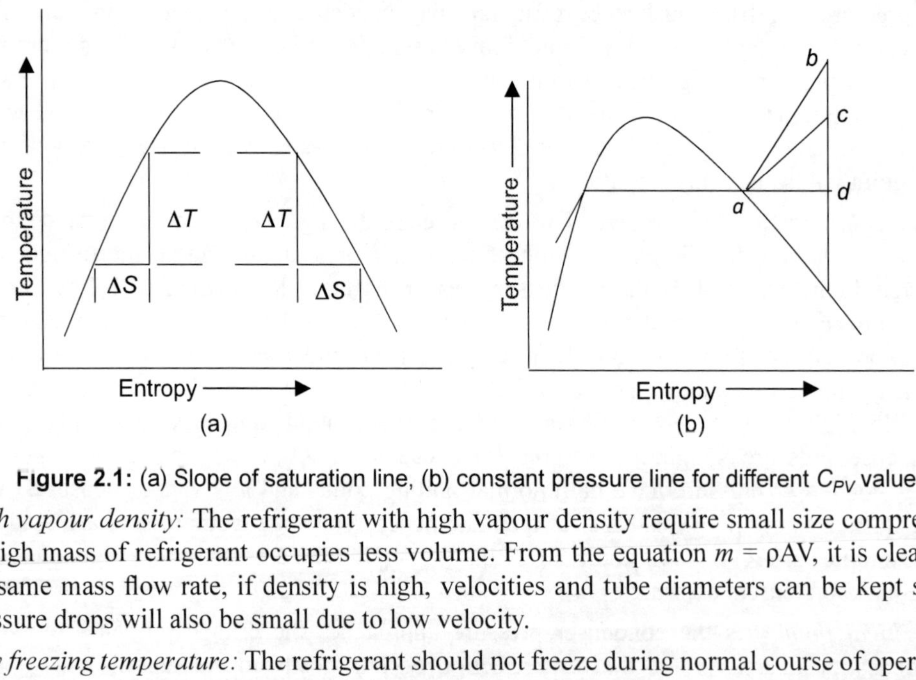

Figure 2.1: (a) Slope of saturation line, (b) constant pressure line for different C_{PV} value

High vapour density: The refrigerant with high vapour density require small size compressors as high mass of refrigerant occupies less volume. From the equation $m = \rho AV$, it is clear that for same mass flow rate, if density is high, velocities and tube diameters can be kept small. Pressure drops will also be small due to low velocity.

Low freezing temperature: The refrigerant should not freeze during normal course of operation; hence, its freezing temperature should be small. Water, for example, cannot be used below 0°C.

High latent heat: High latent heat results in lower mass flow rate of refrigerant. If the vapour density is also high then it will require smaller compressor. Small mass flow rate, sometimes becomes a disadvantage since it is difficult to control small flow rates. For example, NH_3 cannot be used in small refrigeration systems since its mass flow rate is very small of the order of 0.003 kg/s per TR).

High heat transfer characteristics (heat transfer coefficient and viscosity): The heat transfer coefficient depends upon density, specific heat, thermal conductivity, viscosity, surface tension and latent heat or non-dimensional numbers like Reynolds number, Prandtl number and few other parameters for boiling and condensation. Low viscosity and high thermal conductivity for both liquid and vapour are desirable. All of them lead to higher heat transfer coefficients. Higher heat transfer coefficient leads to requirement of smaller area and low viscosity results in lower pressure drop. This makes the system more economical and compact.

Inertness and stability: It should not react with materials used in the refrigeration, i.e., tubes, gaskets, compressor parts, etc. It should be non-corrosive and should not react with water. It should be stable, that is, its chemical composition should not change during its use. Ammonia reacts with copper and cuprous alloys while CH_3Cl reacts with aluminium. Most of refrigerants form acids and bases in presence of water and some of them may react in presence of lubricating oil. CH_3Cl and CCl_2F_2 can form HCl that can dissolve copper from the copper tubes and deposit it on piston.

High dielectric strength of vapours: This is required for their use in hermetically sealed compressors where the refrigerant vapour comes into direct contact with the motor winding (as discussed in Section 5.2.1) and may cause short circuit.

Satisfactory oil solubility with lubricating oil: The lubricating oils are either mineral oils or synthetic oils. Poly butyl silicate is used with R22, alkyl benzenes or poly alkyl glycols are used with R134a.

The lubricating oils may get dissolved in refrigerant or some refrigerant may get dissolved in the oil. This will change properties of refrigerant or the viscosity of oil. Some oil is physically picked up by high velocity refrigerant vapour. The results in a mechanical mixture that may accumulate in some component if it cannot be physically carried by the refrigerant. The adverse effects include reduction in heat transfer coefficient, change in p-T characteristics of refrigerant and adverse effects on lubrication.

The refrigerants are immiscible, miscible or partially miscible with the mineral oils. It is partially a miscible variety which leads to maximum problems.

Oil is completely miscible in refrigerants like in R11, R12, R2l, R113 and R500. It has to be mechanically carried by the refrigerant. The high velocities and the burden of carrying lead to larger pressure drop and more compressor power. If the refrigerant velocity is not sufficient then it cannot carry all the oil back to compressor and there is a possibility that it may accumulate in evaporator. This phenomenon is called oil logging of the evaporator. This must be prevented.

NH_3, SO_2, CO_2, R13, R14 and Rl34a are immiscible with mineral oil. Hence, the oil can be mechanically separated in oil separators right after the compressor. Then it comes back to the compressor.

The refrigerants which are partially miscible pose maximum problem. R22 and R114 are partially miscible in lubricating oil. Above certain temperature depending upon the percentage of lubricating oil, the lubricating oil is completely miscible in refrigerant and partially miscible below this temperature. To avoid these problems either efficient oil separator must be used or synthetic oil that is completely miscible with refrigerant must be used.

Frothing of hermetic compressor: In hermetic compressor the refrigerant comes in direct contact with the lubricant. The refrigerant being a polar substance gets adsorbed on the surface of the lubricating oil. The adsorbed liquid may suddenly boil under adverse conditions and its vapour may carry large quantity of lubricating oil with it as froth. The froth fills the compressor cylinders with lubricating oil depriving its crankcase thereby increasing the wear and tear. This is usually avoided by providing crankcase heaters, which may be part of the condenser coil also.

Low water solubility: All refrigerants form acids or bases in presence of water. These cause corrosion and are harmful to valves, seals and metallic parts. This may also damage the insulation of winding of motors of hermetic compressors.

CO_2, SO_2 and NH_3 dissolve in water, NH_3 particularly dissolves it in all proportions whereas water is practically insoluble in R12, R113 and R114.

If the water is in excess of what can be dissolved, the remaining water remains as free water, which freezes at temperature below 0°C and may choke the flow particularly if it happens in the narrow orifice of the expansion valve. The water expands upon freezing hence if it is in large quantity, it may burst the tubes.

The system should be completely dehydrated before charging refrigerant into the system. The lubricating oil should also be free of moisture. Subsequently, running a system with positive pressure would prevent entry of moisture and moist air. Permanent dryers should be installed in systems. These dryers usually contain silica gel, or anhydrous calcium oxide.

Material compatibility: The refrigerant should not react with the materials commonly used for refrigeration piping, like copper, aluminium, and steel. It should not chemically react with other type of materials used in refrigeration circuit such as seat of compressor valves, hose material, plastics, brazing materials, etc.

Nontoxic: It should be non-poisonous to human beings and foodstuff. The toxicity depends upon the duration of exposure and % of vapours in air. CFCs are generally safe. NH_3 gives an unpalatable flavour to food. SO_2 kills flowers and is a dying agent for textiles.

Non-irritability: It should not irritate the skin, eyes, nose and lungs. NH irritates all mucous membranes. SO_2 also irritates the lungs.

Non-inflammability: It should not support combustion in air or lubricating oil. CFCs, CO_2, SO_2 are non-flammable. Methane, ethane, propane, butane and other hydrocarbons are highly flammable. This is rather unfortunate since properties of propane are similar to NH_3. All chlorofluorocarbons with relatively large number of hydrogen atoms are flammable. NH_3 with 16-25% air makes an explosive mixture.

Easy leak detection: NH_3 and SO_2 can be detected by their characteristic smell. NH_3 has a pleasant odour in small concentrations, which becomes irritating and pungent in strong concentration. SO_2 has an irritating smell that causes sneezing. Strong smell is an advantage in leak detection. Sometimes strong smelling chemical like acrolein may be added to it for detection purpose. Smell just gives the idea of leakage; it does not indicate the location of leakage. NH_3 leakage can be detected by smell however other means are required to find the exact location. NH_3 leaks in water-cooled condenser are very difficult to locate since NH_3 is soluble in water. Leaks may be detected by soap solution. Soap solution cannot detect small leaks.

CFCs can be detected by halide torch. It consists of an alcohol lamp that gives blue flame. The blue flame turns green in presence of CFCs impacting upon copper elements of the burner. This test is based upon Beilstein test for chlorine. SO_2 can be detected with NH_3-H_2O solution as it makes white fumes of ammonia sulphide. Similarly, ammonia can be detected by burning sulphur that will form white fumes of ammonia sulphide. A stick wetted with strong HCI yields white fumes in presence of ammonia. It can also be detected by blue litmus paper.

Electronic detectors are used for detection of Freons based upon the variation of current between two platinum electrodes. All CFCs except R14 can be detected by this method. The presence of alcohol and CO interferes with the detection test.

Rate of leakage is proportional to $1/\sqrt{M}$, pressure ratio, diffusion coefficient, density, kinematic viscosity and capillarity. Leakage of refrigerants may result in accidents, low evaporator pressure and should be detected early and corrected.

Ecofriendly: The two main adverse effects of refrigerants (specifically CFCs and HCFCs) on the environment which are of serious concern are global warming and ozone depletion.

Ozone depletion: All chlorine and bromine containing refrigerants have a high ozone depletion potential. The CFCs are very stable and inert compounds. Once they leak into the atmosphere they stay in the atmosphere for years and decades. These are not soluble in water. Hence, these are not washed down by the rain. These are not wasted by any natural cycle in nature and these are also not biodegradable. Hence, once they leak, they stay in the atmosphere for long. Then they rise upwards very slowly due to atmospheric turbulence and reach stratosphere and ozonosphere. In this region very energetic ultraviolet radiation of small wavelength and high energy from sun is present.

The energy can be expressed from quantum mechanics, $\varepsilon = h\nu$; where ε = energy, h = a constant and ν = frequency of radiation and ($\nu = c/\lambda$, where c = light velocity, and λ = wavelength).

This radiation breaks down the bromine and chlorine bonds and produces a radical and a chlorine atom as follows.

In ozonosphere the following reaction takes place continuously to form ozone layer around the earth so that ultraviolet radiations can be captured.

$O_2 + h\nu \rightarrow O + O$; The oxygen atom combines with oxygen molecule ($O + O_2$) to give ozone (O_3).

But before the reaction occurs to form O_3 following reaction occurs.

$Cl^- + O \rightarrow ClO$ and $ClO + O \rightarrow Cl + O_2$

In this process it destroys two molecules of O_3 to form one chlorine atom.

The reaction may occur in other way also.

$CCl_3F + h\nu \rightarrow CCl_2F^+ + Cl^-$

$Cl^- + O_3 \rightarrow ClO + O_2 ClO + O \rightarrow Cl + O_2$

In this process chain reaction, one ozone atom and one oxygen atom have been destroyed and the culprit chlorine atom is still left behind to destroy more ozone atoms by chain reaction. It is estimated that one chlorine atom can destroy 1,00,000 ozone molecules during its one to two years of lifetime in atmosphere.

Similar reactions occur with NO and Br.

Thus this ability of a chemical to destroy ozone depends upon the halogen type (Cl or Br), the number of halogen atoms it releases, and their resident time in atmosphere.

ODP is a relative index indicating the extent to which a chemical may cause ozone depletion. The reference level of 1 is in most cases the potential of CFC 11 to cause ozone depletion.

The computer models are used to predict the present atmosphere and future atmosphere with historical emissions of CFCs.

Global warming: The other alarming issue is the global warming or the greenhouse effect that refrigerants may cause and lead to global climate changes. Short waves (high frequency) radiations, mostly light waves, from the sun hit the surface of the earth and the seas. These surfaces absorb these radiations and give out long wave radiations (mostly Infrared) which is trapped by a variety of gases which act like an insulating blanket. (The gases have inherent property that they are transparent to certain wavelength of radiation and opaque to another. Thus, these gases allow shorter wavelength of radiations, that is, of light and block the longer wavelength of radiation, that is of heat, give out by the earth's surface.) There is thus a tendency to increase the earth's temperature causing the global warming effect. However, natural concentration of CO_2 and water molecules in the atmosphere are essential without which the average temperature of earth would be as low as –18°C. Hence, the objective is to keep the global warming effect in balance rather than decreasing or eliminating it. CO_2 and CFC molecules are transparent to the solar radiation, which is in the short wavelength range. However, these are opaque to long wavelength radiation, which is emitted by the ground surfaces at temperatures 300 K. These gases block this radiation thereby trapping it in the earth's atmosphere. The main concern in this regard is about increase in the atmosphere of CO to increases use of fossil fuels and deforestation. Increase in concentration of CO_2, CFCs and few other gases would inevitably lead to global warming. It is estimated that the atmospheric temperature of earth may rise by 3 K over the next 100 years.

Global warming potential of a gas is measured from two factors. One is its atmospheric lifetime and the other one is its infrared absorption capability. This atmospheric lifetime for GWP calculation is based upon, how much of a gas remains in atmosphere after 0 year, or 100 years or 500 years. For example, 41% of CO_2 may remain in atmosphere after 100 years whereas by that time all the CFCs might have washed out. Thus, GWP of different gases are mentioned in three ways, i.e., on the basis of 0 year, 100 years or 500 years. However like ODP, GWP is also a relative index indicating the extent to which a chemical may cause global warming. The reference gas considered is CO_2 whose GWP is 1.

Low cost: It is immaterial sometimes since the quantity of the refrigerant required is small. If it is low then people may not take sufficient safeguards to prevent leakages, for example,

NH_3. Prices of some of the refrigerants with high ozone depletion potential, are being hiked to discourage their use.

Table 2.2 Refrigerant properties

Refrigerants	Molecular weight (gm/mol)	$P_{Critical}$ (bar)	$T_{Critical}$ (°C)	NBP (°C)	Atmospheric lifetime (years)	ODP	GWP (100 years)
R11(CCl_3F)	137.38	44.0	198.0	23.7	45	1.0	3500
R12(CCl_2F_2)	120.91	41.2	112.0	–29.8	100	1.0	7300
R13($CClF_3$)	104.46	38.7	28.8	–81.4	640	1.0	8000
R22($CHClF_2$)	86.47	49.8	96.1	–40.8	11.8	0.055	1500
R114($C_2Cl_2F_4$)	170.97	32.6	145.7	3.6	300	0.8	6800
R115(C_2ClF_5)	154.47	31.4	80.0	–39.0	1700	0.6	6900
R134a($C_2H_2F_4$)	102.03	40.6	101.1	–26.5	14.6	1.5×10^{-5}	1200
R290(C_3H_8)	44.1	42.5	96.7	–42.0	—	0.0	3.0
R600a(C_4H_{10})	58.12	36.5	135.0	–11.8	—	0.0	3.0
R717(NH_3)	17.03	113.4	132.3	–33.3	—	0.0	0.0
R744(CO_2)	44.01	73.8	30.95	–78.5	—	0.0	1.0

2.5 ANALYSIS OF REFRIGERANT PROPERTIES IN RESPECT OF THEIR MOLECULAR WEIGHTS

Scientists have pointed out that from thermodynamic point of view, critical temperature and vapour specific heat are the key factors in determining the performance of theoretical vapour compression cycle. It was observed that ratio of normal boiling point to the critical temperature; (Tnb/Tr) is between 0.6 to 0.7 for most fluids.

Higher NBP refrigerants have lower vapour pressure and vapour densities for required cooling temperature, thereby requiring larger compressor displacement/TR. The COP is however better for these. On the other hand, low NBP refrigerants operate at higher pressures and have higher vapour densities, thereby giving higher volumic refrigeration effect. Their critical temperatures are however low leading to lower COP due to excessive compressor superheats and throttling losses.

The relative performance of refrigerants is directly related to its molecular weight also. Two laws of physical chemistry, namely, Avogadro' s law and Troutons rule explain this relation for refrigerants with a similar vapour pressure. Avogadro's law states that number of molecules in a given volume of gas depends only on the gas pressure and temperature. Hence, a given compressor, running at a fixed revolutions per minute, will pump the same number of molecules regardless of molecular weight as long as the pressure and temperature are similar. Trouton's rule states that for a group of similar substances ($M\ h_{fg}\ /T_{NBP}$) is constant, that is, the latent heat on molar basis is same for a group of similar groups of substances. Atwood suggests that taken together, refrigerants with similar vapour pressures should require similar compressor displacements for equivalent molar flow rate and would result in equivalent refrigeration effect on molar basis.

The higher molecular weight refrigerants would have higher mass flow rates. lower latent heats and lower areas of superheat horn and large throttling loss. The 1ower molecular weight refrigerants would have lower mass flow rates, higher latent heat. higher specific heats and higher areas of superheat horn. In hermetic compressors, which are cooled by suction gas, the lower mass flow rate of low molecular weight refrigerant is balanced by their higher specific heat to give equivalent cooling. Experience has shown that CFCs with molecular weights around 120 ± 40 are well suited for hermetic compressors. The optimum molecular weight will depend upon the application.

Normal boiling points of refrigerants: Amongst the hydrocarbons, NBP increases with increase in number of carbon atoms. In one carbon series, that is, methane derivatives, it is observed that NBP increases from –161°C for methane as more and more hydrogen atoms are replaced by chlorine alone and it is 77°C for CCl_4.

Normal boiling point is found to decrease as the chlorine atoms are replaced by fluorine J atoms for fixed carbon atoms, NBP being –128°C for CF_4.

The same pattern exists for ethane derivatives except isomeric forms, which have lower NBP than normal configuration.

Flammability and toxicity: All methane derivatives with up to two hydrogen atoms and ethane derivatives up to three hydrogen atoms are flammable. Toxicity is not amenable to a systematization scheme since it is not a physical property but pertains to interaction with living organisms. Further confusion arises because of variable toxic effects such as "acute effect from a single but massive exposure or chronic effects from multiple low level exposures. Instead of considering TLV, toxicity is classified herein as low, moderate and high.

Hydrocarbons have low toxicity. Fluorinated compounds with no chlorine atoms have low toxicity with the exception of R161. Toxicity increases from moderate to high as the number of chlorine atoms in the CFC increases. HCFCs have lower toxicity.

Lifetimes: The ozone depletion potential and GWP have been mentioned as indicators of environmental friendliness. Atmospheric lifetime, on the other hand, is important for both ODP and GWP to be effective. Fully halogenated CFCs have very long lifetime, which increases with increase in the number of fluorine atoms, reflecting greater stability of carbon-fluorine bond. Carbon tetra fluoride and R16 have atmospheric lifetime exceeding 500 years. R10, R11, R12 and R13 have atmospheric lifetimes of 50, 65, 120 and 400 years, respectively. Similarly, R113, R114, R115 have atmospheric lifetime of 90, 180 and 380 years respectively.

The hydrogen containing HCFCs have lifetimes which are around 1/3 of fully halogenated CFCs due to reaction of hydrogen with compounds present in troposphere.

McLinden and Didion conclude that amongst methane derivatives R22 and R23 and amongst ethane derivatives R132, R133, R134, R123, R124 and R125 and their isomers satisfy the various criteria to serve as alternative refrigerants.

A Comparative study of CFC (R22) and Natural Refrigerant (NH_3)

In the light of Montreal Protocol, CFCs, HCFCs, etc., are to be phased out completely by 2030 as they are having high ODP and GWP. Again Kyoto Protocol emphasized on refrigerants having high GWP to be phased out gradually. Hence, it is high time to search for alternative refrigerants which are environment-friendly. The alternatives, that are being thought of, are natural refrigerants and refrigerant mixtures like azeotropes, zeotropes, etc. (discussed in the next section). Natural refrigerants are hydrocarbons (like isobutane, propane or blend of these two), ammonia, CO_2, water, air, etc. Here a comparative study has been done between R22 (HCFC) and NH_3 (Natural) refrigerants.

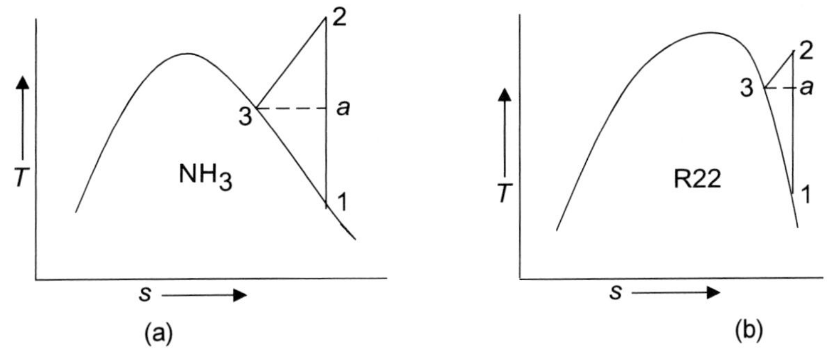

Figure 2.2: Saturation curve, peak temperature and superheat horn for (a) NH_3 and (b) R22

Merits of NH_3 as a refrigerant

Ammonia is a natural refrigerant. The thermodynamic properties are as follows.

Low NBP (–33°C), low freezing point (–78°C), high critical point, high latent heat of vaporisation, high C_P of vapour (2.2 kJ/kg-K), high thermal conductivity, low viscosity, and high density of vapour.

Leak detection is very easy due to its pungent smell. Easily available and low cost.

Demerits of NH_3 in comparison to R22 are as follows.

General public perception is that ammonia is highly flammable and toxic. But its flammability or in its state of application for refrigeration and air conditioning purpose is not significant. Any slightest leakage can be detected easily due to its pungent smell. So precautions can be taken much before it creates any hazard. In this way its toxicity is not significant. Discharge temperature is high because the index of compression is higher (1.32) in comparison to 1.16 of R22. This is evident from Fig. 2.2. Depending upon the slope of the vapour saturation line, which depends upon the specific heat, the compressor discharge temperature and the superheat horn are high in comparison to R22. Incompatible with most commonly used material, copper. Though it has low NBP, ammonia is not preferred to be used for temperature below –20°C. Because as the evaporator temperature decreases, compression ratio increases, and this causes the compressor discharge temperature very high and this high temperature affects the lubricating oil properties.

2.6 SUMMARY OF DESIRABLE PROPERTIES OF A FLUID TO BE USED AS A REFRIGERANT

Thermodynamic Properties

1. The fluid should have low normal boiling point, i.e., the fluid must evaporate at a lower temperature at atmospheric pressure so that evaporator pressure can be maintained at a pressure above atmosphere and atmospheric air will not leak into the refrigeration system.

2. The fluid should condense at moderately low pressure so that the condenser can be made of lightweight equipment and piping.
3. The fluid should have higher critical point than the condensing pressure.
4. The fluid should have higher vapour specific heat and lower liquid specific heat.
5. It should high latent heat of vaporisation.
6. It should have high thermal conductivity for both the liquid and vapour.
7. It should have low freezing temperature.

Physical properties

1. It should have high dielectric strength.
2. It should have high vapour density or low vapour specific volume.
3. It should have low viscosity for both the liquid and vapour.
4. It should have satisfactory oil solubility.
5. It should have lower water solubility.
6. It should be non-flammable.

Chemical Properties

1. It should have material compatibility. Chemically, it should not react with common engineering materials like copper, steel, aluminium, etc. And it should be stable.
2. It should be non-toxic.
3. It should be non-irritable and any leakage can be detected easily.

Besides these properties It should be environment-friendly, i.e., its global warming potential and ozone depletion potential should be very low. It should be easily available at low cost.

2.7 REFRIGERANT MIXTURES

The desirable properties of a refrigerant have been discussed earlier. It is not possible to find a fluid in nature which will meet all the desirable properties so that it can be used as a suitable refrigerant. It is the science that says some properties of a pure fluid can be altered by mixing it with another fluid. Thus, mixing two or three fluids in certain proportion may make the mixture suitable to be used for a particular application where the single component may be found unsuitable. There are three types of refrigerant mixtures: (a) azeotropic mixture, (b) near-azeotropic mixture (NARM), (c) zeotropic mixture. Before discussing the refrigerant mixtures, let us have some idea regarding solutions.

2.7.1 Ideal Solutions

An ideal solution is characterised by the uniformity of intermolecular forces. If A and B are the constituents, the cohesive forces between A-A pair, A-B pair and B-B pair are assumed to be same. Hence, change in volume and enthalpy are zero after mixing. It is only the algebraic sum of the volume and enthalpy of the individual component.

$$h_{\text{Mix}} = h_A + h_B \quad [2.2]$$

and

$$v_{\text{Mix}} = v_A + v_B \quad [2.3]$$

Raoult's law: The partial pressure of a constituent is, in a sense, a measure of its tendency to escape from the liquid surface. In an ideal solution, the partial vapour pressure of A is equal to the mole fraction of A in the solution multiplied by the vapour pressure of the pure A liquid at solution temperature.

$$P_A = x_A P_A^\circ \qquad [2.4]$$

where, P_A° is the vapour pressure of liquid A at solution temperature and x_A is the mole fraction of component A. This is Raoult's law.

Similarly, $$P_B = (1 - x_A) P_B^\circ \qquad [2.5]$$

Hence, the total pressure of the solution

$$P = P_A + P_B \qquad [2.6]$$

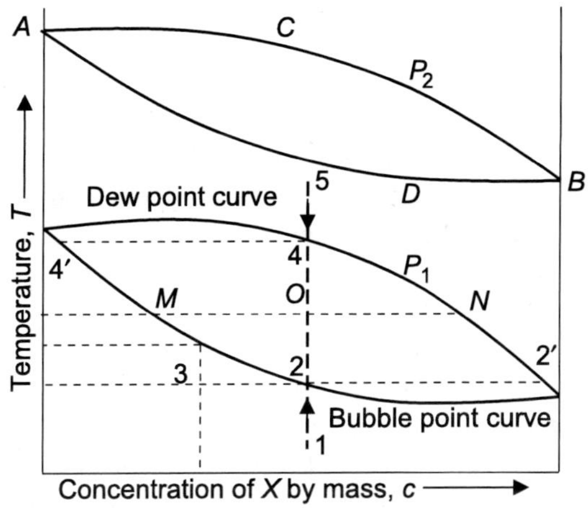

Figure 2.3: Temperature-concentration diagram of an ideal mixture

When a solution of X (solute) and Y (solvent), in a sub-cooled state (1) and at a pressure (P_1), gets heated (Fig. 2.3), boiling starts at certain temperature (T_2) for concentration, c_1. When the concentration changes from c_1 to c_3, boiling temperature also shifts to T_3. Thus, unlike a pure fluid, a solution has different boiling points for different concentration. The locus of these boiling points is called the 'bubble point curve'. Similarly, when the vapour mixture condenses, condensation starts at different temperatures, depending upon the concentration of the vapour mixture. In this figure condensation of vapour at state point 5 (T_5 and c_1) starts condensing at 4. The locus of these condensing points is called 'dew point curve'. This indicates a mixture or a solution boils and condenses over a range of temperatures. M and N are the saturation points for the pure fluids X and Y, respectively. These bubble point (A-D-B) and dew point (A-C-B) curves are different for different pressures (P_2) as shown in above figure.

2.7.2 Non-ideal Solution

Raoult's law, to some extent, is valid for dilute solutions where numbers of solvent molecules are so high in comparison to numbers of solute molecules that there is a negligible effect of solute molecules on solvent molecules on escaping tendency.

When concentration increases beyond a certain limit, there is a repulsive force between unlike molecules which will increase the escaping tendency of the molecules. Hence, their partial vapour pressure will be more than as predicted by Raoult's law. $P_A > x_A P_A°$. Similarly, for component B. So, the total solution pressure is also more than that predicted by Raoult's law. So, the total pressure $P > P_A + P_B$. Such solutions are called 'positive deviation from Raoult's law'. Such solution absorbs heat when formed. So, this heat of mixing is taken as positive, i.e., the enthalpy of the solution is higher than the summation of the enthalpies of the individual components. If there is attractive force between unlike molecules then escaping tendency will decrease. $P_A < x_A P_A^{\circ}$. So, the total pressure $P < P_A + P_B$. This is called 'negative deviation from Raoult's law'. Similarly, the heat of mixing is taken as negative, i.e., the enthalpy of the solution is lower than the summation of the enthalpies of the individual components. Thus, to summarise the characteristics of non-ideal solutions are: (i) showing deviation from Raoult's law, (ii) marked change in volume upon mixing, and (iii) the mixing may be endothermic or exothermic.

The mixtures considered for refrigeration/heat pump applications may be azeotropic, near azeotropic and zeotropic. Azeotropic mixture is the non-ideal mixture which has a single boiling temperature, that is, lower or higher than the boiling points (BP) of individual components. If the boiling temperature is lower than BP of lower boiling component, it is called 'minimum boiling azeotrope' (Fig. 2.4a). The advantage is that the evaporator pressure can be maintained higher than the atmospheric pressure. Due to depression of boiling point of azeotrope, the condenser pressure also decreases accordingly. This leads to low pressure ratio and hence low vapour discharge temperature, less compressor work and higher COP. Azeotrope with boiling point higher than the bp of the higher boiling component is called 'maximum boiling azeotrope' (Fig. 2.4b). It has the advantage of high latent heat of vaporization resulting in a higher refrigerating effect. Mixtures with close normal boiling points, for which only small differences in liquid to vapour compositions exist in equilibrium, are termed **'Near azeotropes' (NARM). 'Zeotropic mixtures'** boil through a wide range of temperatures. **The temperature difference between the beginning of the condensation/boiling and end of it is called the temperature glide** (Fig. 2.5) and the advantage of temperature glide is obtained. An advantage of using zeotropic refrigerant mixture is the possibility of capacity control by varying the concentration of the individual components. Some of the refrigerant mixtures are: R507A (50% of R125 and 50% of R143a), R404A (44% of R125, 52% of R143a and 4% of 134a), R407C (20% of R32, 40% of R125 and 40% of R134a).

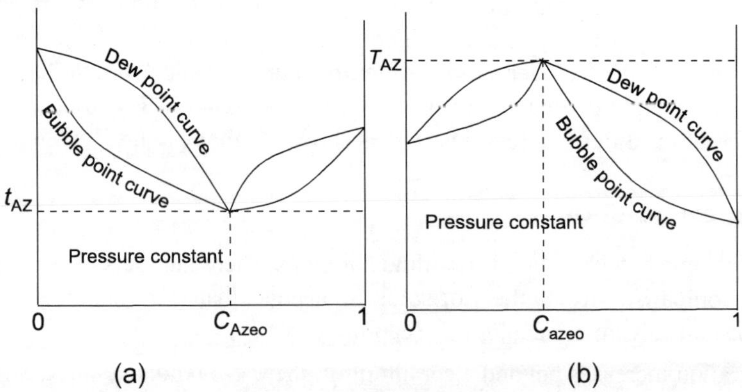

Figure 2.4: (a) Minimum boiling azeotrope (b) Maximum boiling azeotrope

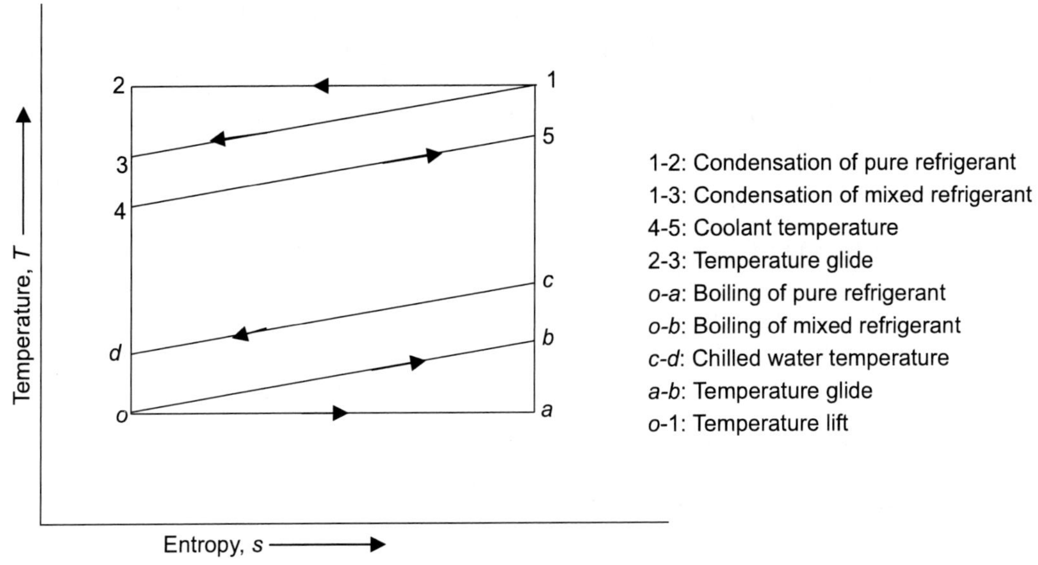

Figure 2.5: Temperature glide and temperature lift in a zeotropic mixture

2.7.3 Vapour Compression System with Zeotropic/Near-Szeotropic Mixture (NARM)

Let us consider the application of these systems to food processing or chemical industries, etc., where the temperatures of both the source and sink vary in space. In such cases, the vapour compression cycle with single component working fluid is not an ideal cycle because the performance of this cycle is limited by reverse Carnot cycle which is not advantageous where the source and sink temperatures vary. The use of zeotropic refrigerant mixtures in vapour compression system has the advantage of boiling or condensation over a range of temperatures, (temperature glide). Hence, for process industries where cooling/heating is carried out over a range of temperature, it gives higher COP. The glide matching with the external fluid reduces the irreversibility during heat transfer resulting in higher COP and lower heat exchanger area. Zeotropes also gives the advantages like low pressure ratio, higher evaporator pressure for low temperature, capacity control by varying the mixture composition only.

It has been studied that the performance of the vapour compression system (VCS) by using near-azeotropic refrigerant mixtures (NARMs) improve when there is correct matching of the temperature glide in the refrigerant and the temperature change in heat transfer fluid. For most of the VCS the heat transfer fluid is air or water whose temperatures do not vary much (<10°C), the performance improvement is not significant. Again, the components of the mixture should have material compatibility and oil solubility. VCS with NARMs are advantageous over single fluid system for medium temperature lift (<70°C) **(the temperature difference between the minimum of the evaporator temperature and the maximum of the condenser temperature)** and small capacity applications. For high temperature lift, suitable NARMs have not been reported. The compressor displacement volume needs to be very large when high refrigerant mass flow rate is required for higher capacity applications. In contrast to the vapour compression systems, the vapour absorption system uses naturally occurring working fluid pair like NH_3-H_2O and H_2O-LiBr. But these working fluid pairs have some disadvantages like toxicity for NH_3-H_2O and limited temperature range for H_2O-LiBr. The gain by using NARMs (temperature glide, volumetric capacity, pressure ratio) increases as the NBP difference of the components becomes higher. This happens with a mixture where one component is a non-volatile absorbent and the other one is refrigerant. Hence, higher temperature glide, higher volumetric capacity and still lower pressure ratio can be achieved with refrigerant absorbent working pair in comparison to NARMs. The pressure of some amount of working fluid rises from an evaporator to condenser in a compressor and for the rest in a solution pump. Hence, compressor size need not be made large. Again, de-superheating of refrigerant vapour of the VCRS is replaced by absorption of vapour, which requires smaller heat exchanger area.

2.8 ALTERNATIVE REFRIGERANTS

In view of the environmental degradation, mainly the ozone layer depletion and global warming, the CFCs and HCFCs are going to be completely phased out by 2030. This is the outcome of Montreal Protocol (1987), Kyoto Protocol (1997) and Kigali Amendment (2016) to phase out HFCs, particularly gases having high GWP.

Hence, the search for alternatives to CFCs and HCHCs and now HFCs are going on by the scientists and researchers. Retrofitting of the new refrigerant with the existing one has become the main challenge. The important factors considered are: (a) the compressor size depending upon the volume flow rate and compression ratio, (b) size of condenser and evaporator depending upon its heat transfer characteristics, (c) size or opening of the expansion device, (d) material compatibility (for tubes as well as gaskets), (e) miscibility with lubricating oil. For example, R134a was tested to have all the desirable properties of a refrigerant and thermodynamically a good replacement for R12. But it took a longer period to get a suitable lubricating oil. Because it is not miscible with the mineral oils (naphthenic and alkyle benzene oils) with which R12 is completely miscible. Then polyol ester or polyalkylene glycol was developed to be used with R134a and R134a got into market.

However, some of the proposed or in use replacement of CFCs and HCFCs are mentioned below.

R12 (CCl_2F_2) is replaced by R134a ($C_2H_2F_4$) whose GWP is 0.27. As per Kigali Amendment this is to be phased out. Hence, the upcoming substitutes are R152a ($C_2H_4F_2$), blends of hydrocarbons (HC290, i.e., propane and HC 600a, i.e., isobutane).

R22 ($CHClF_2$) is popularly used in air conditioning application. This is to be replaced with R410A (zeotropic, more correctly near azeotropic mixture of 50% R32, i.e., CH_2F_2 and 50% of R125, i.e., C_2HF_5). Other alternatives are pure R32, R125, R152a, R143a ($C_2H_3F_3$) and R407c (a tertiary zeotropic blend of R32, R125, and R134a by 20%, 40% and 40%).

R502, an azeotropic mixture of R115 (C_2ClF_5) and R22, mostly used for freezers in supermarket frozen food display cases. This is to be replaced with R507a (an azeotrope of R125, R143a, 50% each), R404a (a tertiary zeotropic blend of R125, R143a and R134a in proportion of 44%, 52% and 4% respectively) and R407c.

Research activities are carried out globally to replace the CFCs, HCFCs and HFCs with refrigerant mixtures (azeotropic, zeotropic or NARM). In the mixture, if there is any component belonging to banned group, the amount is less and hence its environmental impact is very much reduced. Extensive research activities are also being carried out to establish the natural refrigerants like ammonia, hydrocarbons (propane, isobutane), sulphur dioxide, air as the alternative refrigerants.

2.9 SECONDARY REFRIGERANTS

In large establishments like multi-speciality hospitals, malls, multiplexes, etc., conditioned air, directly from the cooling coil, cannot be delivered to the conditioned space. Refrigeration from the cooling coil is carried by fluids like water, brines, glycols and even halocarbons through the ducts to the space where cooling is required. Air is drawn over these ducts to get the chilled air in the space. Thus the fluid that carries the coldness from the refrigerating machine (evaporator coil) to the desired location is called as the secondary refrigerant. The desirable properties of the secondary refrigerants are low freezing point, low viscosity, non-flammability, low vapour pressure, and good stability.

Brines: Brines are formed by dissolving some kind of salts like NaCl, $CaCl_2$, etc., in water. These salts, when dissolved in water, depress the freezing point of water. This freezing point varies with the variation of the concentration of the salt in water. The lowest freezing point temperature that can be attained is called the '*eutectic temperature*' and the corresponding composition called the '*eutectic composition*'. Fig. 2.6 gives the phase diagram and shows the eutectic point of brine solution.

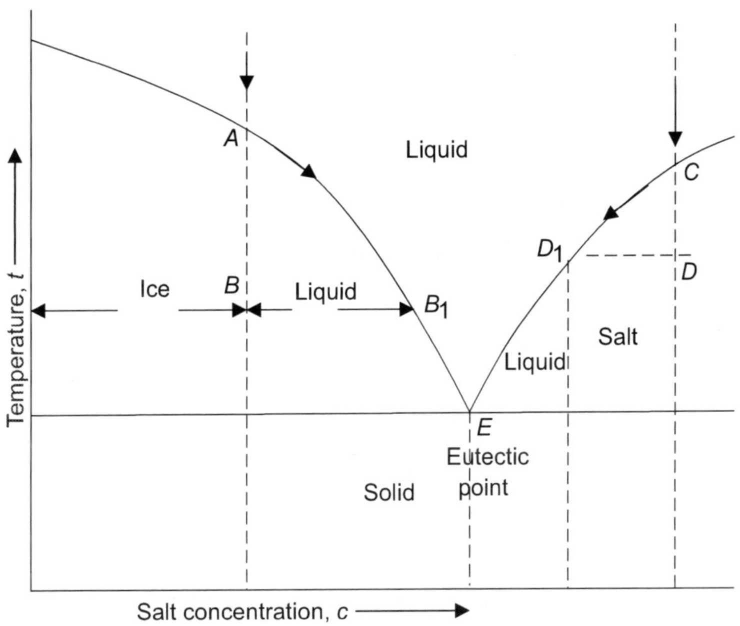

Figure 2.6: Phase diagram of brine solution and eutectic point

If the brine has a salt concentration (*A*) less than the eutectic, then on cooling ice crystals will begin to form at *B* and will separate out of the solution. Therefore, the concentration of salt increases. On further decrease in temperature the salt concentration reaches *D*, the eutectic point. Then the whole solution starts to freeze. Similarly, if the brine has a salt concentration (*E*) more than the eutectic (*D*), then on cooling the salt and solution will be separated until it reaches a salt concentration of *D*. On further cooling the whole solution will freeze. The minimum crystallization temperatures attained with NaCl and $CaCl_2$ are around –15°C and –55°C respectively. In ice plant brine (NaCl) is used to preserve the ice because brine at –15°C preserve ice at 0°C for a longer period.

Ethylene glycol and propylene glycol in solution with water are also used as freezing point depressants. However, all the above secondary refrigerants are used for different applications considering their minimum crystallisation temperature, material compatibility, toxicity for preserved food, etc.

EXERCISE

1. Why dielectric strength is considered to be one of the desirable properties for a refrigerant?
2. Discuss the desirability of lubricating oil solubility with the refrigerant.
3. What are the essential physical properties of a refrigerant?
4. What are the essential chemical properties of a refrigerant?
5. What are the essential thermodynamic properties of a refrigerant?
6. Can air be used as a refrigerant for a domestic refrigerator? Explain.
7. Why the ammonia compressors are usually water cooled while R22 compressors are mostly air cooled ? (any two points)
8. Ammonia an evergreen refrigerant is still popular refrigerant due to its.
9. What do you mean by secondary refrigerant? Where it is used?
10. What do you mean by eutectic point of a solution? What is the value for a NaCl-water solution?
11. Find the chemical formula for R114, R22.
12. What is an azeotropic composition of a mixture? Give an example.
13. What is meant by 'temperature lift'?
14. What is meant by temperature glide and how it helps in increasing the COP of refrigeration system?
15. What is the alternative refrigerant for R22?
16. Why CFCs became most preferred refrigerant among the available refrigerants?
17. What is the significance of 'eutectic point'?
18. Define azeotropic mixture and zeotropic mixture. Give examples of both types of mixtures.
19. How the refrigeration capacity of a system can be altered where zeotropic mixture is used?
20. Differentiate between 'minimum boiling azeotrope' and 'maximum boiling azeotrope'. Give two examples of each.

21. What are the fluids used as the reference fluid for measuring the ozone depletion potential and global warming potential?

22. Write in decrease order in ODP for R11, R12, R13, R21 and R22.

23. Write down the chemical formula of ecofriendly refrigerant R134a.

24. What does halide torch mean for different refrigerants ?

25. State the thermodynamic properties of refrigerant.

26. Suggest substitute for CFC refrigerants from the point of ozone depletion and global warming.

27. The use of R22 for temperature below –300C is not recommended due to its.......................................

28. Environment-friendly refrigerator R134 is used in the new generation domestic refrigerator. Its chemical formula is ....

MULTIPLE–CHOICE QUESTIONS

1. The reference substance for ODP measurement is

(a) R12 (b) R22

(c) R11 (d) R13

2. The reference substance for ODP measurement is

(a) R22 (b) R717

(c) R125 (d) R744

3. Dielectric strength of refrigerant is important for — type compressor.

(a) Scroll (b) Semi-open

(c) Hermetic (d) Open

4. Specific heat of refrigerant vapour and liquid should be — and — respectively.

(a) Low and high (b) Low and low

(c) High and high (d) High and low

5. Which dome on T-s plane is more preferable for an ideal refrigerant?

(a)  (b) 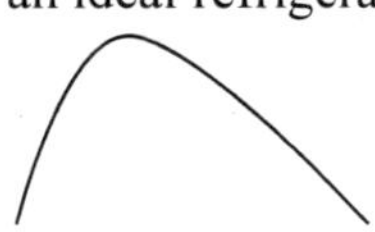

(c)  (d) 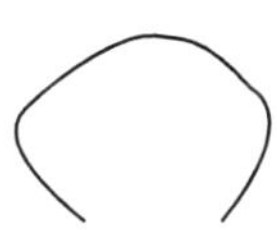

6. Normal boiling point of a hydrocabon ____ if hydrogen is replaced with chlorine.

(a) Increases (b) Decreases

(c) Remains same

7. Of the following which refrigerant has the highest compressor discharge temperature for same CR and initial temperature?

(a) R22 (b) R134a
(c) R717 (d) R12

8. Which is the correct decreasing sequence of GWP?

(a) CFC, HCFC, HC, inorganic (b) HC, inorganic, CFC, HCFC
(c) HCFC, inorganic, HC, CFC (d) HC, inorganic, HCFC, CFC

9. Which is the correct increasing sequence of ODP?

(a) CFC, HCFC, HC, inorganic (b) HC, inorganic, CFC, HCFC
(c) HCFC, inorganic, HC, CFC (d) Inorganic, HC, HCFC, CFC

10. Latent heat of vaporization is ____ for maximum boiling azeotrope.

(a) Low (b) High
(c) Variable

11. Minimum boiling azeotrope has ____ compression ratio.

(a) Low (b) High
(c) Variable

12. Zeotropic mixture has ____ boiling point.

(a) single
(b) a range of
(c) equal to the number of components of the mixture

13. Addition of salt to water ____ its freezing point.

(a) Increases (b) Decreases
(c) Has no effect (d) Depends upon amount of salt

14. Ammonia as a refrigerant

1. has higher compressor discharge temperature compared to fluorocarbons.
2. is toxic to mucous membranes.
3. requires larger displacement per TR compared to fluorocarbons.
4. reacts with copper and its alloys.

Select the correct answer using the codes given below:

(a) 1 and 2 (b) 1, 2 and 3
(c) 1, 2 and 4 (d) 2, 3 and 4

15. An ideal refrigerant should have:

(a) high latent heat of vaporization and low freezing point
(b) high operating pressure and low freezing point
(c) high specific volume and high latent heat of vaporization.
(d) low COP and low freezing point.

16. Which statements are correct?

1. Practically, all common refrigerants have approximately the same COP and power requirement.
2. Ammonia mixes freely with lubricating oil and this helps in lubrication of compressors.
3. Dielectric strength of refrigerants is an important property in hermetically sealed compressorunits.
4. Leakage of ammonia can be detected by halide torch method.

(a) 1, 2 and 4 are correct (b) 2 and 4 are correct
(c) 1, 3 and 4 are correct (d) 1 and 3 are correct

17. The leakage in a R12 can be detected by

(a) oxyacetylene torch (b) halide torch
(c) sulphur torch (d) blue litmus paper

18. Which statements are correct?

1. In Freon 22 system, moisture choking generally does not occur.
2. Freon 11 is mainly used in large capacity air conditioning plants with centrifugal compressor.
3. Presence of lubricating oil in evaporator will increase the heat transfer coefficient.
4. Refrigerants that are completely miscible with oil, do not cause oil choking.

Of these statements:

(a) 2 and 4 (b) 1, 2 and 3
(c) 2, 3 and 4 (d) 1, 3 and 4

19. In ammonia refrigeration systems, oil separator is provided because

1. oil separation in evaporator would lead to reduction in heat transfer coefficient.
2. oil accumulation in the evaporator causes choking of evaporator.
3. oil is partially miscible in the refrigerant.
4. oil causes choking of expansion device.

(a) 1 and 2 are correct (b) 2 and 4 are correct
(c) 2, 3 and 4 are correct (d) 1, 3 and 4 are correct

20. Moisture should be removed from refrigerants to avoid

1. compressor seal failure. 2. freezing at the expansion valve.
3. restriction to refrigerant flow. 4. corrosion of steel parts.

(a) 1, 2, 3 and 4 are correct (b) 1 and 2 are correct
(c) 2, 3 and 4 are correct (d) 1, 3 and 4 are correct.

ANSWERS

1. (c) **2.** (d) **3.** (c) **4.** (d) **5.** (b) **6.** (a) **7.** (c) **8.** (a) **9.** (d) **10.** (b)
11. (a) **12.** (b) **13.** (b) **14.** (b) **15.** (a) **16.** (d) **17.** (b) **18.** (a) **19.** (d) **20.** (c)

3

Air Cycle Refrigeration

[Working Fluid: Air]

Thermodynamic Cycle: Reverse Brayton Cycle

Application: Mostly used in aircraft air conditioning

3.1 INTRODUCTION

Reverse Carnot cycle (described in Section 1.8) seems to be a hypothetical thermodynamic cycle due to the following reasons.

(a) Isothermal heat addition/rejection with zero temperature gradient between the system and source/sink is hardly achievable and also it takes infinite time (it requires very slow movement of piston so that system get sufficient time to transfer the required amount of heat) and isentropic compression/expansion, though possible, requires very less time (it needs very fast movement of piston so that system gets less time to exchange heat with the surrounding). Practically, it is not possible that a compressor runs very slow for first half and very fast for second half of its revolution.

(b) From the thermodynamic cycle diagram, it is clear that there are two consecutive compressions (first is isentropic compression followed by isothermal compression) and two consecutive expansions (first is isentropic expansion followed by isothermal expansion). It seems these processes need a large volume for a reciprocating compressor or two compressors (one isentropic and one isothermal) and two expanders (one isentropic and one isothermal) for the processes.

It is worth mentioning that reverse carnot cycle (the most ideal cycle) is not practically possible. But it is useful to analyse the cycle so that it will be possible to know how much a refrigeration cycle deviates from the ideal one and how much is the irreversibility. Efforts can be made to minimise the deviation and maximise the COP of the refrigeration cycle. On this basis refrigerating efficiency has been defined in Section 1.8, i.e.,

$$\eta_{Ref} = \text{Actual COP} / \text{COP}_{RC} \qquad [3.1]$$

3.2 DEVELOPMENT OF AIR STANDARD REFRIGERATION CYCLE

The above discussion confirms that reverse Carnot cycle is not practically possible. With the availability of high speed motor driven compressors, isentropic processes are feasible. But the processes of isothermal heat addition and rejection of reverse Carnot cycle are replaced with constant pressure heat addition/rejection. Because constant volume heat addition and rejection process is not desirable as it needs an instant of time to transfer the heat (rejection/addition) from/to the system. The new thermodynamic cycle, thus developed, is called **Reverse Brayton cycle** or **Joule's cycle** or **Bell-Coleman cycle** (Fig. 3.1).

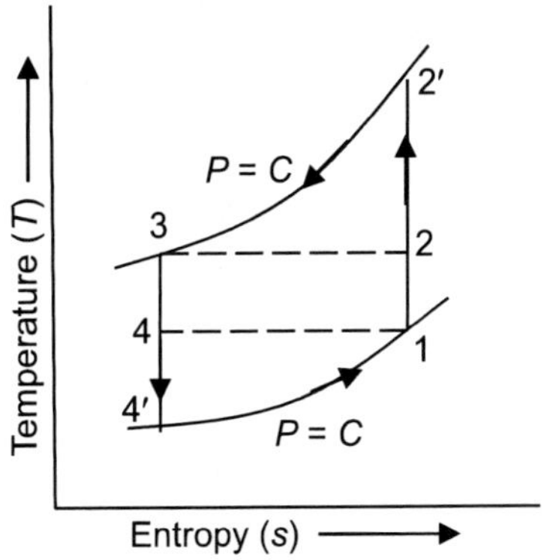

Figure 3.1: 1-2-3-4 RC cycle and 1-2′-3-4′ reverse Brayton

This thermodynamic cycle is used for air cycle refrigeration system. This refrigeration system is so named because the working fluid used here is air or any non-condensable gas. The schematic of this refrigeration system is shown in Fig. 3.2 (a). The cycle diagram on *T-s* and *p-v* planes are also shown in Fig. 3.2 (b) and 3.2 (c) respectively. Process 1-2 is the isentropic compression (external work is done on the system by the compressor), process 2-3 is the isobaric heat rejection (heat transferred from the system to the high temperature sink or environment), process 3-4 is the isentropic expansion (high pressure and high temperature air/gas is expanded to low pressure and low temperature air/gas by an expander, thereby recovering some amount of work) and process 4-1 is the isobaric heat addition (heat transferred from the low temperature heat source or the space to be cooled by the system). Thus, the Joule's cycle is completed. Assumptions made to make this thermodynamic cycle ideal are as follows.

(a) There is no internal friction, no internal heat generation or no heat transfer to or from the system during the processes 1-2 and 3-4. Hence, the compression and expansion processes will be isentropic.

(b) Fluid friction and wall friction during the fluid flow through the tubes are negligible. So isobaric heat rejection /addition processes (2 – 3 and 4 - 1) is possible.

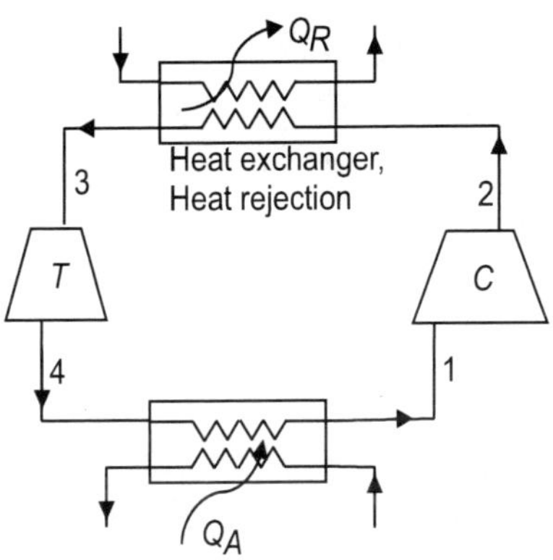

(a)

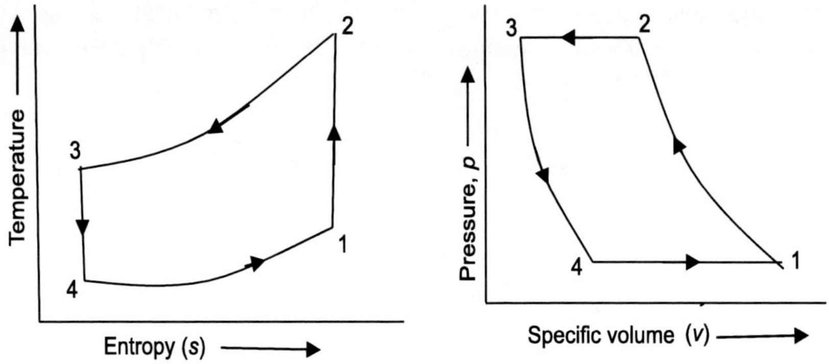

(b) (c)

Figure 3.2: (a) Schematic; (b) diagram on *T-s* plane; (c) diagram on *p-v* plane; of reverse Brayton cycle

The COP of this cycle is calculated as follows.

The refrigerating effect (desired effect),

$$Q_R = m\,C_P\,(T_1 - T_4) \qquad [3.2]$$

The work done due to isentropic compression,

$$W_C = m\,C_P\,(T_2 - T_1) \qquad [3.3]$$

Amount of heat rejected,

$$Q_{Rej} = m\,C_P\,(T_2 - T_3); \qquad [3.4]$$

which is also equals to $Q_R + (W_C - W_E)$

The work recovered due to isentropic expansion,

$$W_E = m\,C_P\,(T_3 - T_4) \qquad [3.5]$$

So COP = Cooling effect / Net work done

$$= m\,C_P\,(T_1 - T_4) / \{m\,C_P\,(T_2 - T_1) - m\,C_P\,(T_3 - T_4)\}$$
$$= (T_1 - T_4) / \{(T_2 - T_1) - (T_3 - T_4)\}$$
$$= (T_1 - T_4) / \{(T_2 - T_3) - (T_1 - T_4)\}$$
$$= 1/\{(T_2 - T_3)/(T_1 - T_4) - 1\}$$

From the perfect gas relationship, for an isentropic process, $T_2/T_1 = (P_2/P_1)^{(\gamma-1)/\gamma} = T_3/T_4$

Again $T_2/T_1 = T_3/T_4 = (T_2 - T_3)/(T_1 - T_4)$ (from equivalent ratio property)

Substituting in the above equation

COP $= 1/\,(T_2/T_1 - 1)$

$= 1/\{(P_2/P_1)^{(\gamma-1)/\gamma} - 1\} = 1/(r^{(\gamma-1)/\gamma} - 1)$

where r is the pressure ratio. [3.6]

[NB: Compression of the gas/vapour by decreasing the volume of the gas and thereby increasing its pressure (like in reciprocating compressor) is expressed by compression ratio. But increasing the pressure by imparting centrifugal force or by converting kinetic energy to pressure energy (as in centrifugal compressors) is expressed by pressure ratio.]

3.2.1 Actual Air Standard Refrigeration Cycle

This calculation is for an ideal reverse Brayton cycle. But the actual cycle deviates from the ideal one as shown in the Fig. 3.3.

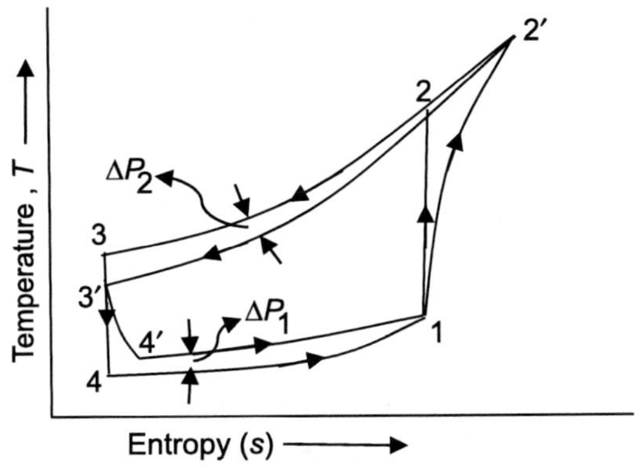

Figure 3.3: 1-2′-3′-4′-1 The actual reverse Brayton cycle

The deviation of the actual cycle from the ideal cycle is as follows.

(a) Isentropic compression (1 – 2) of ideal cycle is not possible in reality because of internal friction and heat transfer to the surroundings. So the compression process becomes non-isentropic and is represented by 1 – 2′.

The isentropic efficiency is expressed by

$$\eta_{Isen} = (T_2 - T_1)/(T_2 - T_1) \qquad [3.7]$$

(b) Isobaric heat rejection (2 – 3) is not possible. As the air/gas flows through a tube there must be drop in pressure (ΔP_2) due to fluid viscosity as well as friction between the fluid and the tube wall. So the process is represented by 2′ – 3′.

(c) Isentropic expansion (3 – 4) is not possible due to the reasons stated above. The expansion of air starts from 3′. The expansion would have continued to a point on the isobaric process (4 – 1). But considering a pressure drop of (ΔP_1) in the cooler tube, the expansion is continued up to 4′ so that the exit air pressure can be at P_1.Thus, the real expansion process is represented by 3′ – 4′.

(d) Isobaric heat addition (4 – 1) is not possible. As the air/gas flows through a tube there must be drop in pressure (ΔP_1) due to fluid viscosity as well as friction between the fluid and the tube wall. So the process is represented by (4′ – 1′).

Example 3.1

A 1 TR air conditioning system is running on Bell-Coleman cycle. Air is the working fluid. Temperature of air coming out of the evaporator is 273 K, and the temperature at the inlet to the turbine is 318 K. Pressure ratio is 4. Find the mass flow rate, compressor work, turbine work recovered, heat rejected at condenser, COP and volume flow rate at the inlet to compressor and exit of turbine.

Ans.: [Ref: Fig. 3.2] Given $T_1 = 273$ K, so $T_2 = T_1\{(P_2/P_1)^{(\gamma-1)/\gamma}\} = 273\ (4)^{0.4/1.4} = 273(1.486)$ = 405.68 K

$T_3 = 318$ K, $T_4 = T_3/(P_3/P_4)^{(\gamma-1)/\gamma}\} = 318/1.486 = 214$ K

Mass flow rate $m = 3.5167 \times TR/\{C_P(T_1 - T_4)\} = 1 \times 3.5167/\{1.005(273 - 214)\} = 0.0593$ kg/s

$Q_C = mC_P(T_2 - T_3) = 0.0593 \times 1.005 \times (405.68 - 318) = 5.226$ kW

$W_C = mC_P(T_2 - T_1) = 0.0593 \times 1.005 \times (405.68 - 273) = 7.908$ kW

$W_T = mC_P(T_3 - T_4) = 0.0593 \times 1.005 \times (318 - 214) = 6.2$ kW

Net work done = $(W_C - W_T) = 7.908 - 6.2 = 1.708$ kW

COP = $1 \times 3.5167/1.708 = 2.059$

Volume flow of air at inlet to compressor,

$$V_1 = mRT_1/p_1 = 0.0593 \times 0.287 \times 273/101.305 = 0.04585 \text{ m}^3/\text{s}$$

Volume flow of air at exit to turbine,

$$V_4 = mRT_4/p_4 = 0.0593 \times 0.287 \times 214/101.325 = 0.03594 \text{ m}^3/\text{s}$$

Example 3.2

With the above data, if the compression isentropic efficiency (η_C) and turbine isentropic efficiency (η_T) are 0.8 and 0.85 respectively, find the above parameters.

Ans.: Refer Fig. 3.3

Isentropic efficiency, $\eta_C = 0.8 = (T_2 - T_1)/(T_2' - T_1)$; $T_2 = 438.85$ K

Similarly, $\eta_T = 0.85 = (T_3 - T_4')/(T_3 - T_4)$; $T_4' = 229.6$ K

Mass flow rate, $m = 3.5167 \times TR/\{C_P(T_1 - T_4')\} = 1 \times 3.5167/\{1.005(273 - 229.6)\}$

$= 0.0806$ kg/s

$Q_C = mC_P(T_2' - T_3) = 0.0806 \times 1.005 \times (438.85 - 318) = 9.79$ kW

$W_C = mC_P(T_2' - T_1) = 0.0806 \times 1.005 \times (438.85 - 273) = 13.434$ kW

$W_T = mC_P(T_3 - T_4') = 0.0806 \times 1.005 \times (318 - 229.6) = 7.16$ kW

Net work done = $(W_C - W_T) = 13.434 - 7.16 = 6.274$ kW

COP = $1 \times 3.5167/6.274 = 0.56$

Volume flow of air at inlet to compressor,

$$V_1 = mRT_1/p_1 = 0.0806 \times 0.287 \times 273/101.305 = 0.06232 \text{ m}^3/\text{s}$$

Volume flow of air at exit to turbine,

$$V_4 = mRT_4/p_4 = 0.0806 \times 0.287 \times 229.6/101.325 = 0.05242 \text{ m}^3/\text{s}$$

Example 3.3

In a Bell-Coleman cycle air enters the compressor at 1.0 bar and comes out at 6 bar. Index of compression (n) and expansion process (m) are 1.25 and 1.3, respectively. Temperature at the beginning of compression and expansion are 7°C and 37°C, respectively. Neglecting the clearance volume of the compressor, calculate the mass flow rate, compressor work, turbine work recovered, heat rejected at condenser, and COP.

Ans.: The net work done = area enclosed by 1-2′-3-4′ as shown in Fig. 3.4.

i.e., $W = W_C - W_T$

Then $W = \{n/(n-1)\}(p_2v_2 - p_1v_1) - \{m/(m-1)\}(p_3v_3 - p_4v_4)$

$= RT_1 n/(n-1)\{(p_2/p_1)^{\{(n-1)/n\}} - 1\} - RT_3\, m/(m-1)\{1 - (p_4/p_3)^{\{(m-1)/m\}}\}$

where, n and m are the indices of compression and expansion, respectively.

Now substituting the respective values, we get

W_C (represented by first part of the equation) = 173.16 kJ/kg of air

W_T (represented by second part of the equation) =130.56 kJ/kg of air

So W = 42.6 kJ/kg of air

$T_2 = T_1 \times (p_2/p_1)^{\{(n-1)/n\}} = 400.67$ K

$T_4 = T_3 /(p_3/p_4)^{\{(m-1)/m\}} = 205$ K

Cooling effect $C_P (T_1 - T_4) = 1.005 (280 - 205) = 75.375$ kJ/kg of air

$\text{COP} = C_P (T_1 - T_4)/W = 75.375/42.6 = 1.77$

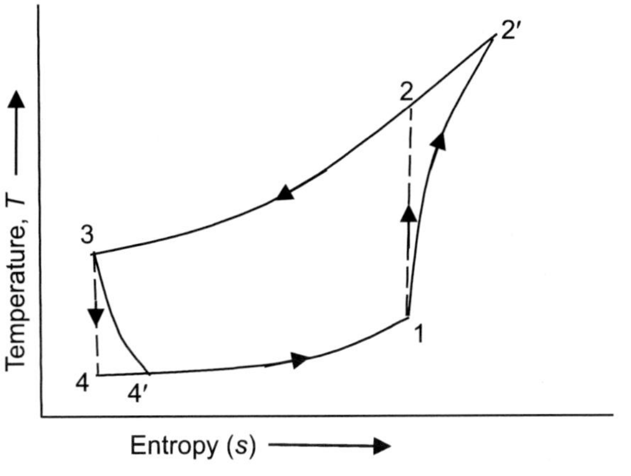

Figure 3.4: 1-2′ and 3-4′ are the non-isentropic compression and expansions, respectively

Note: Here net work done cannot be expressed as $C_P(T_2 - T_1) - C_P(T_3 - T_4)$ because the compression and expansion processes are polytropic processes, not the isentropic one, i.e., index n is not equal to γ. Because when $n = \gamma$, $\{n/(n-1)\}(p_2v_2 - p_1v_1) = C_P (T_2 - T_1)$. This can be proved by substituting the relationship, $C_P - C_V = R$, and $C_P / C_V = \gamma$ in this equation.

Alternately the problem can be solved in the following process.

Here polytropic efficiency (η_P) is defined as the deviation of the polytropic process from isentropic process. $\int vdp$ work may be greater than or smaller than $(h_2 - h_1)$ work.

Polytropic efficiency can be expressed as $\{n/(n-1)\}\ \{(\gamma - 1)/\gamma\}$ for $n > \gamma$ (for centrifugal compressors where there is more frictional heat) and as $\{(n-1)/n\}\{\gamma/(\gamma-1)\}$ for $n < \gamma$ (for reciprocating compressors, i.e., when cooling of compressors is done). (Refer any thermodynamics textbook for the above expression of polytropic efficiency).

Thus, the equivalent temperature, if the process would be isentropic, can be expressed as

$(\eta_P) = (T_2 - T_1)/(T_2' - T_1)$ for compression

and $(\eta_P) = (T_3 - T_4)/(T_3 - T_4')$ for expansion

In this case, it is a reciprocating compressor and $n < \gamma$.

So substituting the values of n and γ for compression, we get $\eta_P = 0.7$ and for expansion $\eta_P = 0.808$

So $T_2' = 452.4$ K and $T_4' = 180.05$ K

Now $W_C = C_P (T_2' - T_1) = 1.005 (452.4 - 280) = 173.262$

$W_T = C_P (T_3 - T_4') = 1.005 (310 - 180.05) = 130.6$

$W = 173.262 - 130.6 = 42.662$ kJ/kg

Refrigerating effect, $C_P (T_1 - T_4) = 75.375$ kJ/kg

COP = 75.375/42.662 = 1.767

Alternatively, this can also be solved from first law of thermodynamics

During compression, $Q_{1-2} = \{(\gamma - n)/(\gamma - 1)\}\{RT_1/(n-1)\}\{(p_2/p_1)^{(n-1)/n} - 1\}$

Substituting the respective values ($n = 1.25$); $Q_{1-2} = -52.0$ kJ/kg of air

Heat rejection, $Q_{2-3} = C_P (T_3 - T_2) = -91.7$ kJ/kg of air

During expansion, $Q_{3-4} = \{(\gamma - n)/(\gamma - 1)\}\{RT_1/(n-1)\}\{(p_2/p_1)^{(n-1)/n} - 1\}$

Substituting the respective values ($n = 1.3$); $Q_{3-4} = 25.12$ kJ/kg of air

Heat absorption, $Q_{4-1} = C_P (T_1 - T_4) = 75.28$ kJ/kg of air

From first law of thermodynamics cyclic $\int (\delta Q)$ = Cyclic $\int (\delta W)$

So net $W = \Sigma \delta Q = -43.22$ kJ/kg of air

Cooling effect, $Q_{4-1} = C_P (T_1 - T_4) = 75.28$ kJ/kg of air

So COP = 75.28/43.22 = 1.742

3.3 AIR AS A REFRIGERANT

Using air as a refrigerant has the main advantage that it is easily and plentily available with practically no cost. It has no adverse effects on environment. Only task is to clean the air before it enters the compressor.

But it has the following demerits.

(i) It has a high specific volume. So it requires high compressor displacement volume for a reciprocating compressor. It is recommended to use dynamic compressors, specifically centrifugal compressor, (refer Section 5.2) for air or gas where the rotational speed determines the displacement volume.

(ii) Since there is no phase change, only sensible heating of air is accounted for the refrigerating capacity. A great mass of air, which is calculated from $Q = mC_P\Delta T$, is required for small refrigerating effect.

(iii) Water vapours in the air freeze for a lower evaporator temperature (< 0°C) and clog the flow. Hence, it may be used for air conditioning system only.

(iv) Large capacity container is required to store the clean air for closed cycle operation. Or open cycle is system is made where clean air is available.

Keeping in view to the above demerits, it is only advantageous to use it as refrigerant in aircraft refrigeration system.

Advantages of air cycle refrigeration system for aircraft cooling

(i) At a high altitude abundant clean atmospheric air is available. So the refrigeration cycle can be an open cycle and the conditioned air after expansion can directly be used in the aircraft cabin. Thus, the use of a heat exchanger, i.e., evaporator can be eliminated.

(ii) A part of the compressed air from the centrifugal compressor used for aircraft engine is used for refrigeration purpose. Thus, the use of an extra compressor is eliminated. Elimination of these two vital components (compressor and evaporator) reduces considerably the weight and cost of the aircraft refrigeration system.

(iii) Thus, cabin pressurisation and air conditioning can be combined into one operation.

(iv) Low temperature air, after ramming effect, is directly used for cooling of compressed air in the cooling chamber. So no other cooling medium is required for heat rejection from the condenser.

3.4 NEED OF AIRCRAFT REFRIGERATION SYSTEM

Aircraft may be considered a closed system. The heat generated inside it due to various factors like passengers' perspiration (human being releases energy nearly 300 kJ/h even at rest), air resistance, electronic devices, solar radiations, etc., increases the temperature and humidity of the cabin. The aircraft also fly at a very high altitude (at about 12,000 m above the sea level) where the air pressure is around 0.15 bar and temperature is –50°C. Air is also quite thin. So our body organs do not function properly at such a low pressure. Bleeding occurs from our internal organs. Hence, the cabin pressure should be maintained close to the standard atmospheric pressure.

3.4.1 Simple Aircraft Refrigeration Cycle

Working Principle: (Refer Fig. 3.5)

Ambient air is available at state point 1. This air enters the aircraft at the speed same as that of the aircraft from opposite direction. Then it is stopped suddenly, which is called *ramming effect.* Thus, it loses all its kinetic energy and this energy is converted to enthalpy, in consequence pressure and temperature of the air increase. Assuming the ramming action to be isentropic, the initial pressure and temperature rise of atmospheric air occurs from state point 1 to 2. This can be expressed, from energy conservation equation

as $$h_2 = h_1 + V^2/2, \text{ or, } C_P \times T_2 = C_P \times T_1 + V^2/2$$

So $$T_2 = T_1 + (V^2/2C_P) \qquad [3.8]$$

Here T_2 is known as the stagnation temperature. This can also be expressed as

$$T_2/T_1 = 1 + \{(\gamma - 1)/2\}M^2 \qquad [3.9]$$

where M is the Mach number of the aircraft and γ is the ratio of specific heats. If the ramming action is non-isentropic, the air reaches at state point 2′ instead of 2. Here the temperature will remain same ($T_2 = T_2'$), but the pressure rise will decrease to P_2'. Thus, the ram efficiency is expressed as

$$\eta_{Ram} = \text{Actual pressure recovery / Ideal pressure recovery}$$

$$= (P_2' - P_1)/(P_2 - P_1) \qquad [3.10]$$

Then the rammed air is isentropically compressed by the centrifugal compressor of the aircraft to state point 3. If the compression is non-isentropic the state point after compression is 3′. A major part of this compressed air is sent to combustion chamber and a little of it is sent for aircraft air conditioning purpose. This air enters the counter flow heat exchanger and gets cooled to state point 4 by the low temperature air available after ramming action. If heat exchanger effectiveness is considered

Effectiveness, $$\varepsilon = (T_3 - T_4)/(T_3 - T_2) \quad [3.11]$$

and $$T_4 = T_3 - \varepsilon\,(T_3 - T_2) \quad [3.12]$$

If the pressure drop (ΔP) in the heat exchanger is considered, then the state point will be at 4′. This low temperature air, after taking the heat of the compressed air, is blown off by the induced fan to the atmosphere. The compressed air at state point 4 is isentropically expanded through the turbine to cabin pressure 5. If the expansion is non-isentropic the state point after expansion is 5′. Here it is directly delivered to the cabin where it takes the cooling load (heat) of the cabin at constant pressure and reaches the state point 6. If the pressure drop (ΔP) in the duct to the cabin is considered, then the air should be expanded to a state point which is ΔP higher than P_5'. However, the pressure of the delivered air to the cabin must yield to the pressure maintained in the cabin.

For simple aircraft refrigeration cycle, cooling effect produced = $C_P\,(T_6 - T_5)$ [3.13]

Net work done by the system = Compression work (2 – 3) – Turbine work (4 – 5) + ram air work (1 – 2) (is provided by the aircraft as it experiences the drag to stop the air flow) + fan work (may be neglected).

So COP = Cooling effect produced/Net work done

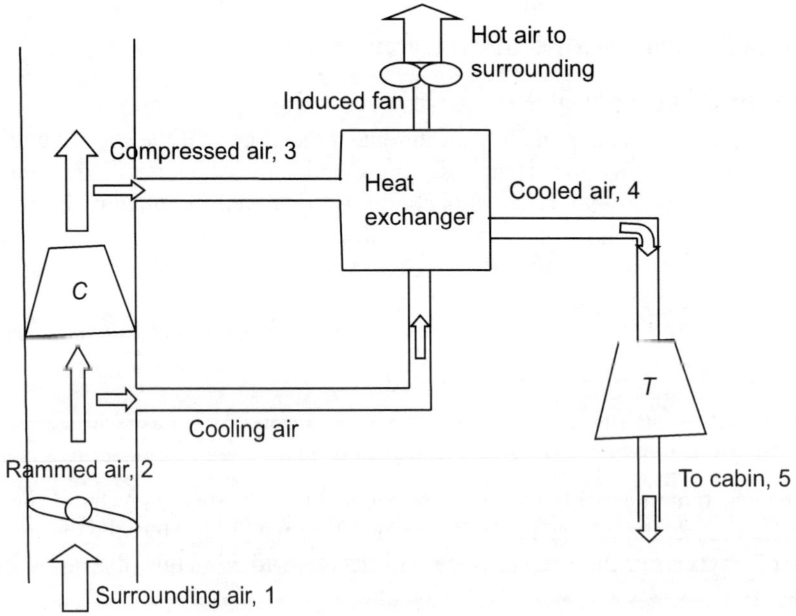

(a)

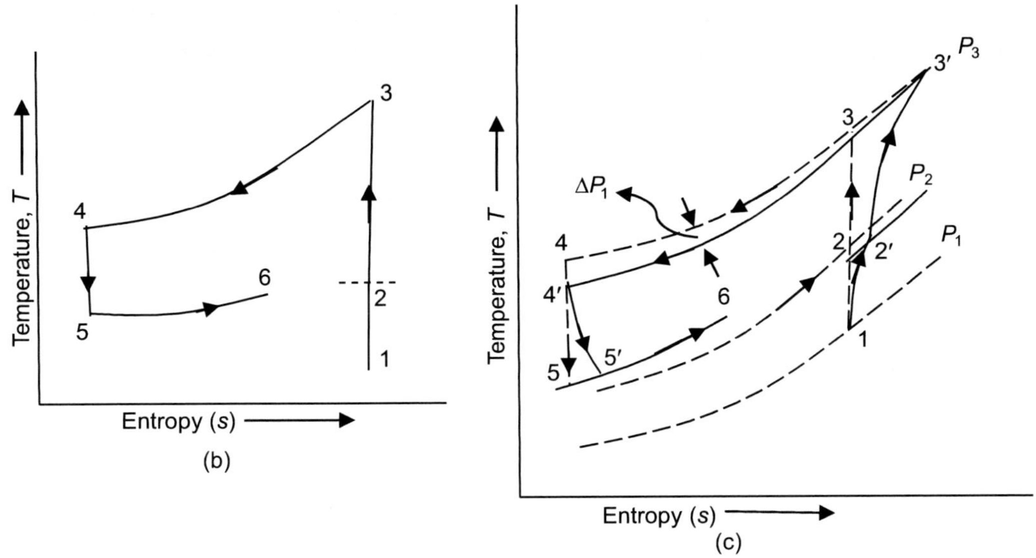

Figure 3.5: (a) Schematic; (b) ideal; (c) actual cycle of simple aircraft refrigeration cycle

Example 3.4

An aircraft is flying at a speed of 800 km/hr. The ambient temperature and pressure are –10°C and 0.5 bar, respectively. The pressure ratio of the compressor is 5. The cabin pressure is 1.06 bar and air leaves the cabin at 25°C. Assuming simple aircraft air conditioning cycle of 1 TR capacity, find the temperature and pressure at different state points, mass flow rate, ram work, compressor work, turbine work, COP, volume flow rates at turbine and compressor exit.

Ans.: $V_{\text{air craft}} = 800 \times 1000/3600 = 222.222$ m/s.

Stagnation temperature, $T_2 = T_1 + (V^2/2C_P) = 263 + (222.222^2/2 \times 1005) = 287.57$ K

$P_2 = P_1\,(T_2 / T_1)^{1.4/0.4} = 0.5\,(287.57/263)^{1.4/0.4} = 0.683$ bar

So $P_3 = 0.683 \times 5 = 3.415$ bar, $T_3 = T_2\,(P_3/P_2)^{0.4/1.4} = 287.57\,(5)^{0.2857} = 455.45$ K

$P_3 = P_4$ and $T_4 = T_2$ (For ideal heat exchanger, i.e., for effectiveness is 1)

$P_5 = P_6 = 1.06$ bar, $T_5 = T_4\,(P_5/P_4)^{0.4/1.4} = 287.57\,(1.06/3.415)^{0.2857} = 205.86$ K

Mass flow rate for 1TR capacity, 3.5167 kW /{1.005 (298 – 205.86)}

$= 0.038$ kg/s $= 136.8$ kg/hr

Ram work = 0.038 (1.005)(287.57 – 263) = 0.94 kW

Compressor work = 0.038 (1.005)(455.45 – 287.57) = 6.411 kW

Turbine work, recovered = 0.038 (1.005)(287.57 – 205.86) = 3.12 kW

So net work = 0.94 + 6.411 – 3.12 = 4.231 kW

COP = 3.5167 / 4.231 = 0.831

Volume flow rates at turbine exit, $mRT_5 / P_5 = 0.038(286)(205.86)/ (1.06 \times 10^5) = 0.0211\ m^3/s$

Volume flow rates at compressor exit, $mRT_3 / P_3 = 0.038(286)(455.45)/ (3.415 \times 10^5)$

$$= 0.0145\ m^3/s$$

Example 3.5

Repeat Example 3.4. Consider the ram efficiency, compressor efficiency and turbine efficiency are 0.85, 0.80 and 0.85, respectively. Effectiveness and pressure drop in the heat exchanger are 0.80 and 0.1 bar. Calculate the parameters of Example 3.4.

Ans. $\eta_R = (P_2 - P_1)/ (P_2 - P_1)$; $0.85 = (P_2 - 0.5)/(0.683 - 0.5)$; So $P_2 = 0.655$ bar

So $P_3 = 5 \times 0.655 = 3.275$ bar

$T_3 = T_2 (P_3/P_2)^{0.4/1.4} = 287.57 (5)^{0.2857} = 455.45$ K

$\eta_C = (T_3 - T_2')/(T_3' - T_2')$; $0.8 = (455.45 - 287.57)/(T_3 - 287.57)$; So $T_3 = 497.42$ K

For heat exchanger effectiveness $0.8 = (T_3' - T_4)/(T_3' - T_2) = (497.42 - T_4)/(497.42 - 287.57)$

So $T_4 = 329.54$ K, $P_4 = P_3 - 0.1 = 3.275 - 0.1 = 3.175$ bar

$T_5 = T_4 (P_5/P_4)^{0.4/1.4} = 329.54 (1.06/3.175)^{0.2857} = 240.87$ K

$\eta_T = (T_4 - T_5')/(T_4 - T_5)$; $0.85 = (329.54 - T_5')/(329.54 - 240.87)$; So $T_5' = 254.17$ K

For 1 TR capacity, mass flow rate, $m = 3.5167/\{1.005(298 - 254.17)\} = 0.08$ kg/s

Ram work = 0.08(1.005)(287.57 – 263) = 1.9754 kW

Compressor work = 0.08 (1.005)(497.42 – 287.57) = 16.872 kW

Turbine work, recovered = 0.08 (1.005)(329.54 – 254.17) = 6.06 kW

So net work done = 1.9754 + 16.872 – 6.06 = 12.7874 kW

So COP = 3.5167/12.7874 = 0.275

3.4.2 Refrigeration for High Speed Aircraft

When an aircraft moves at a very high speed, the speed of air entering the aircraft will be very high. So ram air (which is used as the cooling air as shown in Fig. 3.5a) temperature increases to a comparatively higher value than that obtained for a simple aircraft system described earlier. So cooling of compressed air is incomplete. So the temperature of air entering the turbine remains high. As a result turbine exit temperature will also be high. It is because the turbine has a fixed expansion ratio. Hence, the desired temperature cannot be achieved. So to get the turbine exit air pressure and temperature at the desired state, different methods are adopted. Basing on these, the simple aircraft air conditioning system is modified which is classified into three categories.

1. Bootstrap aircraft air conditioning system
2. Regenerative aircraft air conditioning system
3. Reduced ambient aircraft air conditioning system

3.4.3 Bootstrap System

In bootstrap system (Fig. 3.6) the compressed air is cooled to point 4 as the coolant (ram air) is at higher temperature. If this air is expanded from point 4 to cabin pressure, the temperature will be T_5' which is higher than the desired temperature T_7 at cabin pressure. So it is recompressed by an auxiliary compressor (AC) to 5 and again cooled by the coolant (ram air) to point 6. Then the air is expanded through a larger pressure ratio to obtain the desired temperature (T_7) at cabinet pressure. (If the expansion is done with larger pressure ratio at point 4 to get the desired temperature, the exit pressure will be lower than the cabin pressure.)

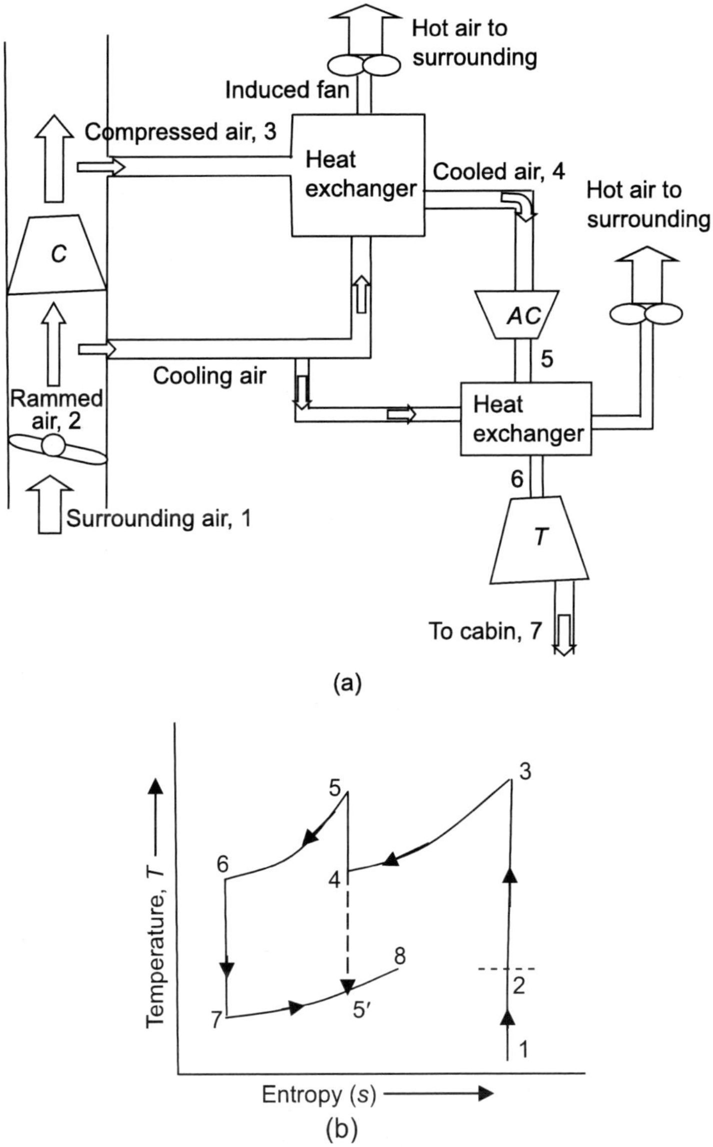

Figure 3.6: (a) Schematic; (b) diagram on *T-s* plane of bootstrap refrigeration system

3.4.4 Regenerative System

In a regenerative system, shown in Fig. 3.7, the compressed air is first cooled to point 4 by the ram air which is at a higher temperature (due to high speed of aircraft). Then a part of the cooled air coming out of the turbine, which is at a lower temperature, is used to cool the compressed air from point 4 to 5 in another heat exchanger.

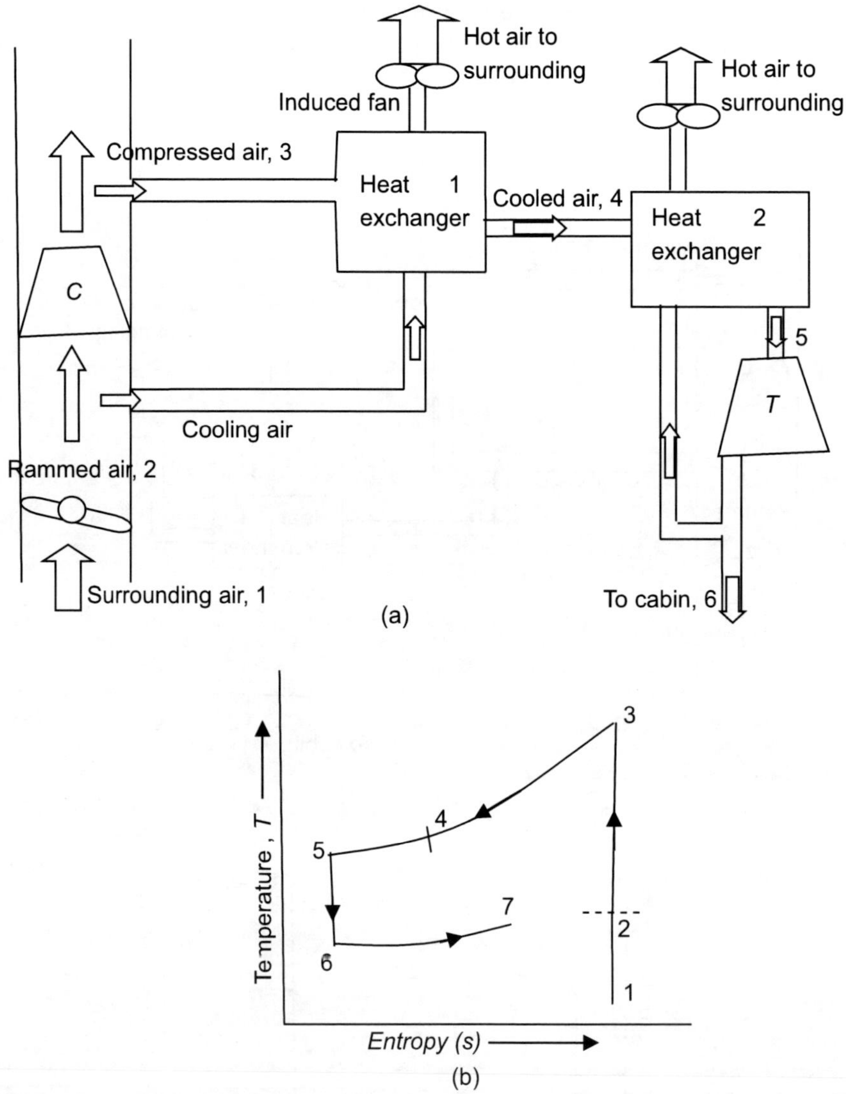

Figure 3.7: (a) Schematic; (b) diagram on *T-s* plane of regenerative refrigeration system

3.4.5 Reduced Ambient System

In reduced ambient system (shown in Fig. 3.8) ram air is first expanded to a lower temperature by an auxiliary turbine. Thus, this chilled ram air can cool the compressed air to the required temperature at the entry of the turbine so that desired temperature at turbine exit can be achieved. The diagram on *T-s* plane will be same as that of the simple aircraft refrigeration system because the state of working fluid (i.e., compressed air) is only shown on the *T-s* plane. In this system the temperature of cooling air is reduced by an auxiliary turbine only.

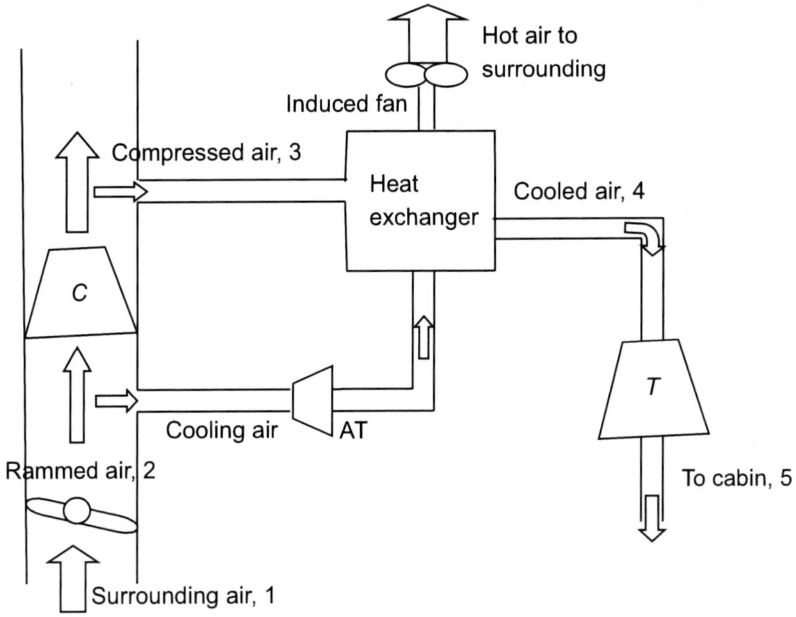

Figure 3.8: Schematic of reduced ambient refrigeration system

It is estimated to go for simple aircraft refrigeration cycle for low Mach number speed aircraft, i.e., sub-sonic aircraft. At high Mach number bootstrap or regenerative cycle is used. For very high Mach number or supersonic aircraft, the reduced ambient cycle is recommended.

EXERCISE

1. What is the need of air conditioning (or pressurizing the cabin) in an aircraft?
2. A cold chamber works on Bell-Coleman cycle. The chamber is maintained at –10°C and 1.0 bar pressure. Air from cold chamber is compressed to 5.0 bar and again cooled to 25°C. Then it is expanded to cold chamber pressure. Assume (a) isentropic compression and expansion, (b) index 1.41, and (c) specific heat 1.007 kJ/kg-K.

   Find net refrigeration effect and COP of the refrigeration system. Also compare the result if the index of compression and expansion is 1.35.
3. Though reverse Carnot cycle is not practically possible, why it is useful to analyse its performance?
4. What is the difference between compression ratio and pressure ratio?
5. For a bootstrap air refrigeration system of an aircraft, flying at an altitude 2500 m (of 0.8 bar and 0°C), the ram air temperature and pressure are 17°C and 1.05 bar, respectively. At

the end of isentropic compression the air is at 4 bar and is cooled to 27°C using ram air. At this temperature air is further compressed in the auxiliary compressor driven by the cooling turbine directly coupled to the compressor. The compressed air from the bootstrap is again cooled to 27°C in another auxiliary heat exchanger. Finally, expansion takes place up to cabin pressure 1 bar and cabin temperature is 25°C. (a) Obtain the maximum pressure in the system and COP if air leaves the cabin at 25°C; (b) repeat the calculation in previous case (a) using compressor efficiencies 0.85 and 0.75 and 0.85 for the main, bootstrap and the cooling turbine respectively.

6. An air refrigerator works between pressure limits of 1 bar and 5 bar. The temperature of the air entering the compressor and expansion cylinder are 10°C and 25°C, respectively. The expansion and compression follows $pv^{1.3}$ = constant. Find the following.
   (i) Theoretical COP.
   (ii) If the load is 10 TR, find the mass flow rate of circulated air assuming actual COP is 50% of theoretical COP.
   (iii) Cylinder dimensions of single acting compressor if it runs at 300 rpm, $\eta_{vol} = 85\%$.
7. An aircraft moving at speed of 1000 km/h uses simple gas refrigeration cycle. The ambient pressure and temperature are 0.35 bar and –10°C, respectively. The pressure ratio of compressor is 4.5. The heat exchanger effectiveness is 0.95. The isentropic efficiencies of the compressor and expander are 0.8 each. The cabin pressure and temperature are 1.06 bar and 25°C. Determine the following.
   (i) Temperature and pressure at each point of cycle.
   (ii) Volume flow rate through compressor and expander for 100 TR.
   (iii) COP of the system.
8. Write down the optimum COP of a gas refrigeration cycle.
9. Sketch the variation COP with pressure ratio for a reversed Brayton cycle.
10. Draw *P-v* and *T-s* diagram for an air refrigerator working on Bell-Colleman cycle and explain.

MULTIPLE–CHOICE QUESTIONS

1. Which thermodynamic cycle is suitable for refrigeration when air/gas is the working fluid?
   (a) S-S S cycle (b) Bell-Coleman cycle
   (c) Brayton cycle (d) Reverse Carnot cycle
2. Which type of compressor should be used for air refrigeration cycle?
   (a) Reciprocating (b) Vane
   (c) Screw (d) Centrifugal
3. Ramming effect means — energy converts to — energy.
   (a) kinetic pressure (b) kinetic potential
   (c) kinetic heat (d) potential kinetic

4. COP of Bell-Coleman cycle — with increase in pressure ratio.

(a) increases (b) decreases

(c) remains unaffected (d) first increases, then remains unaltered

5. COP of Bell-Coleman cycle — with increase in index of compression.

(a) increases (b) decreases

(c) remains unaffected (d) first increases, then remains unaltered

6. Air conditioning in an aircraft is more importantly needed to

(a) maintain the cabin pressure at 1 bar (b) feel chilled

(c) get pure air (d) get more O_2

7. In air refrigeration system high mass of air flow is required because cooling occurs due to

(a) latent heat of vaporisation only (b) sensible heat only

(c) (a) and (b) (d) evaporation

8. The cooling medium used for the condenser of aircraft refrigeration system

(a) atmospheric air (b) rammed air

(c) cabin exhaust (d) compressed air

9. The refrigeration system of an aircraft has the prime aim that the

(a) system has high C.O.P.

(b) H.P./ton is low

(c) weight of refrigerant circulated in the system is low

(d) weight of the refrigeration equipment is low

ANSWERS

1. (b) **2.** (d) **3.** (a) **4.** (b) **5.** (b) **6.** (a) **7.** (b) **8.** (b) **9.** (d)

4

Vapour Compression Refrigeration System

4.1 INTRODUCTION

The Carnot cycle is considered to be the basic thermodynamic cycle for heat engines which gives the maximum cycle efficiency. Similarly, reverse Carnot (RC) cycle (Section 1.8) gives the maximum COP for a refrigeration system. This system is also known as 'mechanical vapour compression cycle', because refrigerant vapour pressure is increased by physical compression by a reciprocating compressor or by the centrifugal action of a centrifugal compressor.

4.1.1 Development of Single-Stage Saturation Cycle

RC cycle is not feasible due to reasons discussed in Section 3.1.

It is made clear that it is not possible for a piston to move very slowly for a moment and very fast in the next moment. But isothermal heat addition/rejection is only be possible in two-phase region where isobaric and isothermal processes coincide with each other. Hence, to have a close approach to RC cycle the refrigeration cycle may operate within two-phase zone of a vapour cycle as shown in Fig. 4.1, so that heat rejection (process 2-3) and heat addition (process 4-1) can be achieved in a two-phase zone as isobaric processes. But the process 1-2 and process 3-4 remain as isentropic compression (from pressure P_e to pressure P_C) and isentropic expansion, respectively. The former two processes replace the two isothermal processes of RC cycle.

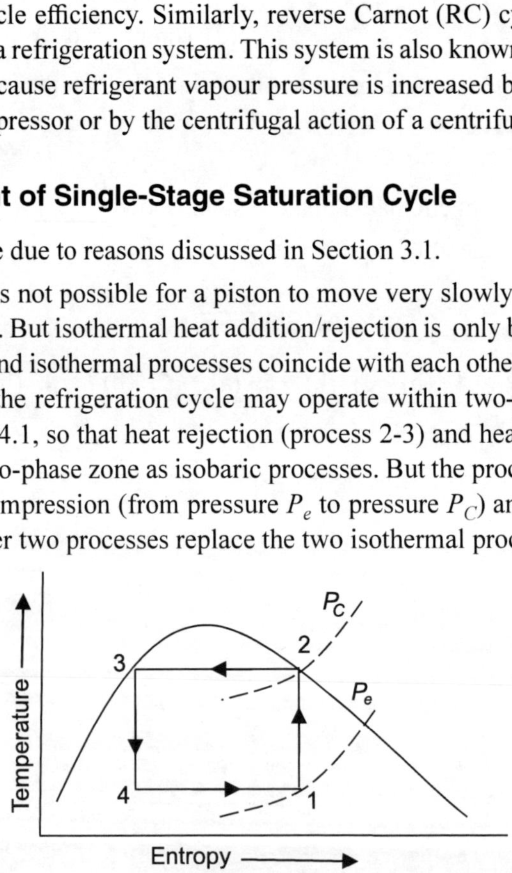

Figure 4.1: Reverse Carnot cycle within two-phase zone

However, the compression in two-phase zone (process 1-2), called wet compression, involves: (a) risk of damage of compressor blade and seat of the valves due to high impact of liquid refrigerant particles (at high velocity liquid refrigerant particles hit the compressor blades like bullets), and (b) carrying away of lubricating oil with liquid refrigerant. Hence, inlet of refrigerant vapour to the compressor is shifted from two-phase zone to dry saturated point. The dry saturated vapour is compressed from 1′ (pressure P_e) to 2′ isentropically (to pressure P). Then to get back to point 2 (pressure P_C) which is required to maintain the RC cycle, the vapour is to be again compressed isothermally from 2′ to 2. Thus, it needs two compressors, one is isentropic compressor and another is isothermal compressor to compress the vapour from P_e to P_C. This is also not possible and also highly expensive. So a single isentropic compressor is employed to compress the vapour from P_e to P_C.

The isentropic expansion of high pressure liquid refrigerant from condenser through the turbine has some demerits like: (a) initial cost increases, (b) maintenance cost added, and (c) system becomes bulky, though it has the advantage of recovering of some amount of work due to expansion. Therefore, turbine is replaced with a throttling device which throttles (reduces the pressure) the refrigerant from condenser pressure to evaporator pressure. This is an irreversible process (hence shown by a dotted line) where enthalpy remains constant. Thus, the basic vapour compression refrigeration cycle, also called single-stage saturation (S-S-S) cycle, is completed. This is so called because the compression process is completed in single-stage compression and the vapour going out of evaporator and entering the compressor is assumed to be saturated vapour. Also the liquid refrigerant coming out of condenser and entering the expansion device is assumed to be saturated liquid. Some also call it ideal vapour compression refrigeration cycle.

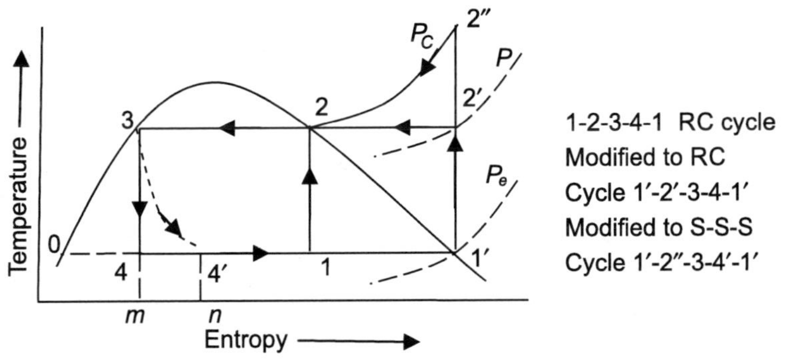

Figure 4.2: Development of S-S-S cycle from RC cycle

When compared S-S-S cycle (1′-2″-3-4′-1) with RC cycle (1′-2′-3-4-1), the differences are as follows.

(a) The extra work done by the isentropic compressor is the area enclosed by 2′-2″-2-2′.
(b) The cooling effect lost amounts to the area under 4-4′-n-m-4
(c) Work recovered due to isentropic expansion (3-4) is lost due to throttling.

However, wet compression is desirable sometimes. It is practicable with a continuous flow machine like a centrifugal compressor or screw compressor with no valves which is an essential part of a reciprocating compressor. It is evident from the diagram that wet compression gives very high COP as it approaches reverse Carnot cycle.

4.2 VAPOUR COMPRESSION REFRIGERATION SYSTEM

Working Principle:

Vapour compression refrigeration system works on the principle of S-S-S cycle. The schematic of a vapour compression system is shown in Fig. 4.3. Low pressure and low temperature refrigerant vapour from the evaporator (state 1) flows to the compressor. It is compressed

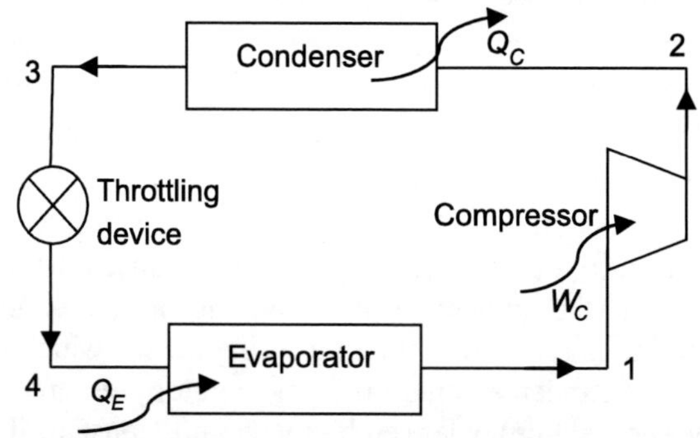

Figure 4.3: Schematic of a vapour compression refrigeration

to the condenser pressure (state point 2) by a compressor utilizing mechanical energy (W_c) from external source. The superheated refrigerant vapour from compressor is condensed to saturated liquid at constant pressure (state point 3) by rejecting heat (Q_C) to the sink. Then the liquid refrigerant is throttled to the evaporator pressure and the refrigerant in two phase state (state point 4), enters the evaporator. It gets vaporised by absorbing heat (Q_E) from the source (refrigerated space) and flows out of the evaporator as refrigerant vapour to the compressor. Thus, the vapour-compression (S-S-S) cycle is completed.

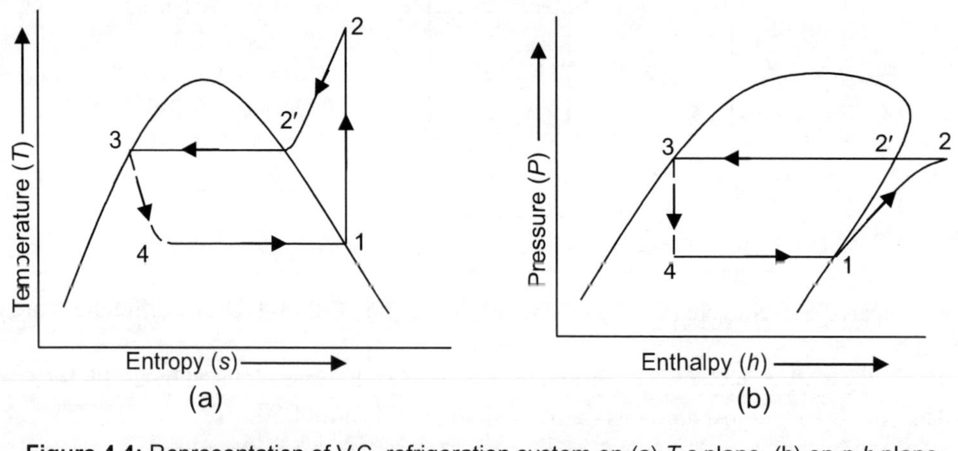

Figure 4.4: Representation of V.C. refrigeration system on (a) *T-s* plane, (b) on *p-h* plane

It looks funny that pressurising and adding the heat to the refrigerants by compression and again depressurising (throttling) to cool. **In an open system if infinite amount of liquid refrigerant is available at atmospheric pressure then refrigerating effect can be achieved just by having an evaporator at atmospheric pressure and having the open cycle as shown.** The temperature of evaporator can be achieved is –40°C because this is the saturation temperature of R22 at atmospheric pressure. But availability of infinite amount of liquid refrigerant is not possible. A closed cycle system must be there so that a fixed amount of refrigerant can be used in a cycle. The vapour coming out of the evaporator should be compressed to high pressure, high temperature so that it can reject heat to the environment (2nd law of thermodynamics) and cooled down to saturated liquid at that pressure. Then the refrigerant liquid at high pressure is brought down to low pressure so that it will boil at the corresponding low saturation temperature.

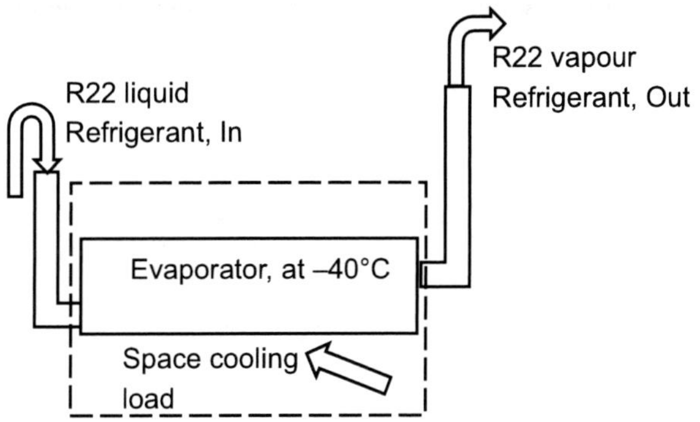

Figure 4.5: Open cooling system

4.2.1 Actual Cycle

The actual vapour compression refrigeration cycle differs a little from S-S-S cycle in the following manner. Refer Fig. 4.6.

(a) The vapour entering the compressor is slightly superheated (1 – 1′) so that it is ensured that the refrigerant vapour entering the compressor is fully dry, i.e., there is no chance of presence of a single liquid refrigerant droplet.

(b) Practically, it is not possible to have isentropic compression. Frictional heat due to movable components in the compressor is inevitable. Entropy increases and compression follows the process 1 – 2′. Heat loss from the compressor to the cooling medium is also essential to prevent the compressor from over-heating. In this case entropy decreases and it follows the process 1 – 2*. Thus, the compression is non-isentropic.

(c) The refrigerant coming out of the condenser must be sub-cooled (process 3 – 3′) so that it is ensured that the liquid refrigerant entering the throttle valve or capillary tube is free from any vapour which would have choked the flow through the narrow passage of the capillary tube. The flow of liquid-vapour mixture at the beginning of the expansion process will decrease the mass flow rate because the density of the mixture is less than that of the pure liquid.

Apart from these there are pressure drops during the flow of refrigerants through (a) condenser tube, and (b) evaporator tube. More pressure drops are also there in the connecting tubes from condenser to expansion device, evaporator to compressor. Considerable amount of pressure drop occurs to overcome the inertia of the inlet and outlet valves of the compressor during suction and discharge of refrigerant vapour. This pressure drop at the valve is called 'wire drawing effect'. This type of frictional pressure drop has a detrimental effect over the system performance. For example, the condensation occurs at X°C at corresponding saturation pressure Y bar. If there is drop of pressure by ΔY amount, then corresponding saturation temperature will decrease by ΔX amount. That means condenser has to be more cooled. Similarly, drop in evaporator pressure has more effect than that in the condenser as far as system performance is considered. Drop in evaporator pressure causes decrease in mass flow rate. This also causes drop in saturation temperature which increases the compression ratio resulting in decrease in COP. Considerable amount of heat transfer/leakage occurs from/to the connecting tube between the condenser and the throttling device and the connecting tube between the evaporator and the compressor to/from the surrounding.

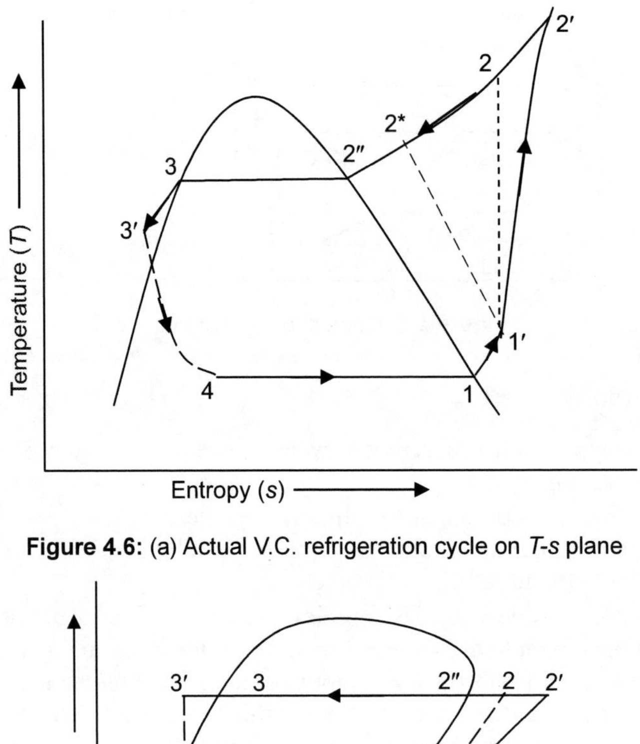

Figure 4.6: (a) Actual V.C. refrigeration cycle on *T-s* plane

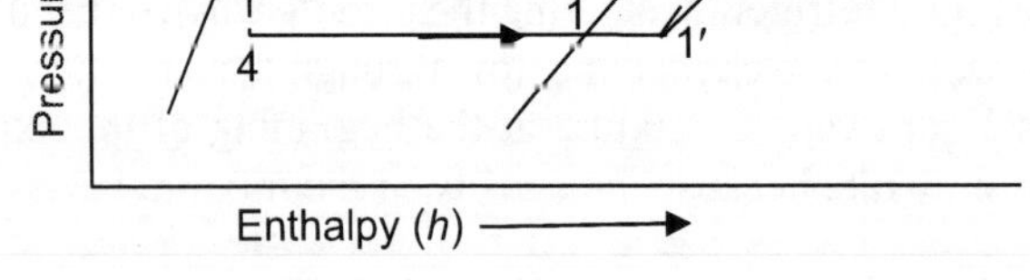

Figure 4.6: (b) Actual V.C. refrigeration cycle on *p-h* plane

4.2.2 Cycle Performance

Vapour compression refrigeration cycle performance is calculated from coefficient of performance (COP) of the system, and expressed as follows. Refer Fig. 4.4 for S-S-S cycle.

Refrigerant flow rate is m kg/s and h is the specific enthalpy of refrigerant.

Compressor work on the system, $W_c = m \times (h_2 - h_1)$ [4.1]

This can also be expressed as $W_c = m \times \int vdp$ (From Tds relationship Eq. 1.23 in Chapter 1) [4.2]

Heat rejection at condenser, $Q_C = m \times (h_2 - h_3) = m \times \{C_{PV} \times (t_2 - t_2) + h_{fg}\}$ [4.3]

Heat absorbed or refrigerating capacity, $Q_e = m \times (h_1 - h_4)$

$Q_e = m \times (h_1 - h_3)$ as $h_4 = h_3$ for isenthalpic process [4.4]

In other way h_4 can be calculated (if dryness fraction x is known at 4)

from $h_4 = h_f + x\, h_{fg}$ (at t_e) [4.5]

COP = Desired effect / Energy input = Refrigerating capacity / Work input

COP = $(h_1 - h_4)/(h_2 - h_1) = (h_1 - h_3)/(h_2 - h_1)$ [m is cancelled] [4.6]

If there is isentropic expansion instead of throttling

Work recovered due to expansion process, $W_T = m \times (h_3 - h_4)$ [4.7]

So net work done, $W_{net} = m \times \{(h_2 - h_1) - (h_3 - h_4)\}$ [4.8]

COP = $(h_1 - h_4)/\{(h_2 - h_1) - (h_3 - h_4)\}$ [4.9]

Detail of the performance calculation is discussed in Example problems.

4.3 PARAMETERS AFFECTING THE PERFORMANCE OF THE VAPOUR COMPRESSION REFRIGERATION SYSTEM

Different operating parameters affect the performance of the refrigeration system. Some of them are superheating of refrigerant vapour before entering the compressor, sub-cooling of refrigerant liquid before entering the expansion device, increase/decrease in evaporator/ condenser pressure, etc. Their effects are discussed below.

4.3.1 Superheating

Superheating of the refrigerant vapour is desirable, as it ensures that the pure vapour is entering the compressor.

Comparing Figs. 4.4 and 4.7, if superheating occurs in the evaporator, the cooling effect is increased from $h_1 - h_4$ to $h_1' - h_4$. If it is superheated by the atmospheric heat which is leaked through the tubing connecting the evaporator to compressor due to improper insulation or the vapour is superheated outside the evaporator, h_1 to h_1' will not be counted towards cooling effect.

Superheating increases the vapour specific volume. This decreases the mass flow rate for volumetric displacement as m = V/v. So it decreases the volumetric efficiency of the compressor and thus cooling capacity decreases.

Superheating increases the compressor specific work from $h_2 - h_1$ to $h_2' - h_1'$ and also increases the compressor discharge temperature. So it decreases the COP.

Due to high compressor discharge temperature the heat rejection from the condenser is increased and more condenser surface area is required.

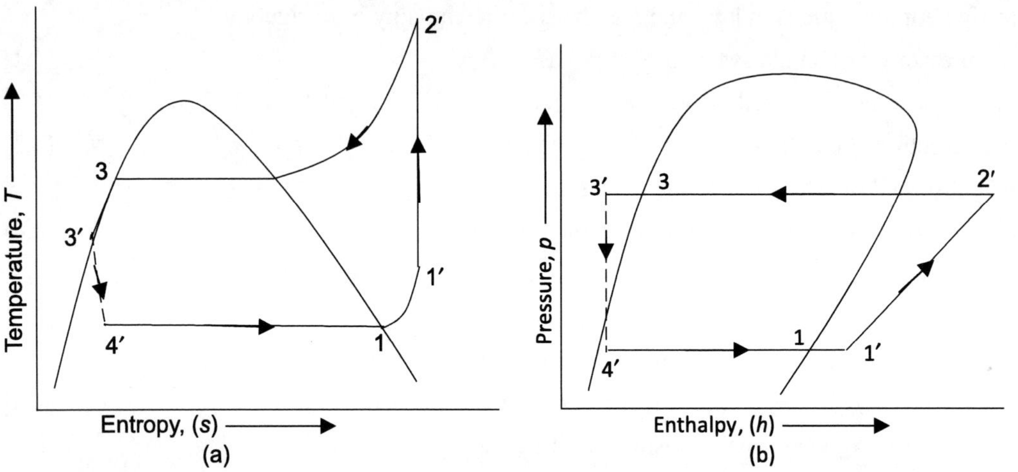

Figure 4.7: Superheating and sub-cooling of refrigerant in V.C. refrigeration system

4.3.2 Sub-cooling

Comparing Figs. 4.4 and 4.7, sub-cooling of the refrigerant from the condenser is desirable as it ensures the refrigerant entering the expansion valve/capillary tube is free from any vapour so that risk of choking will not be there. It also increases the cooling effect from $h_1 - h_4$ to $h_1 - h_4'$. On the other hand, it requires more of the condenser heat transfer surface area for sensible cooling of refrigerant liquid from h_3 to h_3'.

The above discussion shows that superheating and sub-cooling are done more importantly for practical purpose, but not to get an increased COP. However, for refrigerant R12, R22, R134a, etc., the COP increases in some cases but for NH_3 COP decreases due to large superheat area. Superheating and sub-cooling, if done by heat transfer from/or to the environment, put extra thermal load on the system. So internal heat recovery, as shown in Fig. 4.9, system is introduced which increases the COP of the system.

4.3.3 Effect of Evaporator Pressure

As shown in Fig. (4.8 a) P_e is the designed pressure of the evaporator. If the evaporator pressure decreases to P_o' (may be due to more cooling load) the refrigerating effect decreases from $(h_1 - h_4)$ to $(h_1 - h_4)$. Similarly, compressor specific work increases from $(h_2 - h_1)$ to $(h_2 - h_1')$. Hence, the COP decreases. If the evaporator pressure increases, the refrigerating effect increases and the specific compressor work decreases. Thus, COP also increases.

4.3.4 Effect of Condenser Pressure

As shown in Fig. (4.8 b) P_C is the designed pressure. If condenser pressure increases (may be due to leakage of air or non-condensable gases into the system) to P_C, condensation takes place at a higher pressure and corresponding higher saturation temperature. This causes a decrease in refrigerating effect from $(h_1 - h_4)$ to $(h_1 - h_4')$. Specific work of compression increases from

$(h_2 - h_1)$ to $(h_2'' - h_1)$. Hence, the COP decreases. As shown de-superheating zone is increased which requires more condenser heat transfer surface area as the heat transfer coefficient in single phase is much less in comparison to two-phase zone.

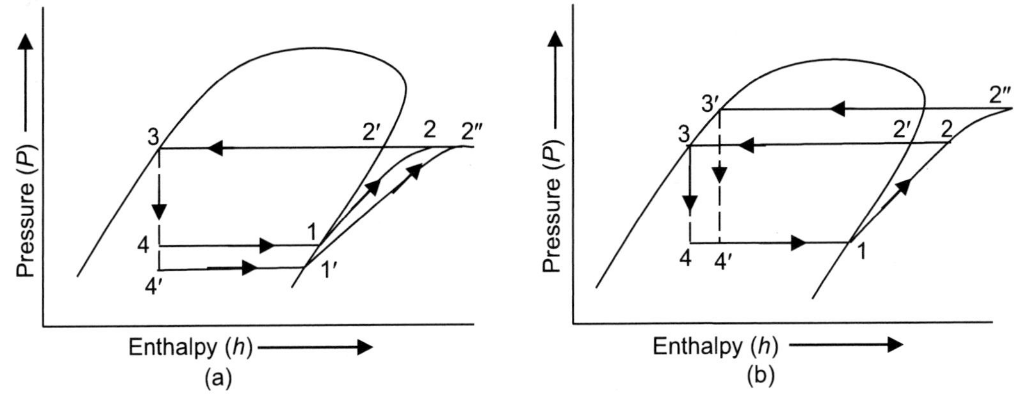

Figure 4.8: (a) Effect of evaporator pressure, (b) effect of condenser pressure

4.3.5 Effect of Presence of Non-condensable Gases in the System

Sometimes the evaporator pressure is maintained at sub-atmospheric pressure to achieve lower evaporator temperature. In this case atmospheric air (a mixture of dry air and water vapour) leaks into the system. Water vapour, being solidified to ice, blocks the small hole of the expansion device. Water vapour may corrode the tube surfaces. Dry air, without being condensed along with the refrigerant in the condenser, comes up to the surface and occupies the surface of the condenser tube. Dry air is a poor conductor of heat. It reduces the heat transfer coefficient. So there will be no complete condensation. Also the condenser pressure becomes the summation of the partial pressures of air and the refrigerant which is higher than the designed condenser pressure. This increases the corresponding saturation temperature and compressor work will also increase due to higher compression ratio. This will reduce the COP.

4.3.6 Internal Heat Recovery by Using Liquid-Gas Heat Exchanger

(Refer Fig. 4.9) Low temperature refrigerant vapour from the evaporator draws the heat from high temperature saturated refrigerant liquid from condenser in a gas to liquid heat exchanger. Thus, the saturated vapour from the evaporator gets superheated and the saturated liquid from condenser gets sub-cooled. The energy balance in the heat exchanger is done from the following equation.

$$C_{PV}(T_1 - T_1) = C_{PL}(T_3 - T_{3'}) \qquad [4.10]$$

The advantage with this internal heat exchanger is that no extra fan power or size of the condenser is required for sub-cooling. The evaporator size need not be made large to get the superheated vapour at the exit of evaporator. Thus, it is advantageous where superheated vapour needs to enter the compressor to avoid any liquid droplet enter the compressor and sub-cooling is needed to avoid any vapour enter the expansion device. Sub-cooling increases the refrigerating effect, but at the same time compression of superheated vapour increases the compressor power consumption. The specific volume of the superheated vapour is high. Hence, the mass flow rate through the compressor decreases. Thus, it is always not true that with internal liquid-gas heat exchanger increases the refrigerating capacity or increases COP. But irreversibility of the system is reduced due to internal heat recovery.

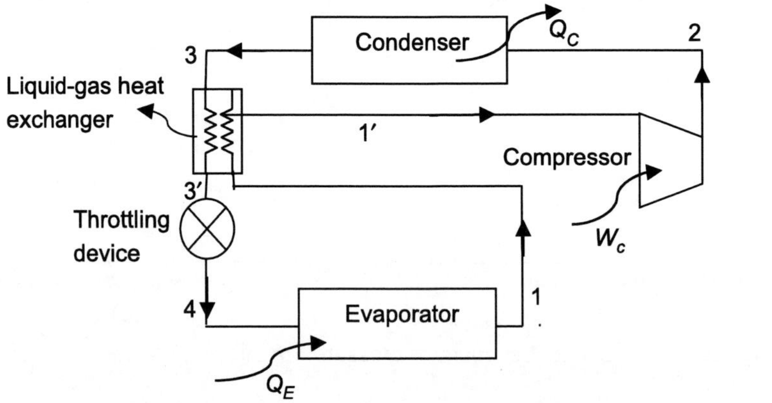

Figure 4.9: Schematic of V.C. refrigeration system with liquid-gas heat exchanger

4.3.7 Sometimes a Question is asked what Happens if the Door of a Refrigerator is Kept Open

Let us discuss it from fundamentals of refrigeration principle. If a refrigerator door is kept open, the cooling load on the evaporator will increase as the outdoor air continuously enters refrigerator. Hence, the compressor goes on running, without stop, as the thermostat cannot reach the pre-set temperature. The exit temperature from evaporator will be highly superheated which will lead to higher compressor discharge temperature. This causes incomplete condensation. Hence, the compression ratio increases. Due to high compression ratio compressor is cut-off by the safety device, i.e., 'high pressure cut-off'. During this process the room temperature will increase, though by very small degrees, due to addition of heat rejected by condenser.

4.4 MULTI-PRESSURE OR MULTI-STAGE REFRIGERATION SYSTEMS

Refer to Fig. (4.8a). When it is desired to achieve lower temperature, corresponding evaporator pressure will be lower. As the evaporator operates at line 4-1 instead of 4-1 the following changes occur.

(a) Compression ratio increases from P_2/P_1 to P_2/P_1' which results in:

(i) increased compressor work and specific work of compression increases by an amount from $h_2 - h_1$ to $h_2' - h_1$. Thus it decreases the COP;

(ii) increased compression ratio causes a decrease in volume of actual mass intake, i.e., from $(V_1 - V_4)$ to $(V_1' - V_4')$. Because the re-expansion of the clearance volume gas occupies more space. So the mass flow rate through the compressor decreases and the cooling effect and COP decrease; and

(iii) higher compressor discharge temperature T_2 to T_2 alters the viscosity and also foaming of lubricating oil occurs. It increases the thermal stress over the compressor blades or seat of the valves. It requires more condenser surface area to reject more amount of heat from $h_2 - h_3$ to $h_2' - h_3$.

(b) Specific volume of refrigerant vapour at lower evaporator pressure increases causing decrease in volumetric efficiency. This decreases the mass flow rate and the cooling capacity. Thus, it decreases the COP.

So to get rid of these demerits multi-pressure refrigeration system is used (normally below –30°C).

Refer to schematic diagram of a multi-pressure refrigeration system and the corresponding thermodynamic cycle on *T-s* plane and *p-h* plane. [Fig. 4.10 (a), 4.10 (b) and 4.10 (c)].

The saturated refrigerant vapour from the evaporator enters the low pressure compressor at 1. It is compressed to an intermediate pressure P_i. This vapour at state point 2, instead of being compressed further to the condenser pressure, it is intercooled to a point very close to the saturation line or to the saturation state at 2′. Then it is compressed by the high pressure compressor to state point 3. **Thus, an amount of compressor work given by area 2-2′-3-2″ is saved and cooling of vapour from 2″ to 3 is eliminated. The peak temperature is also reduced from T_2'' to T_3.** The high pressure and high temperature vapour condense to saturated liquid state at 4. Then the vapour from 4 first expands to the intermediate pressure (P_i) of a flash chamber to point 5. The liquid and vapour are separated to 6 and 2′ (both are at saturated state), respectively.

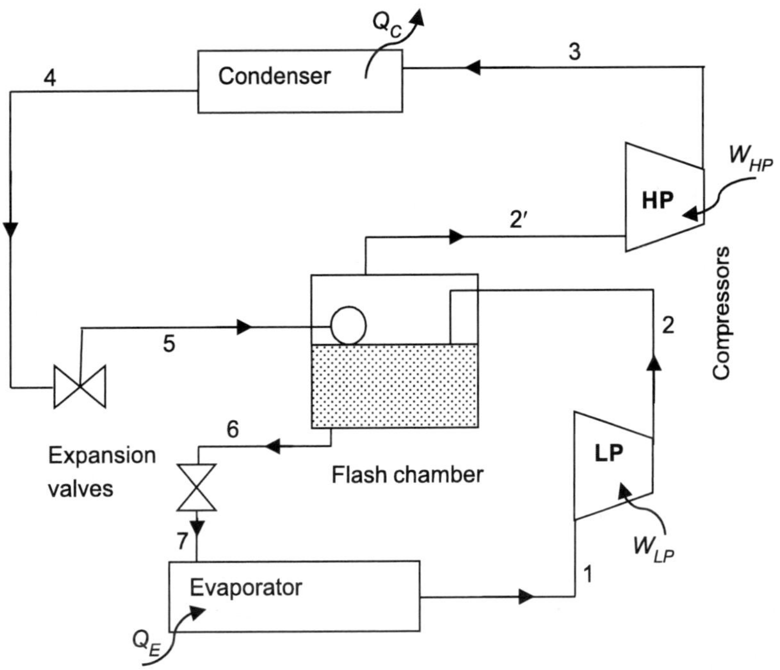

Figure 4.10: (a) Schematic of two-stage V.C. refrigeration system

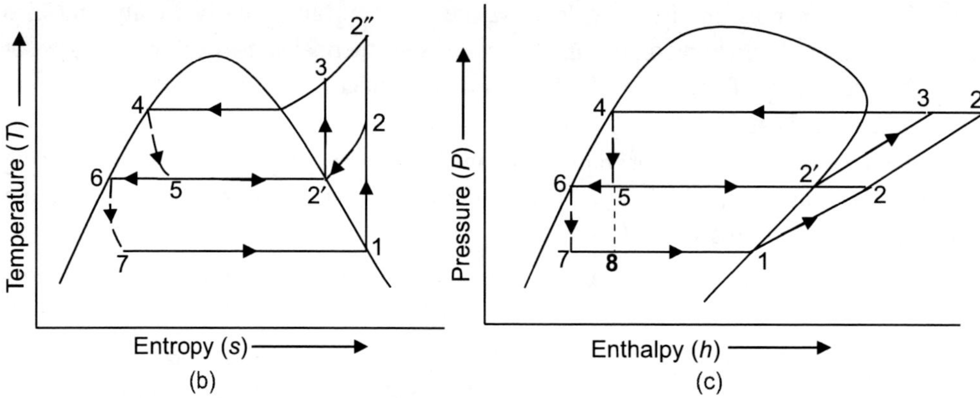

Figure 4.10: Two-stage V.C. refrigeration system on (b) *T-s* plane and (c) on *p-h* plane

The vapour at state 2′ enters the high pressure compressor being mixed with the vapour from low pressure compressor and the saturated liquid at point 6 expands to evaporator pressure at point 7 and enters the evaporator. The space heat is absorbed in the evaporator and the refrigerant comes out of it at 1. Thus, the multi-stage refrigeration cycle is completed.

However, It is observed that as the compression ratio increases, work of compression increases and cooling effect drastically decreases and ultimately decreases the COP. Hence, flash intercooling is desirable where it is advantageous so that cooling effect can be increased from $(h_1 - h_8)$ to $(h_1 - h_7)$. Economic analysis must be done before going for multi-stage refrigeration system. Saving of power must be compared with the cost of the additional equipment required for the system. Refrigerant used, type of compressor, size of the system are also the factors to be compared with single-stage.

4.4.1 Intercooling Methods

In this multi pressure system the important thing to be discussed is the methods of intercooling, its merits and demerits.Because the merits we achieved in multi-staging the refrigeration system is mainly due to intercooling. Intercooling in a refrigeration system can be done by a water-cooled heat exchanger, i.e., water intercoolers or by flash chambers filled with refrigerant. However, a heat exchanger is introduced to cool the high temperature refrigerant vapour from LP compressor to the saturated state.

The discharge temperature for refrigerants like R22, R134a (i.e., CFC group), etc., is not very high. The isentropic compression line runs very close to saturation vapour line. (Refer Figure 2.2 of Chapter 2).This shows superheat horn is small and saving of work is very less. So intercooling in these systems is not of much benefit. Even compressor will need more work, if there is perfect intercooling. Here water intercoolers are used just to reduce the temperature of the refrigerant vapour coming out of the LP compressor and subsequently the temperature of vapour coming out of HP compressor. On the other hand, the compressor discharge temperature for NH_3 is high because its index of compression is high (i.e., 1.31 of NH_3 in comparison to 1.15 of R22). Its compression line is much away from the saturation vapour line. Superheat horn area is more. Hence, intercooling is essential so that there will be more saving of work. It needs flash intercooling as the discharge temperature is much high.

Following are the different methods of intercooling using a flash chamber (in a flash chamber, saturated liquid refrigerant is kept at intermediate pressure P_i and at corresponding saturation temperature).

(a) Flash gas removal

(b) Flash chamber as a liquid sub-cooler

(c) Flash intercooling

(d) Intercooling by flooded type shell and tube heat exchanger (without a flash chamber).

4.4.1.1 Flash gas removal method

In flash gas removal method, high pressure saturated liquid from condenser at 4 expands into a flash chamber at P_i to point 5. The vapour and liquid are separated to 2′ and 6 respectively. The vapour at state 2′ enters the high pressure compressor being mixed with the vapour from low pressure compressor at 2 and attaining a new state point 2′. Thus, the superheated vapour from LP compressor gets inter cooled from point 2 to 2′. The saturated liquid at point 6 expands to evaporator pressure at point 7 and enters the evaporator.

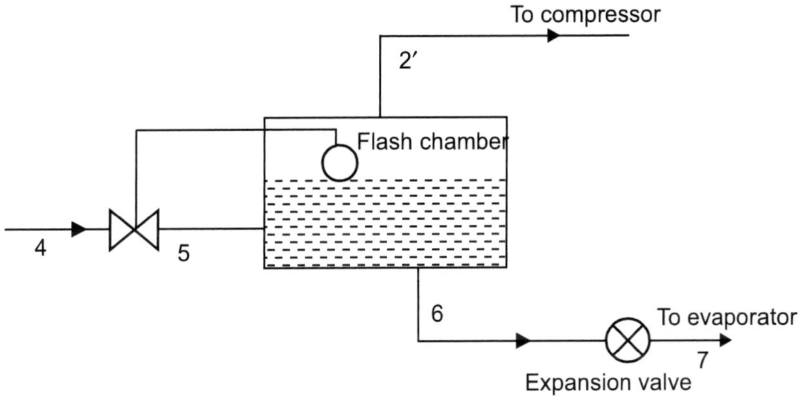

Figure 4.11: Schematic of flash chamber flow circuit

4.4.1.2 Flash chamber as a liquid sub-cooler

High pressure saturated liquid from condenser at 5 is divided into two parts of mass m_1 and m_2. Refrigerant of mass m_2 expands into the flash chamber flashed out as saturated vapour taking the latent heat of vapourisation from the hot refrigerant liquid. Refrigerant mass m_1 is passed through the flash chamber which is at a lower temperature. It rejects heat to the saturated liquid in the flash chamber and gets sub-cooled to 5′. Then this is throttled to the evaporator pressure. The high temperature refrigerant vapour from LP compressor at 2 mixes with low temperature flashed vapour from flash chamber and gets sub-cooled. Then the process continues as discussed above.

4.4.1.3 Flash intercooling

In a flash chamber, saturated liquid refrigerant is kept at intermediate pressure, P_i. When superheated refrigerant vapour at P_i is bubbled through that chamber, some amount of saturated liquid refrigerant flashed to saturated vapour by taking the latent heat of vaporisation from the superheated vapour satisfying the energy equation $m_1C_{PV}(T_2 - T_3) = m_2h_{fg}$. So the amount of vapour going out of the flash chamber is $m_1 + m_2 = m_3$ at saturated vapour state and enters the high pressure compressor. Thus, unnecessary throttling and recompressing the vapour m_2, which is a wasteful energy, can be eliminated as discussed above.

Inter cooling by flooded type shell and tube heat exchanger (without a flash chamber)

Saving of work is also done by having multistaging as explained below.

Saturated liquid refrigerant is throttled to evaporator pressure P_1 to state point 8. At the end more amount of flash vapour is generated which does not contribute to the refrigeration effect. Again this flash vapour is to be recompressed to the condenser pressure. This is a waste of energy. So a suitable intermediate stage (let us take 5) the flashed saturated vapour is separated from saturated liquid. This is done using a flash chamber. The liquid is brought to state 6 and the vapour to state 2′. Then the liquid is throttled to 7 and the vapour along with the vapour from LP compressor is compressed to 3. Thus, the advantages we get: (a) More of refrigerating effect (i.e., 7 to 1 instead of 8 to 1); (b) saving of compressor work for compressing the mass of vapour of an amount $(m - m_1)$ from P_1 to P_i.

A simple energy conservation analysis gives the idea of saving of work due to multistaging of refrigeration system.

For single-stage system: Work done = $m_S(h_2'' - h_1)$; where m_S is the mass flow rate for this system.

For two-stage system: Work done = $m_{LP}(h_2 - h_1) + m(h_3 - h_2')$ [4.11]

So saving of work = $\{m_S(h_2 - h_1)\} - \{m_{LP}(h_2 - h_1) + m(h_3 - h_2')\}$

$$= m_S(h_2'' - h_1) - \{(m - m')(h_2 - h_1) + m(h_3 - h_2'')\}$$

(m' is the mass of vapour separated in the flash chamber; so $m_{LP} = m - m'$)

Rearranging,

Saving of work = $m_S(h_2'' - h_1) - [m(h_2 - h_1) + m(h_3 - h_2'')] + m'(h_2 - h_1)$ [4.12]

The last term (within lunar bracket) represents the work saved due to flash gas removal. This is shown numerically in Example 4.12.

4.4.2 Selection of Intermediate Pressure for Minimum Work Input

Figure 4.12 shows single-stage cycle 1-2-7-a-1 with a pressure limit of P_C and P_e.

With the same pressure limit, the cycle is to operate in two-stage with intermediate pressure P_{i1}.

The two-stage cycle is 1-3-3′-5-7-9-c-1 Saving of work is 2-3-3′-5 and increase in refrigerating effect is area under a-c.

Now the intermediate pressure is changed to (P_{i2}). The new two stages cycle is 1-4-4′-6-7-8-b-1. Saving of work is 2-4-4′-6-2 and decrease in refrigerating effect compared to cycle having intermediate pressure (P_{i1}) by area under *b-c*. It is seen that saving of work depends upon the slope of the saturated vapour line and that of constant pressure line and also effectiveness of intercooling. Similarly, amount of decrease in refrigerating effect depends upon the slope of the saturated liquid line.

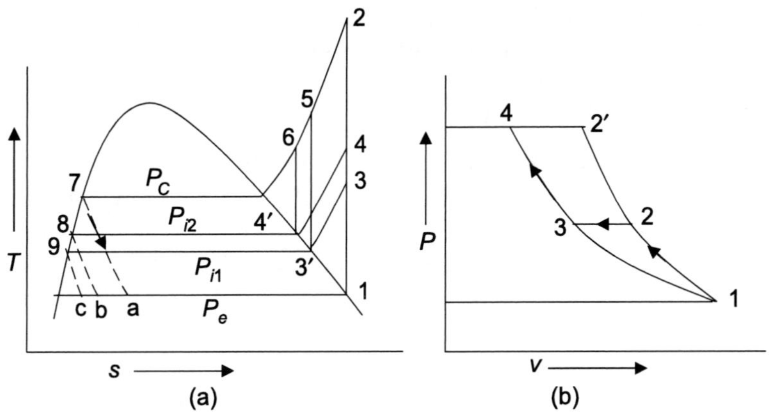

Figure 4.12: (a) Effect of variation of intermediate pressure, and (b) work done in two-stage compression

However, calculation shows that as the intermediate pressure increases saving of work increases first and then decreases. Thus, COP also varies accordingly. Hence, there is an optimum intermediate pressure which gives the maximum COP for that refrigerant, evaporator and condenser pressure.

Refer Fig. 4.12 (Assuming same index of compression)

Work done during the process 1-2, $W_{12} = \frac{n}{n-1} P_1 v_1 \left[\left(\frac{P_2}{P_1} \right)^{\frac{n-1}{n}} - 1 \right]$ [4.13]

Work done during the process 3-4, $W_{34} = \frac{n}{n-1} P_3 v_3 \left[\left(\frac{P_4}{P_3} \right)^{\frac{n-1}{n}} - 1 \right]$ [4.14]

$P_2 = P_3 = P_i$; $P_1 = P_e$; $P_4 = P_C$; and putting the relation $Pv = mRT$ in the above equations and adding both

Net work done, $W_{12} + W_{34} = W_{Net} = \frac{n}{n-1} m_1 R T_1 \left[\left(\frac{P_i}{P_e} \right)^{\frac{n-1}{n}} + \left(\frac{P_C}{P_2} \right)^{\frac{n-1}{n}} - 2 \right]$ [4.15]

For Minimum work, differentiating W_{Net} with respect to the only variable P_i (because P_C and P_e are fixed) and equating it to zero.

Differentiating, $\frac{P_i^{-1/n}}{P_e^{(n-1)/n}} - \frac{P_C^{(n-1)/n}}{P_e^{(2n-1)/n}} = 0$ or $P_i^2 = P_e P_C$

$$P_i = \sqrt{P_C P_e} \qquad [4.16]$$

This shows pressure ratios for the two-stages of compression are same. Hence, work done for each stage of two-stage system is same.

For multi-stage systems, it can be written as

$$\frac{P_{i1}}{P_e} = \frac{P_{i2}}{P_{i1}} = \frac{P_{i3}}{P_{i2}} = \frac{P_C}{P_{i3}} = \left(\frac{P_C}{P_e} \right)^{1/4} \qquad [4.17]$$

Note: Where there is no mention of intermediate pressure or temperature, intermediate pressure should be taken as the geometric mean of the condenser and evaporator pressure.

4.5 CASCADE REFRIGERATION SYSTEM

Multi-pressure refrigeration system has the following limitations.

The minimum temperature attainable by multi-pressure system with low boiling refrigerants like R22, R134a, etc., is around –100°C. For production of still lower temperature (i.e., below –100°C) multi pressure systems will have very low volumetric efficiency or evaporator pressure has to be maintained much below the atmospheric pressure. Hence, for production of very low temperature (normally < –100°C and > – 150°C), the cascade refrigeration system is considered. The study of behaviour of materials and applications of temperature below –150°C, is known as the cryogenic applications. This low temperature is produced by other non-conventional methods which is described in later chapters.

4.5.1 Working Principle of Cascade Refrigeration System

Like binary cycle in heat engine, here one vapour compression cycle operates in conjunction with another. The lower one is called low temperature (LT) cycle and the upper one is called high temperature (HT) cycle as shown in the Fig. (4.13).

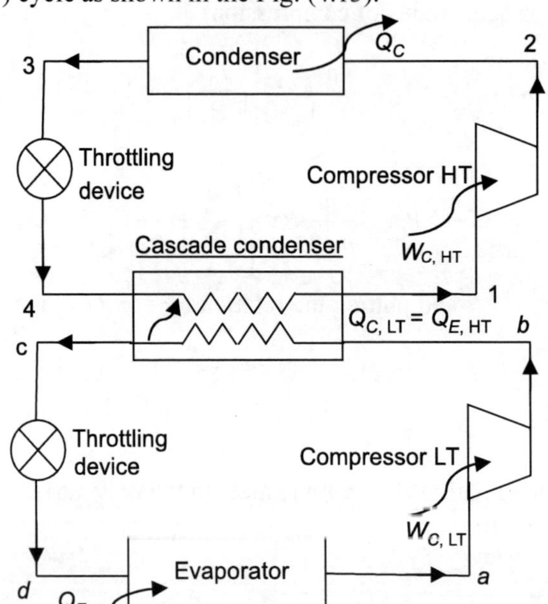

Figure 4.13: Schematic of a cascade refrigeration system

In the LT cycle a very low boiling refrigerant like R13 (NBP –82°C) is the refrigerant. Like the single-stage vapour compression cycle (Fig. 4.14), the low boiling refrigerant (R13, R14, R23, etc.) in the LT cycle is compressed in LT compressor to the condenser pressure, rejects heat to the evaporator of the HP cycle and throttled to the evaporator pressure. There it boils off receiving heat from the refrigerated chamber. Similarly, high boiling refrigerant (R22, NH_3, R134a, etc.) is compressed to high pressure side condenser by HT compressor. Then the saturated liquid refrigerant from the condenser is throttled to HP side evaporator pressure. In the evaporator the two-phase refrigerant receives the heat from the LT side condenser and boils off. Thus, this heat exchanger (Fig. 4.13) is called 'cascade condenser' where the LT side condenser rejects heat to the HP side evaporator. In ideal case the temperatures of LT side condenser and HT side evaporator are assumed to be same and the optimum temperature (at which maximum COP is obtained) is calculated to be $\sqrt{(T_C T_E)}$. In practice a minimum temperature gradient of 5°C is maintained for the heat transfer from LT side condenser to HT side evaporator as shown in Fig. 4.14.

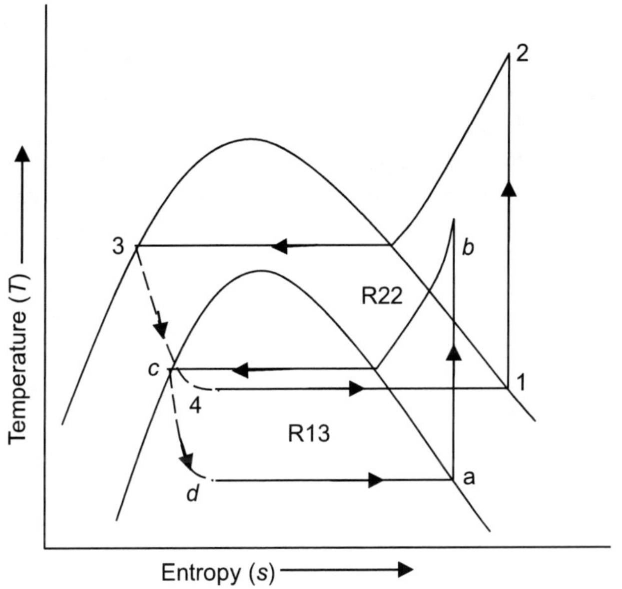

Figure 4.14: Cascade refrigeration system on *T-s* plane

COP Calculation: COP of the Cascade system system is calculated following the similar procedure.

Heat rejected by LT condenser = Heat absorbed by HT evaporator

So $m_{LT} \times (h_b - h_c) = m_{HT} \times (h_1 - h_4)$;

This gives $m_{HT} = m_{LT} \times (h_b - h_c)/(h_1 - h_4)$ [4.18]

Work done by LT and HT compressors $= m_{LT} \times (h_b - h_a) + m_{HT} \times (h_2 - h_1)$ [4.19]

So COP $= (m_{LT}) \times (h_a - h_d)/\{m_{LT} \times (h_b - h_a) + m_{HT} \times (h_2 - h_1)\}$ [4.20]

4.6 PRODUCTION OF SOLID CARBON DIOXIDE OR DRY ICE

The solid state of carbon dioxide is known as dry ice. Before discussing its production process, let us know the relevant thermodynamic properties of CO_2.

The critical point: $P_{critical} = 73.8$ bar, $T_{critical} = 31.0$°C

The triple point: $P = 5.18$ bar, $T = -56.6$°C

CO_2 is available in the atmosphere at 1.0 bar and 30.0°C (say).

The process of producing dry ice is as shown in Fig. 4.15.

The CO_2 at atmospheric pressure (state point 1) is compressed to a pressure near the critical pressure (around 64.4 bar) (state point 2), where the corresponding saturation temperature is 25°C. The highly compressed CO_2 enters the condenser and rejects heat to a cooling medium (temperature must be less than at least 20°C). It comes out as saturated liquid (state point 3). Then it is expanded in a snow chamber or throttled to a state point below triple point where it exists in solid and vapour state (state point 4). Now the solid CO_2 (state 5) and vapour CO_2 (state 6) are separated. This CO_2 mixes with that of the atmosphere (state 7) to be at state 1.

In this process there are following drawbacks.

(a) Compression ratio is very high (around 65). This consumes large amount of power. Hence, multi-pressure compression (two-stage or three-stage) is adopted.

(b) Condenser temperature is very low. So an external cooling medium is required for which cascade system is considered.

(c) Direct expansion from high condenser pressure to atmospheric pressure causes loss of work (as discussed in multi-pressure system) and throttling may also result in choking due to ice formation. So flash intercooler is used and liquid CO_2 is preferably expanded to a snow chamber.

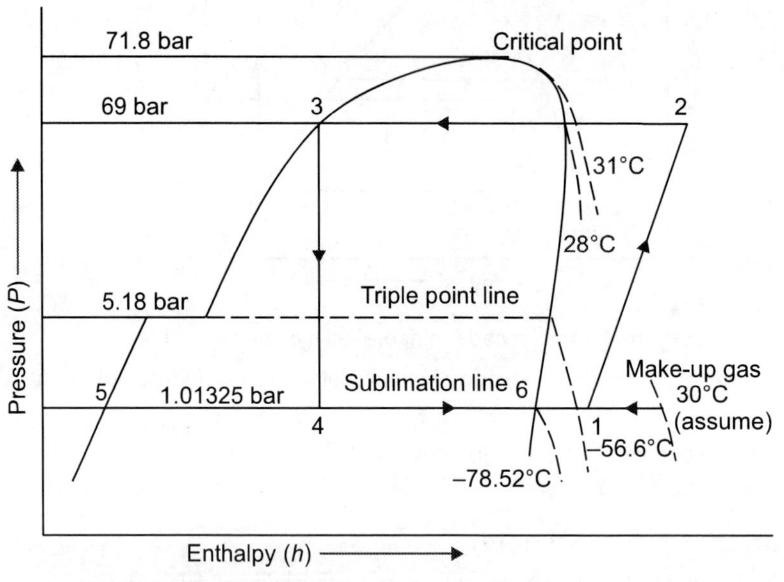

Figure 4.15: Principle of solid CO_2 production on *p-h* plane

The actual production system is explained below and this gives a good example of multi-pressure system and cascade refrigeration system.

Figures 4.16 and 4.17 show the schematic and thermodynamic process on *p-h* plane for the production of dry ice respectively. CO_2 gas at state 1 is compressed by the LP compressor to the intermediate pressure 6.0 bar (state 2). It is intercooled to state 3. Then it is mixed with the CO_2 vapour coming from the flash intercooler at state 9 to attain the state 4. The mixed vapour is again compressed by the HP compressor to a pressure 20.0 bar (state 5). The corresponding saturation temperature is –20°C. This high pressure CO_2 vapour enters the cascade condenser (as it is called in a cascade refrigeration system), where it rejects heat to the evaporator of the HT side (NH_3 is the working fluid) of the cascade system which can be maintained well below at –25°C. Then it comes out of the cascade condenser as saturated liquid at state 6 and expanded through a flash intercooler which is maintained at a pressure of 6.0 bar (little above the triple point pressure so that there is no chance of solidification of CO_2 and blocking of the circuit). Here the liquid and vapour get separated. The vapour at state 9 mixes with vapour at state 3 and the saturated liquid is directly expanded to a snow chamber at atmospheric pressure. Since CO_2 is below its triple point, there will be only solid and vapour state of CO_2 and they are separated which is shown at state 11 and state 12, respectively. The solid state of CO_2 thus formed is like snow. So this is again compressed by a snow compressor to get block of dry ice. The vapour at state 12 mixes with the CO_2 from atmosphere to reach state 1 and thus completes the cycle.

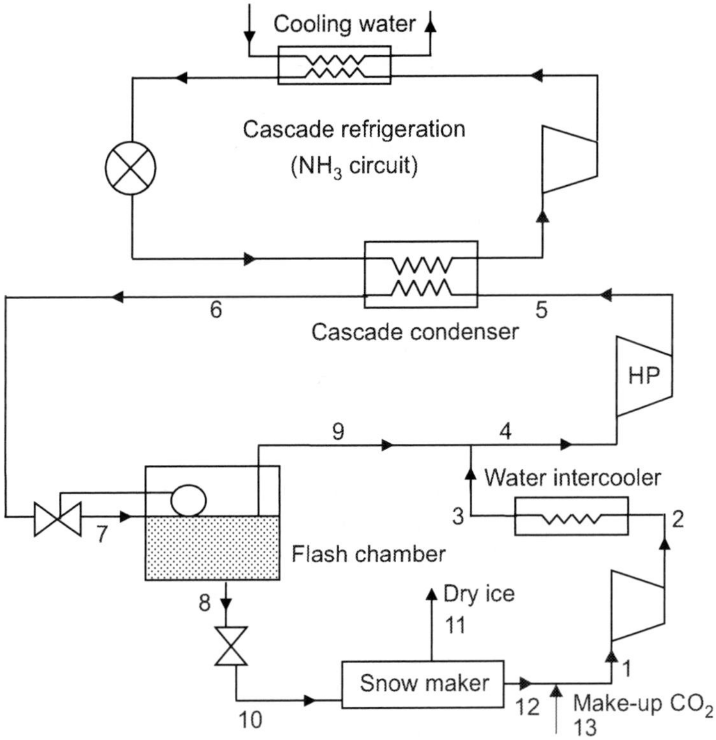

Figure 4.16: Schematic of production of dry ice

The above system is a two-stage compression with a cascade refrigeration system. We may have three-stage compression system where the pressures at different stages may be 1.0 bar, 6.0 bar, 20.0 bar, and 69 bar respectively. In this system the condenser temperature will be around 28°C (saturation temperature at 69.0 bar). So the flash intercoolers are used in intermediate stages. The disadvantages in this case are high cost and difficulty in material selection due to high pressure; also rejection of heat from the condenser which is at 28°C is difficult due to non-availability of a natural cooling medium at such low temperature. However, chilled water from cooling tower can be used as coolant for third stage condenser.

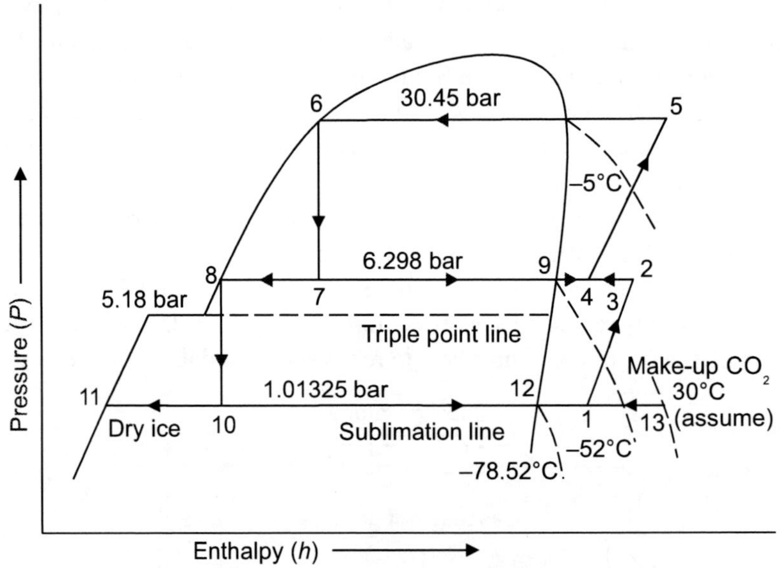

Figure 4.17: Dry ice production process on *p-h* plane

4.6.1 Use of Dry Ice

The most common use of dry ice is to preserve food, biological samples, etc. It is mostly useful when it is required for a short time during transportation of perishable goods. The largest mechanical use of it is 'blast cleaning'. Dry ice pellets are shot from a nozzle with compressed air with the action of the sublimation. This removes the residues from industrial equipment. Besides it is used to loosen asphalt floor tiles, removal and fitting of cylinder liners in large engines. Dry ice chills the liner and it shrinks. So it freely slides into the engine block. Then it is warmed and expanded resulting in interference fit. It is also useful as a cutting fluid. Thus it has many more applications in scientific laboratories, create artificial fog in theatres and in removing warts (surgery), etc., where non-cyclic refrigeration is required. But prolonged exposure to it may cause skin damage (frost bite) and hypercapnia.

4.7 USING OF REFRIGERANT TABLES

Refrigerant property tables are referred to solve the numerical problems on performance and design of a refrigeration system. Some tables are prepared with the reference value at 0°C (See Appendix 5), i.e., the value of liquid specific enthalpy of the refrigerant at 0°C is 200 kJ/kg and liquid specific entropy is 1.00 kJ/kg-K and some others are prepared with reference to –40°C (See Appendix 4), i.e., at –40°C the value of liquid specific enthalpy of the refrigerant is 0.0 kJ/kg and liquid specific entropy is 0.0 kJ/kg-K. Hence, the absolute numerical values of the refrigerant properties will be different at any particular temperature or pressure, but the difference of those numerical values between any two state points will be same. Usually, for refrigeration system designs, the differential values between two state points are considered. So we should not confuse if the tables are prepared with different reference values.

How to use the refrigerant tables: Any fluid or refrigerant thermodynamic properties are tabulated on the basis of either saturated pressure (pressure based table) or saturated temperature (temperature based table). The convenient one is to be selected. Corresponding to saturated temperature or pressure the corresponding values of specific volume (v), specific enthalpy (h), and specific entropy (s) are given. The subscript f, g and fg stand for saturated liquid, saturated vapour and latent heat of vaporisation ($h_g - h_f$). The superheated properties are presented in another table, called superheated table. The properties (specific volume, enthalpy and entropy) of the fluid in superheated state (as of state point 2 in Fig. 4.4) are tabulated.

Interpolation: The properties are tabulated in an interval of different temperatures (2°C, 5°C, etc.). If the value to be estimated falls within this temperature range, the value is estimated more accurately by interpolation (preferably by linear interpolation as the refrigerant properties vary, almost, linearly for a small temperature interval). Example, in the following table, X_1, X_2 and corresponding Y_1, Y_2 values are given. It is required to find out the value of X whose value lies in between X_1 and X_2. Then corresponding Y is calculated from following equation.

$(X - X_1)/(X_2 - X_1) = (Y - Y_1)/(Y_2 - Y_1)$.

X_1	Y_1
X	Y
X_2	Y_2

Similarly, the properties of refrigerant in two-phase zone is estimated by knowing its quality or dryness fraction (x). This is defined as the mass of vapour present in the total mass of liquid vapour mixture. Suppose the pressure (p_{Sat}) and enthalpy (h) of a refrigerant is given. From the saturation table of the refrigerant, corresponding t_{Sat}, h_f, and h_g can be found out. Now dryness fraction (x) can be calculated from $x = (h - h_f)/(h_g - h_f)$.

How to find the state point 2 in the following example

Entropy at 1 is s_1 corresponding to evaporator saturation pressure and temperature (P_e and T_e).

Entropy at 2 is s_2 corresponding to condenser saturation pressure and temperature (P_C and T_C)

For isentropic process, $s_2 = s_1 = x$ (assume).

Since 2 is in superheated state, x is to be found out from the superheated table for T_C and P_C as shown below.

Let's say, x lies in between 15° and 20° superheat.

Assuming a linear variation, the interpolation ratio will be $(x - m)/(m - n) = a/b$

So degree of superheat ΔT can be calculated from $a/b = (\Delta T - 15)/(20 - 15)$

The value of $T_2 = T_{Sat} + \Delta T$

Similarly, h_2 can be calculated.

Superheated table for T_C and P_C (temperature based)

T_{Sat} °C, P_{Sat}	Degree of superheat								
	5	10	15	20	25	30	35	40…….	…….
V									
H									
S			*m*	*n*					

In some data book superheated table is given as below (pressure based).

P, Corresponding T_{Sat} °C	Superheat temperature								
	5	10	15	20	25	30	35	40…….	…….
V									
H									
S			*m*	*n*					

Example Problems

Example 4.1

A room air conditioner of 1.5 TR capacity works on vapour compression cycle. The temperature limits of the cycle are 52°C and 7°C. The refrigerant used is R134a. Calculate (a) the refrigerant flow rate, (b) compressor exit temperature, (c) power consumed by the compressor, (d) compressor swept volume, (e) heat rejected from the air conditioner to the environment, and (f) COP.

Ans:

The state of refrigerant is not mentioned at any point of the cycle. Hence, single-stage saturation (S-S-S) cycle may be assumed. (Refer Fig. 4.4). Now the properties of the refrigerant of the known state points are taken from the Refrigerant Properties Tables. Temperature limits means the temperature of condenser is 52°C and that of evaporator is 7°C for which the system is designed.

The refrigerant at 7°C is coming out of the evaporator in saturated state. The properties of R134a at that state point are: P_{Sat} = 0.3748 MPa, specific volume (v_1) = 0.05462 m^3/kg, specific vapour enthalpy (h_1) = 402.63 kJ/kg, specific vapour entropy (s_1) = 1.7235 kJ/kg-K.

Similarly, properties at state point 3 (saturated liquid): P_{Sat} = 1.3854 MPa, specific liquid enthalpy (h_3) = 274.74 kJ/kg, $h_3 = h_4$ as the process 3-4 is isenthalpic.

Properties at state point 2 can be determined from the superheated table as follows.

Process 1-2 is isentropic compression, so $s_1 = s_2$ = 1.7235 kJ/kg-K.

State point 2 is on the constant pressure line 1.3854 MPa. Now let us check the superheated table corresponding to 1.3854 MPa and at what degree of superheat lies the specific entropy value 1.7235 kJ/kg-K. (If the values are not available directly, interpolation may be done to get more accurate values.) Those values (v, h, s) are the corresponding values of state point 2. Thus, t_2 = 55.5°C, h_2 = 428.01 kJ/kg-K.

(a) Refrigerant mass flow rate, $m = Q_e / (h_1 - h_4) = (1.5 \times 3.5167) / (402.63 - 274.74)$ $= 0.0412$ kg/s

(b) Compressor swept volume, $m \times v_1 = 0.0412 \times 0.05462 = 2.25 \times 10^{-3}$ m^3/s

(c) Compressor discharge temperature, $t_2 = 55.5$°C

(d) Power consumed by compressor, $m (h_2 - h_1) = 0.0412 (428.01 - 402.63) = 1.0456$ kW

(e) Condenser heat rejection, $m (h_2 - h_3) = 0.0412 \times (428.01 - 274.74) = 6.315$ kW

(f) COP = cooling effect/power consumed = $(1.5 \times 3.5167)/1.0456 = 5.045$

(g) $(\text{COP})_{RC} = (7°\text{C} + 273)/(52°\text{C} - 7°\text{C}) = 6.222$

[If superheated table is not available, the properties of state point 2 can be calculated taking suitable average vapour specific heat value between state point 2 and 2′].

Thus, t_2 can be calculated from the relation $s_2 - s_2' = C_{P,\text{vap}} \ln(T_2 - T_2')$ (for isobaric process) $(1.7235 - 1.7064) = 1.32 \ln (T_2/325)$ ($C_{P,\text{vap}}$ is taken 1.32 kJ/kg-K).

This gives,

$$T_2 = 329.2 \text{ K} = 56.2°\text{C and } h_2 = h_2 + C_{P,\text{vap}} (t_2 - t_2') = 424.15 + 1.32\ (4.2) = 429.7 \text{ kJ/kg}$$

The minor difference of values in t_2 and h_2 are due to unreliable prediction of specific heat values of superheated vapour.

To check the correctness of calculation energy balance is done on the system.

Energy In = $W_c + Q_e = (1.0456 + 5.28)$ kW = 6.325 kW

Energy Out = $Q_C = 6.315$ kW

Energy In ≈ Energy Out

Example 4.2

The data of Example 4.1 remains same. In place of R134a, R22 is considered.

Ans:

We proceed in the same manner as in Example 4.1.

R22 properties are taken from the table where the reference is $h_f = 0.0$ kJ/kg and $s_f = 0.0$ kJ/kg-K

R22 properties at 7°C : $P_{\text{sat}} = 0.6211$ MPa $= P_1$, $v_g = 0.0380$ m^3/kg $= v_1$, $h_g = 253.5$ kJ/kg $= h_1$, $s_g = 0.9215$ kJ/kg-K $= s_1$

At 52°C: $P_{\text{sat}} = 2.0227$ MPa $= P_2$, $h_f = 110.83$ kJ/kg $= h_3 = h_4$, $h_g = 262.37$ kJ/kg $= h_2$, $s_g = 0.8596 = s_2$

Properties at state point 2 are calculated from R22 superheated table corresponding to 2.0227 MPa.

(Since values up to 1.9327 MPa (50°C) are given, extrapolations of the values are done.)

Compressor discharge temperature, $t_2 = 52 + 20.66 = 72.66$°C (20.66 is the degree of superheat), $h_2 = 283.123$ kJ/kg

(a) Refrigerant mass flow rate, $m = Q_e / (h_4 - h_1) = (1.5 \times 3.5167) / (253.5 - 110.83) = 0.037$ kg/s

(b) Compressor swept volume, $m \times v_1 = 0.037 \times 0.0380 = 1.406 \times 10^{-3}$ m³/s

(c) Compressor discharge temperature, $t_2 = 72.66$°C

(d) Power consumed by compressor, $m\,(h_2 - h_1) = 0.037\,(283.123 - 253.5) = 1.096$ kW

(e) Condenser heat rejection, $m\,(h_2 - h_3) = 0.037 \times (283.123 - 110.83) = 6.375$ kW

(f) COP = cooling effect/ power consumed = (1.5 × 3.5167)/1.096 = 4.813

(g) $(\text{COP})_{RC}$ = (7°C + 273) / (52°C – 7°C) = 6.222

[If superheated table is not available, the properties of state point 2 can be calculated taking suitable average vapour specific heat value between state point 2 and 2′.

Thus, t_2 can be calculated from the relation $s_2 - s_2' = C_{P,\text{vap}} \ln(T_2 - T_2')$ (for isobaric process) $(0.9215 - 0.8596) = 1.02 \ln (T_2/325)$ ($C_{P,\text{vap}}$ is taken 1.02 kJ/kg-K).

This gives $T_2 = 345.34$ K, $t_2 = 72.34$°C and $h_2 = h_2' + C_{P,\text{vap}}\,(t_2 - t_2') = 262.37 + 1.02\,(20.34) = 283.12$ kJ/kg

The minor difference of values in t_2 and h_2 are due to unreliable prediction of specific heat values superheated vapour]

To check the correctness of calculation energy balance is done on the system.

Energy In = $W_c + Q_e$ = (1.096 + 5.275) kW = 6.371 kW

Energy Out = Q_C = 6.375 kW

Energy In ≈ Energy Out

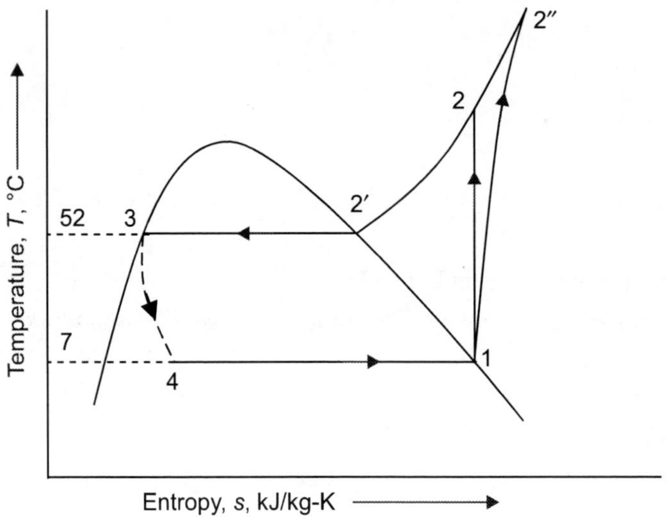

Figure 4.18

If non-isentropic compression is considered

Assume the compressor isentropic efficiency is 85%. (Fig. 4.18)

So $0.85 = (t_2 - t_1) / (t_2'' - t_1) = (72.66 - 7) / (t_2'' - 7)$

This gives $t_2 = 84.25$°C

So degree of superheat is 32.25°C and $h_2 = 293.65$ kJ/kg

Power consumed by compressor = 0.037 × (293.65 – 253.5) = 1.486 kW

COP = 1.5 × 3.5167/ 1.486 = 3.55

Example 4.3

A cold storage of 100 TR capacity uses NH_3 as refrigerant and works within a temperature range of –10°C and 45°C. The refrigerant comes out of the evaporator at 0°C and leaves the condenser at 38°C. Calculate (a) the refrigerant flow rate, (b) compressor exit temperature, (c) power consumed by the compressor, (d) compressor swept volume, (e) heat rejected from the air conditioner to the environment, and (f) COP.

Answer: Refer Fig. 4.19

Refrigerant coming out of the evaporator is being superheated at state point 1′. So the properties at state point 1 are evaluated from superheated table at corresponding saturated pressure 0.291 MPa. Specific volume, $v_1' = 0.435$ m^3/kg, $\;' = 1457.5$ kJ/kg, $s_1 = 5.56$ kJ/kg-K.

After isentropic compression, properties at state point 2 are evaluated from superheated table at corresponding saturation pressure 1.782 MPa. $t_2 = 138°C$, $h_2 = 1746.5$ kJ/kg.

The liquid refrigerant coming out of the condenser is sub-cooled by 7°C, i.e., the liquid temperature is 38°C. Liquid enthalpy at 38°C is calculated from $(h_{f45} - C_{Pf} \times 7°C)$. Or enthalpy of saturated liquid at 38°C $(h_{f,\,38})$ can also be taken without much error because in liquid phase the constant pressure line runs very close to saturation line. Here $h_{f\,38}$ is taken, i.e., 361.9 kJ/kg. Now $h_3' = h_4$

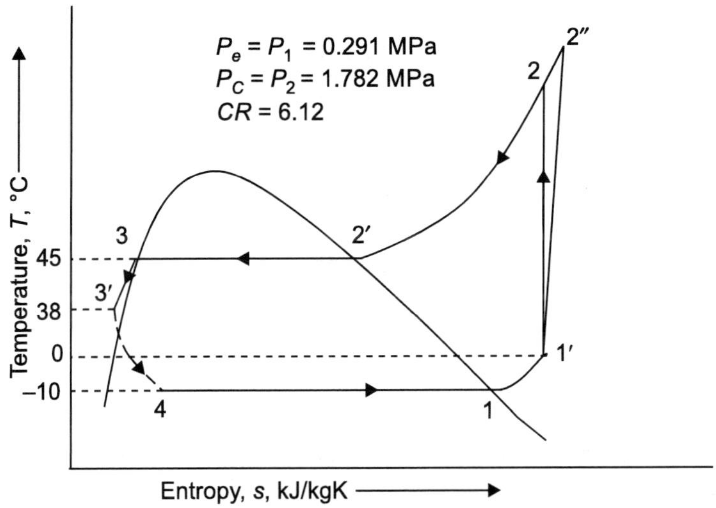

Figure 4.19

(a) Refrigerant mass flow rate, $m = Q_e/(h_1' - h_4) = (100 \times 3.5167)/(1457.5 - 361.9) = 0.32$ kg/s

(b) Compressor swept volume, $m \times v_1 = 0.32 \times 0.435 = 0.14$ m^3/s

(c) Compressor discharge temperature, $t_2 = 138.0°C$

(d) Power consumed by compressor, $m\,(h_2 - h_1') = 0.32\,(1746.5 - 1457.5) = 92.48$ kW

(e) Condenser heat rejection, $m\,(h_2 - h_3') = 0.32 \times (1746.5 - 361.9) = 443.072$ kW

(f) COP = cooling effect/ power consumed = $(100 \times 3.5167)/\,92.48 = 3.8$

(g) $(COP)_{RC} = (-10°C + 273)\,/\,(45°C + 10°C) = 4.78$

Energy balance of the system gives 351.67 kW + 92.48 kW = 444.15 kW ≈ 443.072 kW

Example 4.4

Sometimes refrigeration systems are subjected to pressure limits depending upon the tube material, refrigerants, sealing and working conditions.

A R22 vapour compression refrigeration system of 3 TR capacity works within a pressure range of 0.25 MPa to 1.5 MPa. The refrigerant vapour enters the compressor at –5°C and leaves the condenser at 35°C. Calculate: (a) COP, (b) swept volume of the compressor, assuming 80% volumetric efficiency, if the compressor runs at 600 rpm.

Ans: Refer Fig. 4.20

From R22 refrigerant table, the saturation temperature corresponding to 0.25 MPa is –19.5°C and 1.5 MPa is 39°C.

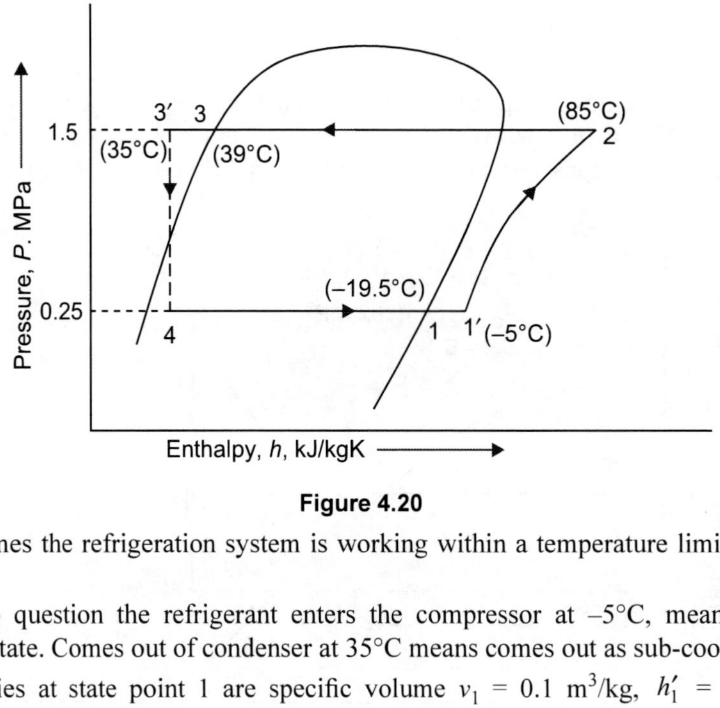

Figure 4.20

Now it becomes the refrigeration system is working within a temperature limit –19.5°C and 39°C.

According to question the refrigerant enters the compressor at –5°C, means it enters in superheated state. Comes out of condenser at 35°C means comes out as sub-cooled liquid.

Now properties at state point 1 are specific volume v_1 = 0.1 m^3/kg, h_1' = 251.61 kJ/kg, s_1 = 0.9941 kJ/kg–K

State point 2: t_2 =85°C, h_2 = 301.1 kJ/kg

State point 3′, h_3 = 88.7 kJ/kg, i.e., h_f at 35°C. $h_3' = h_4$.

Mass flow of refrigerant, m = (3 × 3.5167) / (251.61 – 88.7) = 0.065 kg/s

Work done by compressor, W_c = 0.065 × (301.1–251.61) = 3.217 kW

COP = 3 × 3.5167/ 3.217 = 3.28

Volume swept by the compressor / second, mv_1 / η_{vol} = (0.065 × 0.1 / 0.8) = 8.125 × 10^{-3} m^3/s

Cylinder swept volume, N/60 × (8.125 × 10^{-3}) = 600/60 × (8.125 × 10^{-3}) = 8.125 × 10^{-4} m^3.

Example 4.5

A refrigeration system of 1.5 TR capacity works on vapour compression cycle. The temperature limits of the cycle are 52°C and 7°C. The refrigerant used is R134a. The system uses a liquid-gas heat exchanger to improve the performance of the system. The saturated refrigerant vapour coming out of the evaporator gets superheated to 15°C in the heat exchanger. Calculate (a) the refrigerant flow rate, (b) compressor exit temperature, (c) power consumed by the compressor, (d) compressor swept volume, (e) heat rejected from the air conditioner to the environment, and (f) COP. Compare the performance of the system with that given in Example 4.1.

Ans: Refer Fig. 4.8 for schematic of the system. Figure 4.19 for diagram on *T-s* plane (data are not same).

The saturation pressures corresponding to 7°C and 52°C are $P_1 = 0.377$ MPA and 1.39 MPa.

Refrigerant properties at state point 1′ are calculated by interpolation from superheated table 0.377 MPa and 15°C.

Specific volume $v_1' = 0.056$ m^3/kg, $h_1' = 409.5$ kJ/kg, $s_1' = 1.75$ kJ/kg-K

At state point 2: $t_2 = 64$°C and $h_2 = 438.9$ kJ/kg.

Since this is an internal heat exchanger, amount of rejected heat by the liquid refrigerant from condenser is equal to amount of heat gained by the refrigerant vapour from evaporator.

So $h_1' - h_1 = C_{P,f} \times (t_3 - t_3')$; $(409.5 - 402.63) = 1.582 \times (52 - t_3')$; So $t_3' = 47.66$°C.

Now $h_3' = h_{f,47.66} = 268.0$ kJ/kg $= h_4$

(a) Refrigerant mass flow rate, $m = Q_e/(h_1 - h_4) = (1.5 \times 3.5167) / (402.63 - 268.0) = 0.0392$ kg/s. Here h_1 is considered instead of h_1' because superheating occurs outside of the evaporator. So it does not contribute to cooling effect.

(b) Compressor swept volume, $m \times v_1 = 0.0392 \times 0.056 = 2.195 \times 10^{-3}$ m^3/s

(c) Compressor discharge temperature, $t_2 = 64.0$°C

(d) Power consumed by compressor, $m\,(h_2 - h_1') = 0.0392 \times (438.9 - 409.5) = 1.1525$ kW

(e) Condenser heat rejection, $m\,(h_2' - h_3) = 0.0392 \times (438.9 - 274.74) = 6.435$ kW

Here h_3 is considered instead of h_3' because sub-cooling occurs outside of the condenser.

(f) COP = cooling effect/ power consumed = $(1.5 \times 3.5167)/ 1.1525 = 4.577$

Energy balance gives [1.1525 (W_c) + 5.27505 (Q_e)] kW = 6.42755 kW ≈ 6.435 kW (Q_C)

When compared with the performance of the refrigeration system without HEx. of Example 4.1, for the same cooling capacity of 1.5 TR, the mass flow rate is decreased by $(0.0412 - 0.0392)/0.0412 = 0.0485$, i.e., 4.85% (marginal).

Compressor swept volume and thus the size of compressor is reduced by $(2.25 - 2.195)/2.25 = 2.44\%$ (marginal).

But the COP decreased by $(6.222 - 4.577)/6.222 = 0.2644$, i.e., 26.44%

It is due to the increase in work of compressing the superheated vapour. Discussed in Section 4.3.1.

But internal heat recovery reduces the irreversibility of the system.

Example 4.6

In a vapour compression refrigeration system test rig, the working fluid being R22, temperatures recorded at compressor inlet, compressor outlet, condenser outlet are 12°C, 84°C and 37°C respectively. The pressures recorded are 5.0 bar and 15.0 bar. Work of compression, as recorded by the energy meter, 5.5 kW. Calculate (a) mass flow rate, (b) compressor isentropic efficiency, and (c) cooling effect and COP of the system.

Ans: Refer Fig. 4.19.

Saturation temperature corresponding to 5.0 bar evaporator pressure is 0°C and that corresponding to 15.0 bar condenser pressure is 39°C.

Temperature readings show that the refrigerant vapour enters the compressor in superheated state and liquid refrigerant comes out of condenser in sub-cooled state.

Properties of refrigerant vapour at state point 1 (superheated table at 5.0 bar and 12°C) are

$v_1' = 0.05$ m^3/kg, $h_1' = 258.3$ kJ/kg, $s_1' = 0.9566$ kJ/kg-K.

After isentropic compression, properties of refrigerant vapour at state point 2 are $t_2 = 72$°C and $h_2 = 290.5$ kJ/kg. But temperature recorded at compressor outlet (t_2'') is 84°C. Hence, it is a non-isentropic compression.

So compressor $\eta_{isen} = (72 - 12) / (84 - 12) = 0.8333$, i.e., 83.33%

Hence, $h_2'' = h_2' / \eta_{isen} = 290.5/0.8333 = 348.6$ kJ/kg

Enthalpy of sub-cooled liquid refrigerant at the exit of condenser ($h_{f,37}$) $h_3 = 90.3$ kJ/kg $= h_4$

So mass flow rate of the refrigerant, $m = W_c / (h_2'' - h_1') = 5.5/(348.6 - 258.3) = 0.061$ kg/s

Compressor isentropic efficiency is 83.33%

Refrigerating effect, $m \times (h_1' - h_4) = 0.061 \times (258.3 - 90.3) = 10.25$ kW

COP = 10.25/5.5 = 1.86

Example 4.7: (Effect of decrease in evaporator pressure and increase in condenser pressure) A vapour compression refrigeration system of 10 TR capacity works between a temperature range of 5°C and 40°C uses R22 as refrigerant. Find the required mass flow rate, compressor swept volume and work done, heat rejection from condenser and COP.

(a) Calculate the above parameters if evaporator temperature is decreased by 10°C.

(b) Calculate the above parameters if condenser temperature is increased by 10°C.

Ans: Refer Fig. 4.7 a and 4.7 b.

Here, assumption is made that the vapour compression refrigeration system works on S-S-S cycle.

Properties of saturated refrigerant vapour at state point 1 are

$P_{sat} = 0.584$ MPa, $v_1 = 0.0404$ m^3/kg, $h_1 = 252.84$ kJ/kg, $s_1 = 0.9244$ kJ/kg-K

State point 2 is located in 40°C superheat table with $s_1 = s_2 = 0.9244$ kJ/kg-K

$P_{sat} = P_C = 1.5267$ MPa

By interpolation it is found that $t_2 = 57$°C, $h_2 = 276.9$ kJ/kg

At state point 3, $h_{f,40} = 95.4$ kJ/kg $= h_4$

Mass flow rate, $m = 10 \times 3.5167/(252.84 - 95.4) = 0.2234$ kg/s

Compressor swept volume, $mv_1 = 0.2234 \times 0.0404 = 9.024 \times 10^{-3}$ m^3/s

Compressor work, $W_c = 0.2234 \times (276.9 - 252.84) = 5.375$ kW

Condenser heat rejection, $Q_C = 0.2234 \times (276.9 - 95.4) = 40.547$ kW

COP = 35.167 / 5.375 = 6.543

Energy balance: 5.375 + 35.167 = 40.542 kW ≈ 40.547 kW

Now for part (a) if the evaporator temperature decreased by 10°C, t_e = –5°C and corresponding saturation pressure P_1 = 0.422 MPa, v_1 = 0.0554 m^3/kg, h_1 = 249.29 kJ/kg, s_1 = 0.9393 kJ/kg-K

State point 2 is located in 40°C superheat table with $s_1 = s_2$ = 0.9393 kJ/kg-K

$P_{sat} = P_C$ = 1.5267 MPa

By interpolation it is found that t_2 = 62°C, h_2 = 283.3 kJ/kg

At state point 3, $h_{f,40}$ = 95.4 kJ/kg = h_4

Now mass flow rate, m = 10 × 3.5167/ (249.29 – 95.4) = 0.2285 kg/s

Compressor swept volume, mv_1 = 0.2285 × 0.0554 = 12.66 × 10^{-3} m^3/s

Compressor work, W_c = 0.2285 × (283.3 – 249.29) = 7.771 kW

Condenser heat rejection, Q_C = 0.2285 × (283.3 – 95.4) = 42.93 kW

COP = 35.167 / 7.771 = 4.525

Energy balance: 7.771 + 35.167 = 42.938 kW ≈ 42.93 kW

Thus, there is an increase of 40% in compressor size and decrease of 31% in COP.

If we take the same swept volume of the compressor as 9.024 × 10^{-3} m^3/s

Then mass flow rate will be 9.024 × 10^{-3} / 0.0554 = 0.163 kg/s

So compressor work, W_c = 5.544 kW, Q_e = 25.0841 kW = 7.133 TR and Q_C = 40.55 kW

Thus, the condenser will become oversized.

(Students may analyse if the condenser is oversized then there will be sub-cooling. This will increase the refrigerating effect and hence COP).

For part (b), if the condenser temperature is increased to 50°C, corresponding saturation pressure, P_C, becomes 2.033 MPa.

State point 1 will be same as for saturated condition at 5°C.

State point 2 is located in 50°C superheat table with $s_1 = s_2$ = 0.9244 kJ/kg-K

By interpolation it is found that t_2 = 71°C, h_2 = 283.1 kJ/kg

State point 3, $h_{f,50}$ = 108.22 kJ/kg = h_4

The mass flow rate, m, is calculated as 0.2432 kg/s

Swept volume of compressor is 9.8253 m^3/s

Compressor work, W_c = 7.36 kW and condenser heat rejection Q_C = 42.531 kW

COP = 4.78

Thus there is an increase of 8.8% in compressor size and 26.9% decrease in COP.

Hence decrease in evaporator pressure/temperature has more adverse effect on the refrigerating system than increase in condenser pressure/temperature.

Example 4.8

Evaporator pressure of a R134a refrigeration system is maintained at 120 kPa. The refrigerant enters the evaporator with a quality of 0.31 and leaves the compressor at 60°C. The compressor consumes 475 W of power. Assuming Saturated states at evaporator and condenser exit, calculate (a) mass flow rate, (b) the condenser pressure, and (c) COP of the refrigeration system.

Ans: Refer Fig. 4.21.
From R134a table; at 120 kPa, t_1 = –22.32°C, h_g = 236.9, h_f = 22.4, s_g = 0.9478
Specific enthalpy at 4, $h_4 = xh_g + (1 - x)h_f$ (at 120 kPa)
= 0.31 × 236.9 + 0.69 × 22.4 = 88.89 kJ/kg which is same as h_3.
Saturated liquid enthalpy 88.89 kJ/kg corresponds to saturated pressure of 700 kPa.
So the condenser pressure is 700 kPa.
From superheated table of 700 kPa, t_2 = 35°C,
h_2 = 273.5 kJ/kg
So compressor η_{isen} = {35 – (–22.32)}/{60 – (–22.32)}
= 0.6963 = $(h_2 - h_1)/(h_2' - h_1)$
So h_2 = 289.46 kJ/kg
So power input to compressor = $m\,(h_2' - h_1)$;
475/1000 = m (289.46 – 236.9); m = 0.00904 kg/s = Mass flow rate through the system
COP = $(h_1 - h_3)/(h_2' - h_1)$ = (236.9 – 88.89)/(289.46 – 236.9) = 2.82

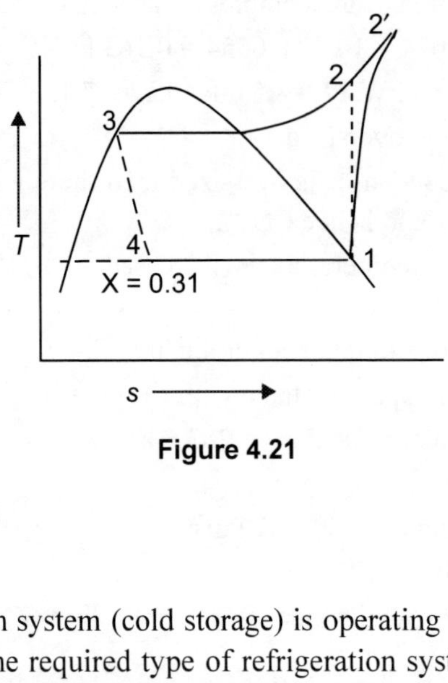

Figure 4.21

Multi-Pressure systems

Example 4.9

A 100 TR, NH_3 refrigeration system (cold storage) is operating between temperature limit of –30°C and 40°C. Suggest the required type of refrigeration system, power consumption and COP of the system.

Ans: Since the evaporator temperature is as low as –30°C, two-stage refrigeration system is suggested. The refrigerant used is NH_3. So intercooling with flash gas removal is suggested. Figures 4.9 and 4.10 may be referred to solve the problem.
The intermediate pressure is taken as geometric mean of condenser pressure and evaporator pressure for optimum compressor power.
Condenser pressure (P_C) is saturation pressure corresponding to 40°C = 1555.0 kPa
Evaporator pressure (P_e) is saturation pressure corresponding to –30°C = 119.5 kPa

So intermediate pressure (P_i) = $\sqrt{(1555.0 \times 119.5)}$ = 431.071 kPa ≈ 430.0 kPa
Corresponding saturation temperature (from NH_3 table) is 0°C.
Required refrigerant flow rate through the evaporator,
$-m_1 = 100 \times 3.5167/(h_{g(-30)} - h_{f(0)}) = 351.67/\ (1403.11 - 180.88) = 0.2877$ kg/s
From energy balance, taking the C.V. around the flash intercooler
$m_2h_2 + m_5h_5 = m_2' h_2' + m_6h_6$;
From the mass balance, $m_2 = m_6 = m_1 = m_{LP}$; and $m_2' = m_5 = m_{HP}$
From NH_3 saturation and superheat tables, h_2 = 1577 kJ/kg, h_5 = 371.47 kJ/kg, h_2' = 1443.34 kJ/kg, and h_6 = 180.88 kJ/kg

Substituting the values in energy balance equation, $m_2 = m_{HP}$ = 0.3747 kg/s

Again h_3 = 1625.0 kJ/kg

LP compressor work = 0.2877 × (1577 – 1403.11) = 50.028 kW

HP compressor work = 0.3747 × (1625 –1443.34) = 68.068 kW

Total compressor work = $W_{HP} + W_{LP}$ = 50.028 + 68.068 = 118.096 kW

COP = 351.67/118.096 = 2.98

Example 4.10

A two-stage NH_3 system designed for a fish freezing plant uses flash gas removal for inter-cooling. The condensing temperature is 35°C. The high temperature evaporator is 0°C and its capacity is 90 kW. The low temperature evaporator is –40°C and its capacity is 170 kW. Determine refrigerant flow rate, the power requirement and COP.

Ans: Refer Figs. 4.22 and 4.23 (a)

The properties of NH_3 at –40°C: h_1 = 1387.15, s_1 = 5.9518; v_1 = 1.55, P_1 = 0.717 bar

The refrigerant vapour should be compressed to the saturation pressure corresponding to 0°C, i.e., 4.294 bar as the evaporator 2 is maintained at 0°C.
Now the properties of state point 2 are evaluated as follows.
$s_2 = s_1 = 5.9518$;
Interpolating in the superheated table of 4.294 bar we get t_2 = 79°C, h_2 = 1635.2; v_2 = 0.39
At the exit of the flash intercooler, t_3 = 0°C; h_3 = 1443.34, s_3 = 5.3368
By interpolating $s_3 = s_4$ in the superheat tale of 35°C, t_4 = 35 + 46.5 = 81.5, h_4 = 1604.5
After compression of refrigerant vapour in HP compressor
Properties of state point 5, $h_5 = 346.89 = h_f$ at 35°C
Enthalpy, $h_8 = h_7$ = 181.1 kJ/kg = h_f at 0°C
Mass flow rate through low temperature evaporator m_1 = 170/ (1387.15 – 181.1) = 0.141 kg/s
$m_1 = m_7 = m_2 = m_8$. Also $m_5 = m_3$
Taking the control volume around flash chamber and high temperature evaporator
Energy balance gives $m_5h_5 + 90 + m_2h_2 = m_3h_3 + m_7h_7$
Substituting the values $m_5346.89 + 90 + 0.141 \times 1635.2 = m_31443.34 + 0.141 \times 181.1$

So $m_3 = m_5 = (90 + 205.03)/1096.45 = 0.27$ kg/s

Power requirement $= m_1(h_2 - h_1) + m_3(h_4 - h_3) = 0.141 \times 248.05 + 0.27 \times 161.16 = 78.488$ kW

COP $= (170 + 90)/78.488 = 3.3126$

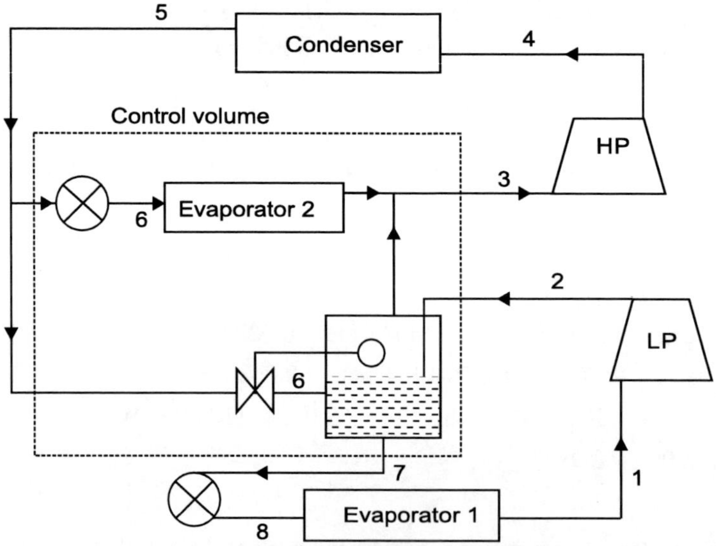

Figure 4.22

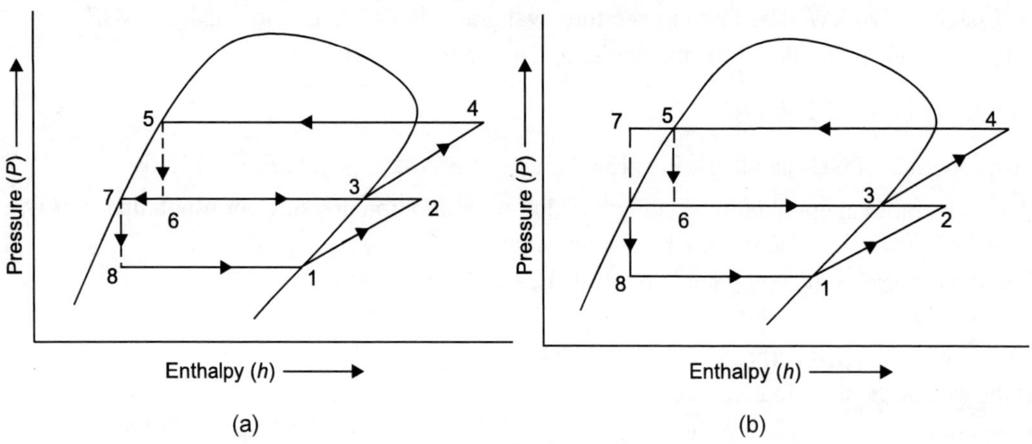

Figure 4.23 (a) and (b)

Example 4.11

A R134a, two-stage refrigeration system works between pressure limit of 800 kPa and 140 kPa. The flash chamber is at 400 kPa. The refrigerant vapour from the low pressure compressor mixes with the flashed vapour from the flash chamber and enters the high pressure compressor. Assume refrigerant leaves the evaporator and condenser in saturated state and unit mass flow rate (1.0 kg/s) through condenser, find (a) the mass of refrigerant being flashed out of the flash chamber, (b) refrigeration capacity and compressor work, and (c) COP.

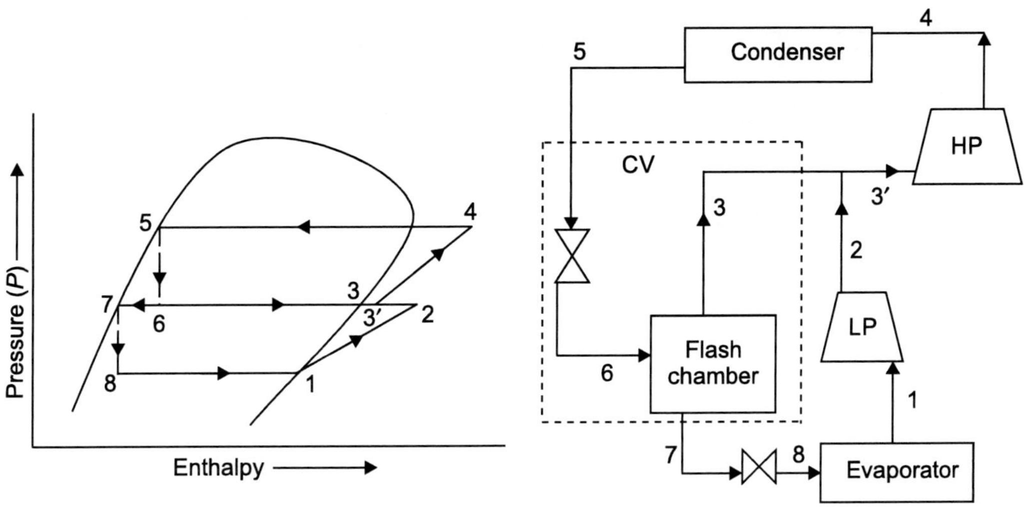

Figure 4.24: (a), (b)

Ans:

Refer Fig. 4.24(a) and (b) for diagram on *T-s* plane and schematic, respectively.

Mass balance in the control volume, $m_5 = m_3 + m_7$; $1.0 = m_3 + m_7$

Energy balance in the control volume gives $m_5h_5 = m_7h_7 + m_3h_3$

$1 \times 95.47 = m_7 \times 63.94 + m_3 \times 255.55$; So $m_7 = m_2 = m_{LP}$; m_3 is the mass flashed.

$95.47 = (1 - m_3) \times 63.94 + m_3 \times 255.55$; $m_3 = 0.165$ kg/s

So mass flow rate through the evaporator, $m_7 = m_1 = 1.0 - 0.165 = 0.835$ kg/s

Cooling capacity = $0.835 \times (h_g$ (at 140 kPa) $- h_f$ (at 400 kPa)) = $0.835 \times (239.16 - 63.94)$ = 146.31 kW

Enthalpy h_2 and enthalpy h_4 from superheated table are 261.0 and 275.5 kJ/kg respectively.

Enthalpy at 3 = $h'_3 = (0.165 \times 255.55 + 0.835 \times 261.0)/1.0 = 260.1$ kJ/kg

Compressor work = $0.835\ (h_2 - h_1) + 1\ (h_4 - h_3) = 0.835\ (261 - 239.16) + 1\ (275.5 - 260.1)$ = 33.64 kW

COP = 146.31/33.64 = 4.35

Example 4.12

A 10 TR capacity refrigerated space is to be maintained at –40°C. Condenser temperature is 45°C. The refrigerant used is R22. The intermediate temperature to be maintained at 0°C. Determine the important parameters of the system.

Ans: The temperature to be maintained is below –30°C, a two-stage refrigeration system is suggested. Intercooling of refrigerant vapour from LP compressor is intercooled to saturation temperature and the liquid refrigerant from condenser is sub-cooled to 5°C. [Refer Fig. 4.25 (a) and (b)]

$P_1 = 1.053$ bar, $h_1 = 234.17$ kJ/kg, $s_1 = 1.0044$ kJ/kg-K,

For intermediate temperature of 0°C

$P_i = 4.98$ bar, $h_3 = 251.2$, $s_3 = 0.9317$, $v_3 = 0.0472$

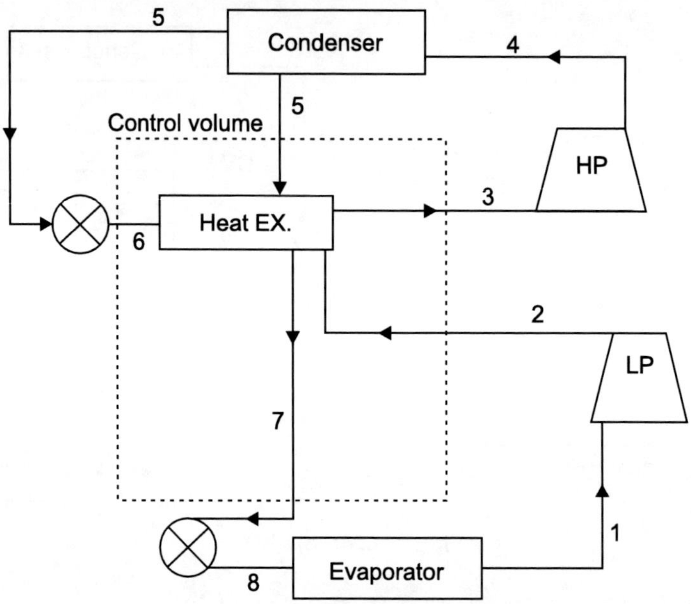

Figure 4.25: (a)

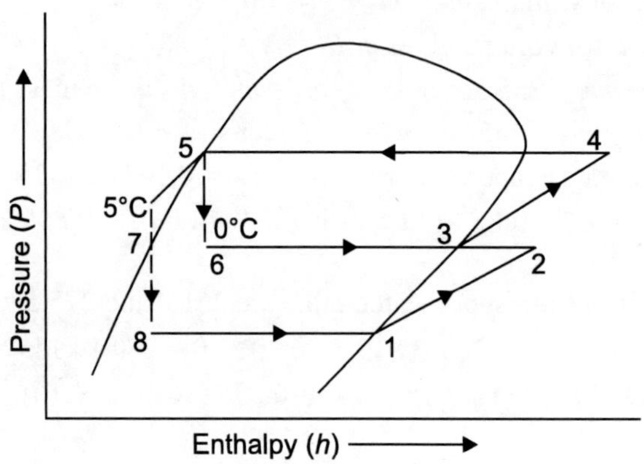

Figure 4.25: (b)

Interpolating in 0°C superheated table for $s_2 = s_1 = 1.0044$, we get

$$t_2 = 9°C, h_2 = 271.8, v_2 = 0.0564$$

After compression in HP compressor, state point 4 is evaluated by interpolating $s_3 = s_4 = 0.9317$ in 45°C (17.209 bar) superheated table $t_4 = 45 + 24 = 69°C$, $h_4 = 284.9$ kJ/kg, $v_4 = 0.0156$ m^3/kg

Now the mass of saturated liquid refrigerant from condenser is divided into two parts. One part is used for intercooling and sub-cooling process and another part goes to evaporator.

State point 5; $h_5 = 101.76 = h_6$; $h_{7(5)} = 52.23 = h_8$

Mass flow rate through evaporator $m_{LP} = 10 \times 3.5167/(234.17 - 52.23) = 0.1933$ kg/s

Taking control volume around HEx;

Energy balance gives $m_2h_2 + m_5h_5 + m_6h_6 = m_3h_3 + m_5h_7$

But $m_{LP} = m_2 = m_5; m_3 = m_6 + m_5 = m_{HP}$

Now the energy equation is $m_{LP}h_2 + m_{LP}h_5 + (m_{HP} - m_{LP})h_6 = m_{HP}h_3 + m_{LP}h_7$

$M_{HP} = m_{LP}(h_2 + h_5 - h_6 - h_7)/(h_3 - h_6)$, but $h_5 = h_6$;

Substituting the values, $m_{HP} = 0.2842$ kg/s

$\eta_{LP} = 1.0 + \varepsilon - \varepsilon(v_1/v_2) = 1.04 - 0.04\ (0.206/0.0564) = 0.894$

$\eta_{LP} = 1.0 + \varepsilon - \varepsilon(v_3/v_4) = 1.04 - 0.04\ (0.0472/0.0.0156) = 0.919$

$V_{S\,LP} = m_{LP}v_1/\eta_{LP} = 0.0445\ m^3/s$

$V_{S,\,HP} = m_{HP}v_3/\eta_{HP} = 0.0146\ m^3/s$

$W_{LP} = m_{LP}(h_2 - h_1) = 7.274$ kW

$W_{HP} = m_{HP}(h_4 - h_3) = 9.578$ kW

$Q_C = m_{HP}(h_4 - h_5) = 52.0484$ kW

COP = 10 × 3.5167/ (7.274 + 9.578) = 2.084

Considering single-stage

State point 1 is same as above.

Properties of state point 2 is calculated by interpolating in 45°C superheated table at $s_1 = s_2 =$ 1.0044.

So $t_2 = 45 + 52 = 97°C$, $h_2 = 308.9$ kJ/kg, $v_2 = 0.018\ m^3/kg$

Mass flow rate, $m = 10 \times 3.5167/(234.17 - 101.76) = 0.2656$ kg/s

$\eta = 1.0 + \varepsilon - \varepsilon(v_1/v_2) = 1.04 - 0.04\ (0.206/0.0.018) = 0.582$

$V_S = mv_1/\eta = 0.0943\ m^3/s$

$W = m(h_2 - h_1) = 19.852$ kW

So saving of work = 19.852 – (9.578 + 7.274) = 3.00 kW

Note: Saving of work can be segregated as mentioned in Eq. 4.12.

Work saved due to flash intercooling and thereby avoiding recompressing of vapour in LP compressor is 3.42 kW. But the excess work of 0.42 kW is done due to unequal mass flow rate and specific work on high pressure side which is compensated and the net saving of work is 3.0 kW).

COP = 10 × 3.5167/ 19.852 = 1.772

Improvement in COP over single-stage = (2.084 – 1.772)/1.772 = 0.176, i.e., 17.6%

For NH_3 system

Two-stage system: $t_4 = 103°C$, $m_{LP} = 0.0297$ kg/s, $m_{HP} = 0.0406$ kg/s

$V_{LP} = 0.0523\ m^3/s$, $V_{HP} = 0.0128\ m^3/s$, $W_{LP} = 7.342$ kW, $W_{HP} = 8.461$, total work = 15.803 kW

COP = 2.225

Single-stage system, $t_2 = 201°C$, $m = 0.0353$ kg/s

Volumetric efficiency, = $\eta = 0.548$, $V_S = 0.10\ m^3/s$

$W_c = 18.472$ kW, COP = 1.904

So saving of work = 18.472 – 15.803 = 2.669 kW

Improvement in COP over single-stage = (2.225 – 1.904)/1.904 = 0.1686, i.e., 16.86%

Analysis: The above examples on two-stage and single-stage systems and R22 and NH_3 systems show that

1. The exit temperature of refrigerant from HP compressor is considerably low for NH_3 system in comparison to single-stage system. But for R22 system the difference is not appreciable.
2. Required compressor size is considerably large for both NH_3 and R22 in single-stage in comparison to two-stage due to decrease in volumetric efficiency for same refrigerating capacity.
3. Work saving and improvement in COP in two-stage system in comparison to single-stage system for both R22 and NH_3 system are not so high in comparison to expenses for the number of equipment required for two-stage system.

Specific work requirement is less for two-stage system in comparison to single-stage system.

Example 4.13

Temperatures of a cold storage and its office are to be maintained at –40°C and 0°C, respectively. The required cooling capacities are 10 TR and 5 TR, respectively. It is assumed that vapour leaves both the evaporators at saturated state. Condenser temperature is 40°C and the condensate is sub-cooled to 5°C in a sub-cooling HEX before entering the low temperature evaporator. Find the required mass flow rate, swept volume, Work requirement and COP.

Answer: For a cold storage NH_3 is the preferred refrigerant. And the above problem can be solved by selecting a two-stage refrigeration system having two compressors and two evaporators. (Refer Figs. 4.22 and 4.23 b)

(Compare this with Example 4.10 where there is no sub-cooling of the condensate in a HEx passing through the flash chamber.)

First the properties of different state points are to be evaluated.

For evaporator 1

At –40°C; h_1 = 1387.15, s_1 = 5.9518; v_1 = 1.55, P_1 = 0.717 bar

The refrigerant vapour should be compressed to the saturation pressure corresponding to 0°C, i.e., 4.294 bar as the evaporator 2 is maintained at 0°C.

Now the properties of state point 2 are evaluated as follows.

$s_2 = s_1$ = 5.9518;

Interpolating in the superheated table of 4.294 bar

we get t_2 = 79°C, h_2 = 1635.2; v_2 = 0.39

State point 3; h_3 = 1443.34; v_3 = 0.29; s_3 = 5.3368

After compression of refrigerant vapour in HP compressor, properties of state point 4 are:

t_4 = (40 + 52) = 92°C; h_4 = 1626.4; v_4 = 0.1058

Enthalpy, $h_8 = h_{7,f(5)}$ = 204.7 kJ/kg; $h_6 = h_{f(40)}$ = 371.47 kJ/kg

Mass flow rate through LP compressor, m_{LP} = 10 × 3.5167/(1387.15 – 204.07) = 0.02972 kg/s

Volumetric efficiency of LP compressor, $\eta_{LP} = 1.0 + \varepsilon - \varepsilon\,(v_1/v_2)$

$= 1.04 - 0.04\,(1.55/0.39) = 0.881$

Volumetric efficiency of LP compressor, $\eta_{HP} = 1.0 + \varepsilon - \varepsilon\,(v_4/v_5)$

$= 1.04 - 0.04\,(0.290/0.1058) = 0.9303$

To determine the mass flow rate through the HP compressor, let us take the control volume around the HEx, evaporator 2.

The energy balance equation gives $m_{HP}h_6 + Q_{e,2} + m_{LP}h_2 = m_{LP}h_8 + m_{HP}h_3$

Substituting the respective values m_{HP} = 0.05607 kg/s

$$W_{HP} = 0.05607\,(h_4 - h_3) = 10.264 \text{ kW}$$

$$W_{LP} = 0.02972\,(h_2 - h_1) = 7.372 \text{ kW}$$

$$\text{Total work} = 10.264 + 7.372 = 17.636 \text{ kW}$$

$$\text{COP} = (10 + 5) \times 3.5167/17.636 = 2.99$$

Cascade Refrigeration System

Example 4.14

In a cascade refrigeration system R13 is used as LT side refrigerant and R134a is used as HT side refrigerant. The system capacity is 1 TR and works in temperature limit of 193 K and 318 K. Find the work done by LT and HT compressors and COP of the system.

Ans: The intermediate temperature to be selected as the sqrt ($T_{LT}\,T_{HT}$) = 247 K = –26°C and this temperature should be the evaporator temperature of HT side because this temperature is just above the nbp of R134a. So the condenser temperature of LT side should be 247 + 5 = 252 K (A minimum 5 K temperature gradient should be maintained to transfer the heat). (Refer Figs. 4.13 and 4.14).

So now for LT side T_e = 193 K, T_C = 252 K

Since the superheated table for R13 is not available, the required values are calculated from relationship $\Delta s = C_P \ln (T_2/T_1)$

$$S_a = s_b = 0.5889,\ s_e = 0.5210.$$

Average C_P for R13 in the working range is taken to be 0.754 kJ/kg-K

$$T_b = T_b \exp\{s_a - s_b\}/C_P = 252 \exp\{(0.5889 - 0.5210)/0.754\} = 275.73 \text{ K}$$

$$H_b = h'_b + C_P\,(T_b - T'_b) = 130.56 + 0.754\,(275.73 - 252) = 148.45$$

$$H_c = 20.23 \text{ and } h_a = 110.27$$

So

$$m_{LT} = 3.5167 / (110.27 - 20.23) = 0.039 \text{ kg/s}$$

$$W_{LT} = 0.039\,(148.45 - 110.27) = 1.489 \text{ kW}$$

$$Q_{C\,LT} = 0.039\,(148.45 - 20.23) = 5.00 \text{ kW}$$

For HT side

From R134a table, At –26°C, $h_1 = 382.82$, $s_1 = 1.7471 = s_2$

Interpolating in R134a superheated table at 45°C and $s = 1.7413$

State point 2, $t_2 = 45 + 57 = 102$, $h_2 = 433.6$, $h_3 = h_4 = 263.9$

Now taking the C.V. around cascade condenser, from energy conservation

$$m_{HT}(h_1 - h_4) = m_{LT}(h_b - h_c) = 5.002$$

$$\text{So } m_{HT} = 5.002/(382.82 - 263.9) = 0.04206 \text{ kg/s}$$

$$W_{HT} = 0.04206\,(433.6 - 382.82) = 2.136 \text{ kW}$$

$$Q_{C\,HT} = 0.04206\,(433.6 - 263.9) = 7.1376 \text{ kW}$$

$$\text{COP} = 3.5167 / (W_{HT} + W_{LT}) = 3.5167 / (2.136 + 1.489) = 0.97$$

EXERCISE

1. What are the processes of RC cycle not feasible practically? And how they are replaced with to develop the S-S-S cycle?

2. Compare the isentropic expansion process with throttling process.

3. How does the performance of refrigeration system is affected if evaporator pressure is below atmospheric?

4. Discuss the effect of sub-cooling at the exit of condenser on the performance of a refrigeration system.

5. Discuss the effect of superheating at the exit of the evaporator on the performance of a refrigeration system.

6. What is wire drawing effect?

7. What are the advantages of using a flash chamber in multi-stage compression system?

8. A two-TR air conditioner rejects 8.57 kW of heat to the environment from the condenser. What is the compressor power consumption?

9. (a) Compare the compression work and compressor discharge temperature for isentropic compression of R22 from evaporator pressure of 110 kPa to condenser pressure of 1250 kPa in single-stage and two-stage with perfect intercooling. The mass flow rate is 2.0 kg/s.

(b) Compare the compression and discharge temperature of 2.0 kg/s of NH_3 from evaporator pressure of 110 kPa to 1250 kPa in single-stage and two-stage with perfect intercooling.

(c) Compare the results of (a) and (b) and suggest in which case two-stage compression with intercooling is more beneficial.

10. A two-stage NH_3 refrigeration system with flash gas removal and intercooling operates with two evaporators. One is operating at 2°C with a refrigerating capacity of 150 kW and the other one is operating at –30°C having the capacity of 250 kW. Condensing temperature is 40°C. Calculate the mass flow rate through the high pressure compressor.

11. Derive an expression for the inter-stage pressure for a two-stage system so that least work is done.

12. A R22 two-stage refrigeration system has a capacity of 200 kW at an evaporating temperature of –30°C and condensing pressure of 1.5 MPa. Calculate the total power required by the two compressors in the system, as shown in figure below, where there is no intercooling but there is flash gas removal at 650 kPa.

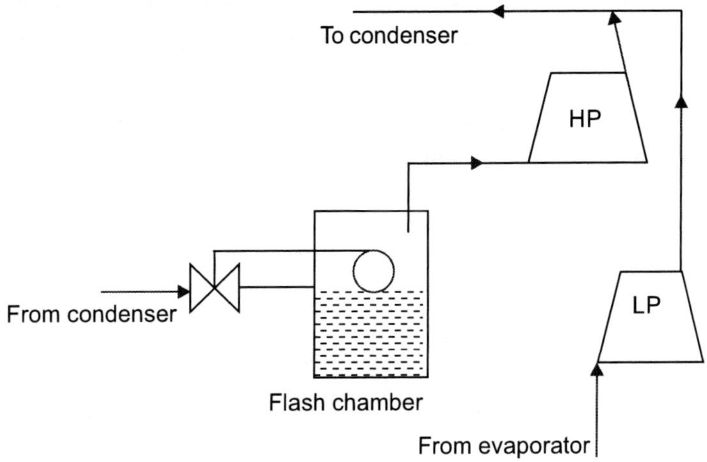

13. A R22 refrigeration system comprises one compressor and three evaporators having capacities 10 TR, 20 TR and 30 TR. The temperatures to be maintained are 10°C, 5°C and –10°C respectively. Each evaporator is fitted with an individual expansion valve. Condenser pressure is 9.6 bar. The exit conditions from evaporator is dry saturated vapour and from condenser is liquid sub-cooled by 10°C. Determine (a) mass flow rate in each evaporator, (b) refrigerating effect of each evaporator, (c) compressor power, and (d) COP of the system.

14. The particulars of a 20 TR R22 food freezer are as follows.

Evaporator temperature –30°C, condenser temperature 40°C, temperature of vapour leaving the evaporator –25°C, sub-cooling of liquid at condenser exit is by 5°C. Find the mass flow rate, compressor work and COP.

15. The size of LP evaporator is more than the HP evaporator in VCR systems. Is this statement true/false? (justify it)

16. What will happen in the VCR system if the R22 is used in R12 refrigerator? (any two points)

17. Why is superheating considered to be good in certain cases?

18. Sketch *T-S* and *p-h* diagram for an actual and theoretical VCR cycle.

19. In multistage ammonia plants, when liquid refrigerant is used for intercooling, will the total power requirement increase/decrease? Prove it.

20. State the purpose of (i) accumulator, and (ii) flash chamber

21. A refrigeration system using R12 as refrigerant consists of 3 evaporators of capacity 20 TR at 10°C, 30 TR at 5°C, and 10 TR at –5°C. The vapours leaving the evaporator at dry and saturated. The system is provided with multiple expansion valves and back pressure valves. The condenser temperature is 40°C and the liquid refrigerant leaving the condenser is sub-cooled to 30°C. Assuming isentropic compression for each compressor, find out (a) the mass flow rate through each evaporator, (b) power consumption for each compressor in kW, and (c) COP of the system.

22. A two-stage refrigeration systems uses LP and HP compressors both having bore = 100 mm and stroke 90 mm and rpm = 1500 mm.

Refrigeration used	: R22
Condenser temperature	: 40°C
Evaporator temperature	: –20°C
Volumetric efficiency	: 100%
Efficiency of LP and HP compressor	: 80%
Flash intercooler temperature	: 5°C

Determine (a) tonnage of LP evaporator, (b) refrigeration obtained in flash intercooler, (c) power to compressors, (d) COP, and (e) heat rejection to condenser.

23. A 15 TR refrigeration machine working on simple vapour compression cycle with R12 as refrigerant, operating between condenser temperature of 40°C and evaporator temperature of –5°C. Determine the following using refrigeration table:

(i) refrigeration effect

(ii) mass flow work of refrigerant in kg/min

(iii) compressor specific work

(iv) COP_R, COP_{Carnot} and refrigeration efficiency

(v) piston displacement and compressor dimension for 1450 rpm, double cylinder with square size of $\eta_{vol} = 0.65$

Compare the various air cooling systems used for aircraft:

24. A simple VCR m/c, with R12 as refrigerant, has a capacity of 20 TR operating between –28°C and 26°C. The refrigerant is sub-cooled by 4°C before entering expansion valve and the vapour is superheated by 5°C before leaving the evaporator. The m/c has a six cylinder single acting compressor with L = 1.5D. It has *c* clearance of 3% of stroke length. Determine the following using refrigeration table/ pH chart.

(i) Power required to run the compressor.

(ii) COP of the cycle

(iii) Volumetric efficiency

(iv) Bore and stroke of cylinder if the speed of compressor is 1000 rpm.

25. In a 15 TR ammonia plant, compression is carried out in two stages with flash intercooler. The details of the plants are as follows:

Condenser pressure	: 12 bar
Evaporator pressure	: 3 bar
Flash intercooler pressure	: 6 bar
The temperature for intercooing and subcooling	: 20°C

Estimate the following ***using pH chart.***

(i) COP of the plant

(ii) Power required for each compressor

(iii) Swept volume of each compressor if volumetric efficiency is 80%.

26. Whether large subcooling is desirable? Justify your answer.

27. List the assumptions made in the determination of the optimum interstage pressure.

28. COP of a reversd Carnot refrigerator is higher in winter than that in summer. Justify.

29. Draw *P-v* and *T-s* diagram for a VCR system and explain it.

30. Superheating always increases COP. (True/False)

31. How is refrigeration efficiency is defined?

MULTIPLE–CHOICE QUESTIONS

1. SSS cycle resembles — thermodynamic cycle

(a) Reverse Carnot (b) Reverse Brayton

(c) Reverse Rankine (d) Reverse Erricson

2. The isentropic expansion of RC cycle is replaced with throttling process in SSS cycle to

(a) Save energy (b) Increase cooling effect

(c) Reduce the initial cost (d) Increase the mass flow

3. Wire drawing effect in vapour compression refrigeration system means pressure drop in

(a) evaporator tube (b) throttle valve

(c) opening the compressor inlet valves (d) condenser tube

4. Sub-cooling of liquid refrigerant at the condenser exit

(a) increases the cooling effect

(b) decreases the possibility of throttle valve chocking

(c) increases the COP

(d) All of the above

5. Superheating of refrigerant vapour at the compressor inlet

(a) Increases compressor work (b) Decreases cooling capacity

(c) Cause discharge temperature high (d) all of the above

6. Increase in evaporator pressure for a VCR system

(a) increases COP (b) decreases COP

(c) has no effect on COP (d) decrease the cooling effect

7. Increase in condenser pressure in a VCR system

(a) COP decreases (b) condenser size increases

(c) compressor work increases (d) all of the above

8. Lower is the evaporator pressure, — will be the volumetric efficiency of the compressor.

(a) higher (b) lower (c) no effect

9. Increased compression ratio causes — mass flow rate through compressor.

(a) higher (b) lower (c) no effect

10. For maximization of COP of a multi-pressure system, the intermediate pressure is considered

(a) $(P_1 + P_2)/2$ (b) $\sqrt{(P_1 P_2)}$

(c) $P_2 - P_1/2$ (d) $P_1 + P_2/2$

11. For maximization of COP of a cascade system, the intermediate temperature is considered

(a) $(T_1 + T_2)/2$ (b) $\sqrt{(T_1 T_2)}$

(c) $T_2 - T_1/2$ (d) $T_1 + T_2/2$

12. Dry ice is the other name of

(a) ice below –30°C (b) snow

(c) solid carbon dioxide (d) alum

13. Which is correct for vapour compression refrigeration system?

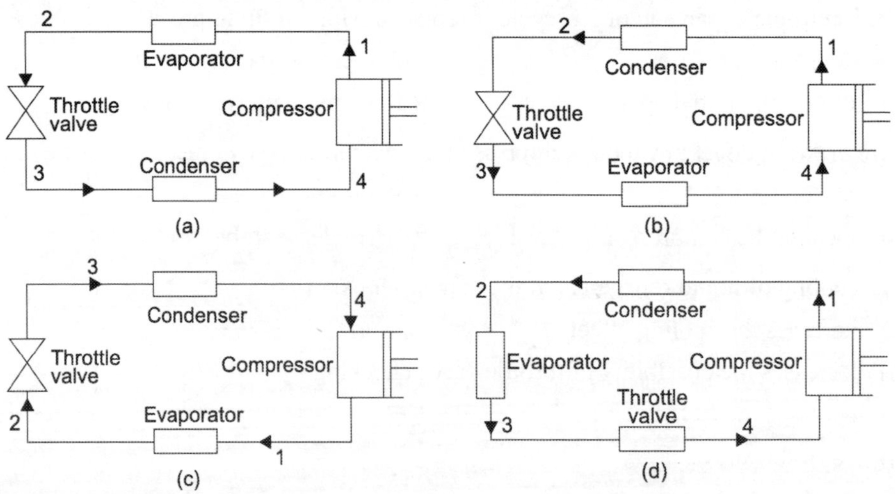

14. In a VCR plant, the enthalpy at exit of the evaporator, compressor and the condenser are 350 kJ/kg, 375 kJ/kg and 225 kJ/kg respectively. The refrigerating efficiency is 80%. Find the power required per kW of cooling to be produced.

(a) 0.25 kW (b) 4.0 kW
(c) 12.5 kW (d) 11 kW

15. A heat pump is working on VCR cycle, enthalpy at the end of heat addition process, at the end of compression process, at the end of heat rejection process, and at the end of isenthalpic expansion process are 195 kJ/kg, 210 kJ/kg, and 90 kJ/kg, respectively. The mass flow rate is 0.5 kg/s. Then the heating capacity of heat pump is, nearly

(a) 7.5 kW (b) 45 kW
(c) 52.2 kW (d) 60 kW

16. In a VCR system, a thermometer kept in the liquid line can indicate whether the

1. refrigerant flow is too low.
2. water circulation is adequate.
3. condenser is fouled.
4. pump is functioning properly.

Of these statements

(a) 1, 2 and 3 are correct (b) 1, 2 and 4 are correct
(c) 1, 3 and 4 are correct (d) 2, 3 and 4 are correct

17. In a VCR system, a throttle valve is used in place of an expander because

(a) it considerably reduces the system weight.
(b) it improves the COP, as the condenser is small.
(c) the positive work in isentropic expansion of liquid is very small.
(d) it leads to significant cost reduction.

18. The refrigerant, immediately after expansion valve in a VCR cycle is

(a) saturated liquid (b) subcooled liquid
(c) dry vapour (d) wet vapour

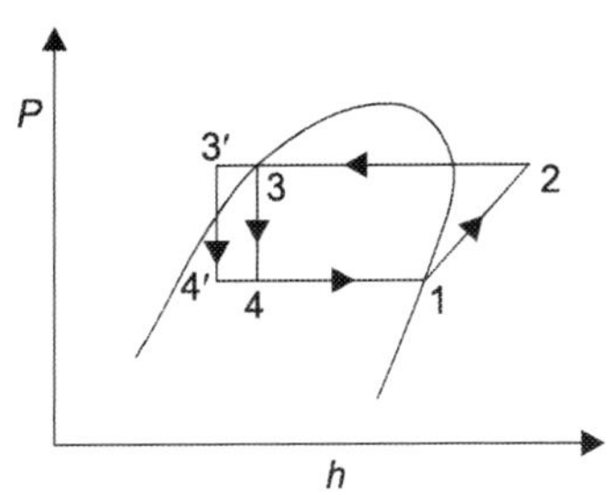

19. In a VCR system, liquid to suction heat exchanger is used to

(a) keep the COP constant
(b) prevent the liquid refrigerant from entering the compressor
(c) subcool the liquid refrigerant leaving the condenser
(d) subcool the vapour refrigerant from the evaporator

20. High condenser pressure in a refrigeration system can occur because

1. the water flow rate is less than the desired value.
2. non-condensable gases are present in the system
3. of accumulation of lubricating oil in condenser
4. of low charge of refrigerant in the system.

Of these statements

(a) 1, 3 and 4 are correct (b) 1, 2 and 3 are correct
(c) 1, 2 and 4 correct (d) 2, 3 and 4 are correct

21. High pressure drop in liquid line in a refrigerating system causes

(a) high condenser pressure (b) flashing of the liquid refrigerant
(c) higher evaporator pressure (d) under cooling of the liquid refrigerant

22. In a VCR system A vapour is superheated by 5°C in the evaporator while in another VCR system B vapour is superheated by 5°C in a liquid vapour regenerative heat exchanger, other conditions being the same. Then

(a) COP of A = COP of B
(b) COP of both A and B > COP of reversed Carnot cycle
(c) COP of A > COP of B (d) COP of A < COP of B

ANSWERS

1. (c) **2.** (c) **3.** (c) **4.** (d) **5.** (d) **6.** (a) **7.** (d) **8.** (b) **9.** (b) **10.** (b)
11. (b) **12.** (c) **13.** (b) **14.** (a) **15.** (d) **16.** (a) **17.** (a) **18.** (d) **19.** (c) **20.** (b)
21. (b) **22.** (c)

5

Components of a Vapour Compression Refrigeration System

5.1 INTRODUCTION

The working principle of a vapour compression refrigeration system has been discussed. The schematic of this system, as shown in Fig. 4.3, shows different components of the system. The main components of a vapour compression refrigeration system are: (a) compressor, (b) condenser, (c) expansion device, and (d) evaporator. Besides, there are flow control and safety devices. In this chapter a brief discussion on all these components is presented.

5.2 COMPRESSORS

Compressor is said to be the heart of the vapour compression refrigeration system. There are mainly two types of compressors used in refrigeration systems basing upon the principle of operation: (a) positive displacement type which compresses low pressure gas (at p_1) physically, reducing the volume of gas causing an increase in pressure (to P_2). The ratio of P_2 and p_1 is called **compression ratio**; (b) dynamic type in which kinetic energy is imparted to the fluid by the rotating blades. This kinetic energy is converted to pressure energy (head) as the fluid flows through a diverging passage and the ratio of high and low pressure is called **pressure ratio**.

Positive displacement type compressors are divided into two groups; one is reciprocating type and another is rotary type. Rotary type compressors are: (a) vane type, (b) scroll type, (c) rolling piston type, and (d) screw type. Dynamic type compressors are grouped into (a) turbo type, and (b) ejector (jet) type. Turbo compressors are of two types. One is centrifugal compressor, the other one is axial type compressor. The commonly used dynamic compressor for large establishments is centrifugal compressor.

It is obvious that for positive displacement type compressors, amount of compression is expressed in compression ratio and for dynamic compressors, it is expressed in pressure ratio as in case of air refrigeration system.

5.2.1 Reciprocating Compressor

Figure (5.1) shows the operating principle of a reciprocating compressor. The extreme positions of the piston are known as top dead centre (TDC) and bottom dead centre (BDC).

As the piston moves from TDC (V_3) to BDC (V_1) low pressure (less than the external pressure) is created inside the cylinder. The gas is sucked at pressure P_1 into the cylinder. The volume of the cylinder from TDC to BDC is called the *swept volume*. The volume of gas sucked is from V_4 to V_1. Volume V_{Cl} ($V_3 = V_{Cl}$ = Clearance volume) to V_4 is the volume occupied due to re-expansion of the gas entrapped in the clearance volume, V_{Cl}. The gas inside the cylinder is compressed to a higher pressure P_2, as the piston moves from V_1 to V_2. This compression process is assumed to be isentropic. As the gas pressure reaches the preset value of P_2, the delivery valve opens and the high pressure gas is delivered at constant pressure, as the piston moves from V_2 to V_3. At this point the forward motion of the piston ends and it starts moving backward from V_3. Then the piston starts moving from V_3 (TDC). The gas pressure inside the cylinder decreases from P_2 to P_1 as it reaches V_4. The gas in the clearance volume expands from V_3 to V_4 and this expansion process is assumed to be isentropic. When the pressure decreases to P_1 the inlet valve opens and the gas is sucked at constant pressure P_1 into the cylinder. Thus, the thermodynamic cycle is completed.

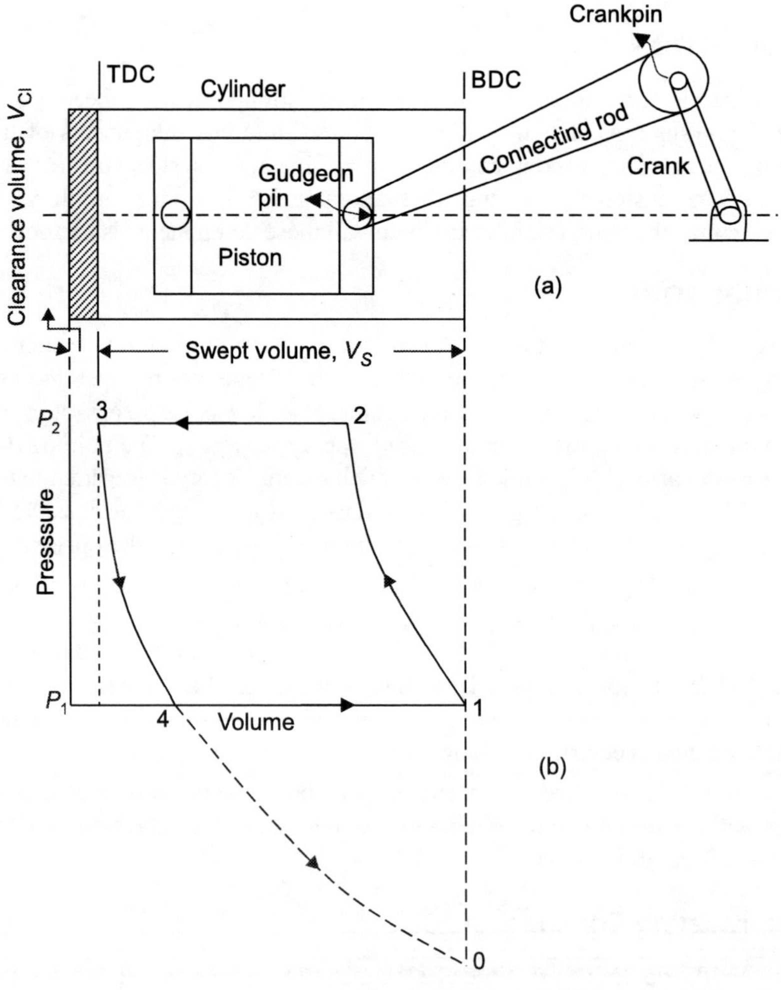

Figure 5.1: (a) Schematic of a reciprocating compressor, (b) Its performance on *P-v* plane

Work done by the compressor to compress the gas from pressure P_1 to pressure P_2 is calculated as follows.

Compressor is considered an open or flow system. For a flow system the enthalpy of the fluid per unit mass is expressed as $h = u + pv$

So $dh = du + pdv + vdp$;

From *Tds* relationship, for isentropic compression, $Tds = du + pdv = 0$ (as $ds = 0$)

So $dh = vdp$

Work done per unit mass during the compression from 1 to 2 for the compression process $pv^{\gamma} = C$

$h_2 - h_1 = \int Vdp = \gamma\,(p_1V_1 - p_2V_2)/(\gamma - 1) = \gamma\,(p_1V_1)\{(p_2/p_1)^{(\gamma-1)/\gamma} - 1\}\,/\,(\gamma - 1)$ [5.1]

Thus, compressor work is also expressed as $h_2 - h_1 = \gamma\, mR(T_1 - T_2)\,/\,(\gamma - 1)$ [5.2]

If γ is replaced by *n* for polytropic process, the work done is also expressed by substituting n in place of γ in the above expression.

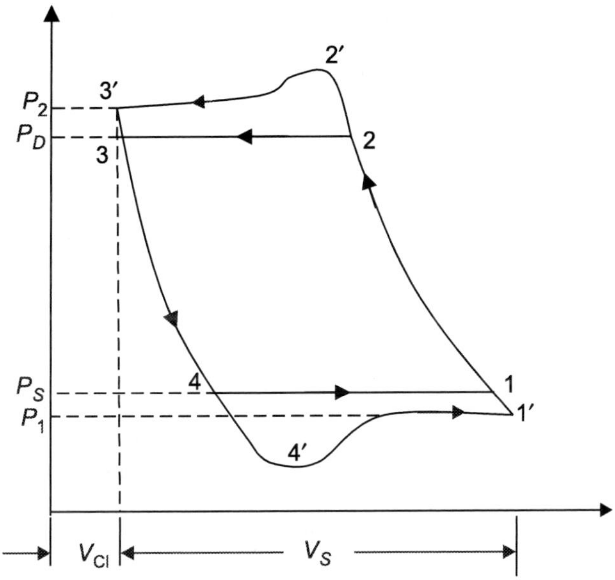

Figure 5.2: Performance of reciprocating compressor on *P-v* plane considering the pressure drops

Actual cycle of a reciprocating compressor: Refer Fig. 5.2. The manifold pressure (P_S) before the suction valve should be little higher than the pressure (P_1) maintained inside the cylinder so that air can enter the cylinder during suction stroke. Again, more pressure drop occurs in overcoming the inertia of the valves while opening and is represented by 4′. Similarly, there is pressure drop, represented by point 2′, in overcoming the inertia of the delivery valve while opening during starting of the delivery stroke and then it maintains a pressure of (P_2) inside the cylinder which is little more than P_D maintained in the delivery chamber so that compressed air can be delivered easily. Hence, for actual compressor work calculation pressure range should be considered P_1 to P_2 instead of P_S to P_D. However, if not mentioned, pressure range is considered P_1 to P_2 in rest of the text.

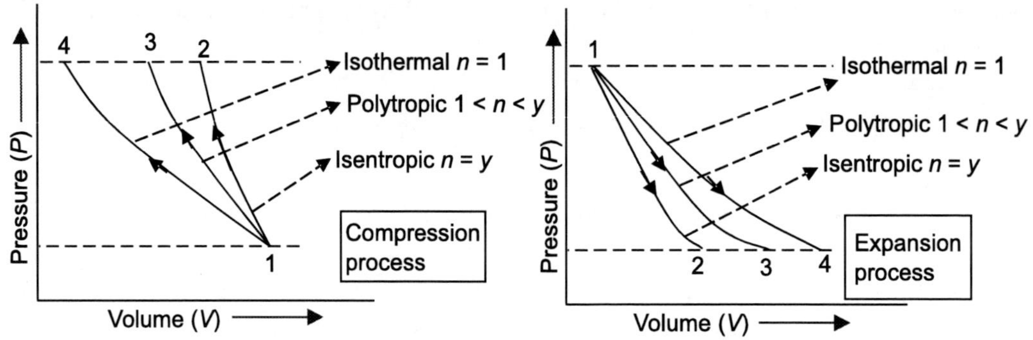

Figure 5.3: Work during different compression and expansion processes

Thermodynamically, minimum work is done by isothermal compression process. Refer Fig. 5.3. But isothermal process is possible when piston moves very slow so that heat generated due to compression gets sufficient time to be transferred to environment and constant temperature can be maintained. This is not possible. Similarly, maximum work output occurs during isothermal expansion. But this process again requires slow compressor movement.

Reciprocating compressors are also classified as: (1) hermetic (2) semi-hermetic, and (3) open types. Hermetic compressors are sealed compressors containing compressor and driving motor enclosed inside the outer casing. The driving motor gets cooled by the refrigerant vapour coming from the evaporator. The refrigerant vapour from evaporator first flows over the windings of the driving motor and receives heat from the winding. So the refrigerant vapour gets superheated. Then it enters the compressor. That is why dielectric strength of the refrigerant vapour should be high so that there will be no danger of short-circuit of the motor winding. As shown in Fig. 5.4, refrigerant vapour from evaporator is sucked through the suction tube at the lowermost part and discharged through tube placed at higher level. Suction tube diameter is larger than the discharge one. Because the specific volume of the refrigerant vapour at suction pressure is more, so the vapour needs higher diameter tube to maintain the required velocity in the suction tube. Above the discharge tube there is a tube connection to charge the compressor with refrigerant when there is loss of it due to leakage or any other reason. This is kept pinched when not in use. These compressors are usually of smaller capacity and available up to 2.0 kW.

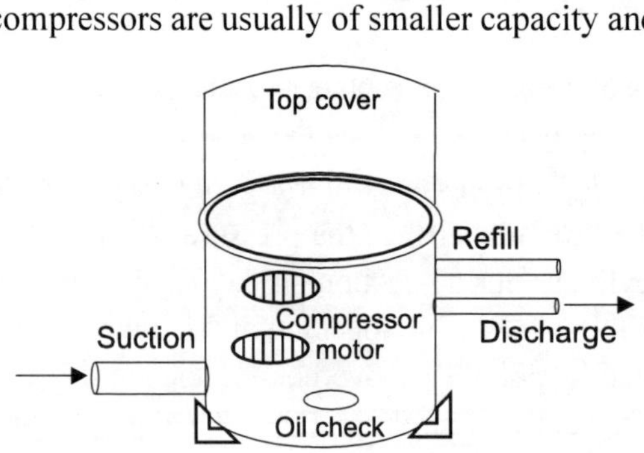

Figure 5.4: View of a hermetic compressor

In semi-hermetic type driving motor is kept outside of the sealed casing. So the driving motor can be detached from the compressor-motor assembly for repair of the motor without breaking the seal. This saves time and money for repair of the motor and also prevents the leakage of the refrigerant from the compressor. In the open type, motor is connected to compressor using a coupling or belt. Neither the compressor nor the motor is sealed. So it is easy for the maintenance of the compressor-motor assembly without disturbing the refrigeration system. But there is chance of leakage of lubricant and refrigerant from the system. The motors get cooled by providing fins for air cooled compressors and water jackets for water cooled compressors.

Let us discuss some important components of a reciprocating compressor.

A reciprocating compressor basically consists of a cylinder, a piston, a connecting rod and a crank. (Fig. 5.1a). In refrigeration vertical type compressors are used. Piston moves up and down inside the cylinder during compression and suction stroke respectively. Cylinders are made up of closed-grained cast iron. Pistons are made up of cast iron or aluminium. Two rings are mounted on the piston. (Fig. 5.5). One is slightly tapered and placed first. This is called compression ring. It prevents the leakage of gas from cylinder to crankcase during compression. Behind this the second ring, called oil ring, is mounted. This spread the lubricating oil between the piston and cylinder surface so that friction is reduced. The small end of the connecting rod is connected to piston by a pin called gudgeon pin and big end of it is connected to crank via crankpin. Crank is connected to crankshaft which placed in the crankcase. Crankcase contains the lubricating oil from which it splashes off into the cylinder.Thus, the reciprocating motion of the piston is transferred to rotary motion of the crankshaft through connecting rod and crank.

Valve plate assembly: In reciprocating compressor of a refrigeration system the valves are to be most carefully designed so that pressure drop across the valves (also termed **wire drawing effect**) is minimum. Generally, small weight reed valves and disc type valves are used for the purpose.These valves are fixed on the valve plates which are placed just above the cylinder block. They open or close due to the pressure difference across the cylinder and the manifold. To facilitate the opening/closing of valves within a short time, light spring (made up of spring steel or titanium) are attached to the valves.

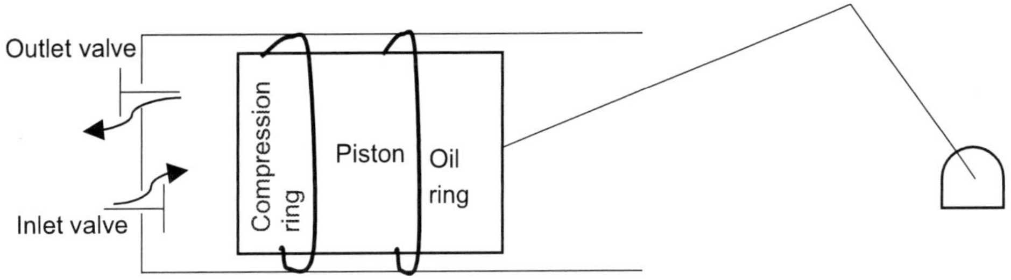

Figure 5.5: View of the piston rings and valves positions

Bearings: Sleeve type bearings are mounted in the journals of the crankshaft.

Shaft seal: In the open type compressors, crankshaft comes out of compressor body for connecting it to the prime mover. It must be properly sealed at the point where it comes out to prevent any leakage of refrigerant gas and lube oil from the crankcase or air from outside to crankcase.

5.2.1.1 *Volumetric efficiency of reciprocating compressor*

This is defined as the actual mass of air delivered by the compressor to the mass of air would have been occupied in the swept volume of the compressor.

Volumetric efficiency is affected by the following factors.

1. Clearance volume
2. Drop in pressure at suction and delivery valve
3. Heat transfer from the gas to cylinder wall and vice versa
4. Superheating of gas

Leakage (past the delivery valve and piston)

Clearance volume and volumetric efficiency

The reciprocating compressors are provided with clearance volume as shown in Fig. (5.1) so that impact between the piston head and the cylinder head can be avoided. The gas occupied in the clearance volume, after the delivery stroke, is re-expanded during the expansion stroke from V_3 to V_4 and attains the suction pressure P_1. After this fresh gas is inducted from V_4 to V_1.

So clearance volumetric efficiency,

$$\eta_{ClVol} = m_{act}/m_{swept} = \rho_{act.} V_{act.}/\rho_{ideal.} V_{swept} = \rho_{act} (V_1 - V_4)/\rho_{ideal} V_S$$

If $\rho_{act.}$ and ρ_{ideal} of the gas are assumed same (incompressible gas), then

$$\eta_{ClVol} = (V_1 - V_4)/V_S$$

$$\eta_{ClVol} = (V_{Cl} + V_S - V_4)/V_S = V_{Cl}/V_S + 1 - V_4/V_s$$

The re-expansion from 3 to 4 is a polytropic process and n is the index of expansion.

So $$V_4/V_3 = (P_2/P_1)^{1/n},$$

As ($V_3 = V_{Cl}$), and clearance ratio (ε) = V_{Cl}/V_S, $\eta_{ClVol} = \varepsilon + 1 - V_4/V_s$

$$V_4/V_s = (V_{Cl}/V_S)(V_4/V_{Cl}) = \varepsilon\,(V_4/V_3) = \varepsilon(P_2/P_1)^{1/n}$$

So $$\eta_{ClVol} = \varepsilon + 1 - \varepsilon(P_2/P_1)^{1/n} = 1 - \varepsilon\,((P_2/P_1)^{1/n} - 1) \quad [5.3]$$

The clearance ratio is set by the compressor manufacturer and the value normally varies from 0.03 to 0.05. If not specifically mentioned, the value may be considered to be 0.045.

Clearance volumetric efficiency, $\eta_{ClVol} = 0$, when $\varepsilon\,((P_2/P_1)^{1/n} - 1) = 1$

or, $$(P_2/P_1) = \{1 + (1/\varepsilon)\}^n \quad [5.4]$$

Above this pressure ratio, reciprocating pump acts as a vacuum pump

If a reciprocating compressor is used as a vacuum pump, the minimum pressure that can be achieved (from Eq. 5.4), is $P_{1(Min)} = P_{atm}/\{1 + (1/\varepsilon)\}^n$. In vacuum pump upper pressure limit is the atmospheric pressure. So in place of P_2 in Equation 5.4, P_{atm} is considered.

Drop in pressure across the suction and discharge valves

(Figure 5.2) In reality air is sucked into compressor when the cylinder pressure is lower than the suction manifold pressure. Similarly, air is discharged when the cylinder pressure is higher than the discharge manifold pressure. This pressure difference is needed to open the valves and make the air flow. Thus, the actual pressure ratio is higher than the ideal pressure ratio. Drop in pressure across the suction line and the compressor inlet which is sufficient to overcome the spring tension, valve weight and valve inertia is called 'wire drawing effect'.This causes decrease in volumetric efficiency. Again due to inertia of the valves, it remains open even after the suction stroke and it closes just after the compression stroke begins (i.e., it closes at point 'a' instead of 1′). So some amount of gas expelled out (from 1′ to a) reducing the volumetric efficiency.

Variation of index of compression/expansion

Higher the index of expansion steeper will be the expansion curve (Fig. 5.3). This causes an increase in aspirated volume. So the volumetric efficiency increases. NH_3 has index of expansion 1.3 whereas Freons have 1.1. So the volumetric efficiency is greater for NH_3 than the Freons and that is why the Freon compressors have low clearance volume.

Superheating of gas/vapours

As the vapour from evaporator flows through the ducts to the compressor, it gets superheated absorbing the environmental heat. The specific volume of the superheated vapour is higher. So the mass occupied by the vapour in the same aspirated volume will be less. This decreases the volumetric efficiency.

Leakage of gas/vapour

If the compression ratio is high the vapour, during discharge stroke, leaks past the piston ring to the crank case. So the amount vapour actually discharged is reduced. This causes a decrease in volumetric efficiency.

Overall volumetric efficiency is defined as the ratio of mass delivered by the compressor per cycle to the mass of vapour in swept volume at the suction flange condition.

So mathematically, this can be expressed, considering all the factors discussed above as.

So $$\eta_{\text{Vol,Overall}} = [(1 + \varepsilon)(P_S/P_1)^{1/n} - \varepsilon(P_D/P_1)^{1/m}](v_1/v_a) - 0.01(P_D/P_S) \qquad [5.5]$$

However, overall volumetric efficiency is considered to be same as clearance volumetric efficiency for easy and fast calculation with minimum error.

So $$\eta_{\text{ClVol}} = 1 - \varepsilon\,((P_2/P_1)^{1/n} - 1)$$

Similarly, for the compressor dimensions, commonly used L/D ratios of a reciprocating compressor are as follows. Vacuum pump and high-speed compressors $L/D < 0.5$; CFC compressors $L/D = 0.8$; and ammonia compressors $L/D = 1.0$. The mean piston speed is given by $V_P = 2NL/60$ where, L is the stroke and N is the rpm. V_P is approximately 2 m/s for slow speed compressors and 3 m/s or sometimes 4 m/s for high speed and modern compressors.

Example 5.1

A R22, 2 TR capacity, twin cylinder hermetic compressor with 4.5% clearance volume works between temperature limit of 50°C and 6°C. Compression index is 1.15. Stroke-to-bore ratio is 0.8 and rotates at 1400 rpm. Neglecting the pressure drop determine: (i) power consumption by the compressor and COP, (ii) volumetric efficiency and dimensions of compressor. If evaporator temperature is reduced to 4°C determine power consumption and COP of the system.

Ans: Assume S-S-S cycle. Refer Fig. 4.4.

Corresponding to evaporator temperature 6°C,

$P_e = 6.023$ bar, $v_1 = 0.0392$ m^3/kg, $h_1 = 253.18$ kJ/kg, $s_1 = 0.9230$ kJ/kg-K

State point 2 is fixed from $s_1 = s_2$; $p_2 = 19.327$ bar

And from superheated table at 50°C $t_2 = 71$°C, $h_2 = 283.1$ kJ/kg, For state point 3, $h_3 = h_4 = 108.22$ kJ/kg

So mass flow rate = 2 × 3.5167/(253.18 – 108.22) = 0.04852 kg/s

Work of compression = 0.04852 (283.1 – 253.18) = 1.45 kW; and COP = 4.844

Work of compression can also be calculated from $m \times n\,(p_1V_1)\{(p_2/p_1)^{(n-1)/n} - 1\}/(n-1) = 1.443$ kW

Volumetric efficiency,

$$\eta_{Cl} = 1 - \varepsilon\,((P_2/P_1)^{1/m} - 1) = 1 - 0.045\,\{(19.327/6.023)^{(1/1.15)} - 1\} = 0.921$$

Piston displacement,

$$V_S = ½\,\{(mv_1)/\eta_{Cl}\} = (0.04852 \times 0.0392)/(2 \times 0.921) = 1.03256 \times 10^{-3}\ \text{m}^3/\text{s}$$

Piston displacement,

$$V_S = \pi D^2/4 \times L \times N = \pi D^2/4 \times 0.8D \times 1400/60;\ \text{So } D = 0.0413\ m = 41.3\ \text{mm}$$

Stroke length,

$$L = 0.8 \times 0.0413 = 0.03304\ \text{m} = 33.04\ \text{mm}$$

If evaporator temperature is reduced to 4°C

$$P_1' = 5.659\ \text{bar},\ v_1' = 0.0417\ \text{m}^3/\text{kg}$$

So specific work done,

$$n(p_1V_1)\{(p_2/p_1)^{(n-1)/n} - 1\}/(n-1)$$
$$= 1.15(5.659 \times 10^2 \times 0.0417)\{(19.327/5.659)^{0.15/1.15} - 1\}$$
$$= 31.4353\ \text{kJ/kg}$$

Volumetric efficiency, $= \eta_{Cl} = 1 - \varepsilon\,((P_2/P_1)^{1/m} - 1)$

$$= 1 - 0.045\,\{(19.327/5.659)^{(1/1.15)} - 1\} = 0.914$$

Mass flow rate, $m' = 2 \times V_S \times \eta_{Cl}/v_1' = 2 \times 1.03256 \times 10^{-3} \times 0.914/0.0417 = 0.04526$ kg/s

Cooling capacity = 0.04526 × (252.51 – 108.22) = 6.53 kW = 1.857 TR;

Power consumed = 0.04526 × 31.4353 = 1.423 kW; COP = 4.59

Note: Piston displacement (V_S) will remain same. So the mass flow rate will change as per the volumetric efficiency.

Example 5.2

A R22 compressor having 4 cylinders runs at 750 rpm and works between the temperature limits of –10°C and 45°C. Bore and stroke of each cylinder are 10 cm and 12 cm, respectively. Clearance volume ratio is 0.045. For isentropic compression, calculate (i) index of compression, (ii) volumetric efficiency, (iii) swept flow rate, (iv) mass flow rate, and (v) refrigerating capacity.

Ans: For –10°C, p_1 = 3.550 bar, v_1 = 0.0654 m^3/kg, s_1 = 0.9473 kJ/kg-K,

For 45°C, p_2 = 17.209 bar, from superheated table, for state point 2, v_2 = 0.0159 m^3/kg,

t_2 = 72°C

For isentropic compression, $p_1 v_1^n = p_2 v_2^n$, substituting the values n = 1.116

Volumetric efficiency = 1 + 0.045 – 0.045(17.209/3.550)$^{1/1.116}$ = 0.86

Swept flow rate, $V_S = V_D$ × No. of cylinders × N/60 = π/4 D^2L × 4 × 750/60 = 0.0471 m^3/s

So mass flow rate = $V_S \eta_{CI} / v_1$ = 0.0471 × 0.86/0.0654 = 0.6184 kg/s

Refrigerating effect, $m (h_1 - h_3)$ = 0.6184 × (247.37 – 101.76) = 90.045 kW = 25.61 TR

Example 5.3

A 8-cylinder NH_3 compressor having *L/D* ratio 1.0 has an average piston speed of 4.0 m/s and 1400 rpm. If it operates between temperature limit of –20°C and 40°C, find its maximum cooling capacity. Assume its volumetric efficiency as 0.75.

Ans: For evaporator temperature of –20°C, v_1 = 0.624 m^3/kg, h_1 = 1417.97 kJ/kg

For condenser temperature of 40°C, h_3 = 371.47 kJ/kg

Average piston speed, $V_P = 2NL$; 4 = 2 × (1400 /60) × L; So L = 0.0857 m and D will also be = 0.0857 m

Swept flow rate, $V_S = V_D$ × No. of cylinders × N/60 = π/4 D^2L × 8 ×1400/60 = 0.0923 m^3/s

Mass flow rate, $V_S \eta_{CI} / v_1$ = 0.0923 × 0.75/0.624 = 0.111 kg/s

Refrigerating capacity = 0.111 (1417.97 – 371.47) = 116.1615 kW = 33.03 TR

5.2.2 Rotary Compressors

All rotary type compressors run on the same principle as is described below. There lie constructional differences only. They are in a pair. When the pair rotates, the space between the pairing components, near to the inlet, starts enlarging, creating low pressure. So the vapour enters the rotor casing. As it further rotates, the space between the pairing components decreases and the vapour is compressed. As it continues to rotate, the pair starts disengaging from each other near the outlet. At this moment outlet opens and the compressed vapour goes out.

5.2.2.1 Vane type compressor

Refer to Fig. 5.6. It consists of a fixed outer cylinder. Inside it, a roller, being mounted on an eccentric shaft, rotates. The shaft and the cylinder have the same axis. The maximum annular space available in this position,

$$V_D = \Pi/4\,(d_C^2 - d_R^2)L \quad [5.6]$$

where d_C and d_R are the diameters of cylinder and roller, respectively and L is the length of the cylinder. V_D is also called the displacement volume. So the swept volume per second is

$$V_S = \Pi/4\,(d_C^2 - d_R^2)L * N/60 \quad [5.7]$$

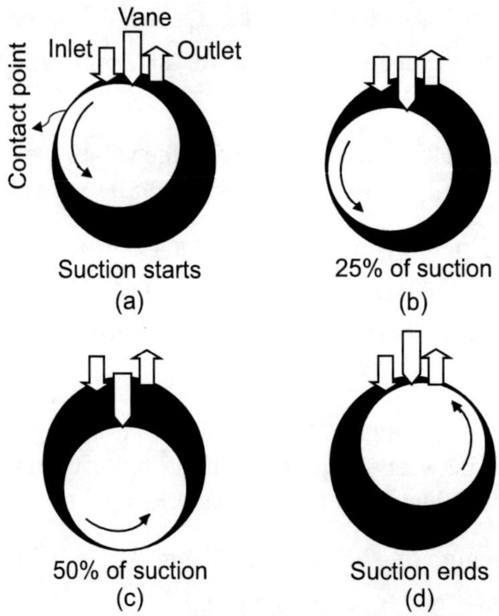

Figure 5.6: Schematic of working principle of rotary (vane type) compressor

where N is the rpm. (Fig. 5.6 a). Due to eccentricity of the roller shaft, as the roller rotates, the area between the roller and the cylinder wall goes on increasing behind the contact point causing a low pressure on that side. Hence, the refrigerant vapour is sucked to that space. Thus, the annular space goes on increasing. At the same time the annular space, i.e., ahead of contact point, goes on decreasing. Thus, the roller starts compressing the vapour and at the end of compression process the compressed vapour is released through the outlet port to the refrigerant line.

In sliding vane or rolling piston type rotary compressor the suction port and the delivery port are separated by a spring loaded vane mounted on a slot in the cylinder wall. This vane moves in or out as the roller comes closer to or away from it. The roller rotates inside the cylinder as if it rolls on the inner side of the cylinder wall. Thus, at any time the surface of roller is in contact with the cylinder wall at a point. In rotating vane or multiple vane type compressor two or more sliding vanes are mounted on eccentrically mounted slotted rotor. The sliding vanes are held against the cylinder wall by the action of centrifugal force or by spring force to obtain a better seal against the cylinder wall. The volumes between the eccentric rotor and sliding vanes vary with angular position. Larger models may have eight or more vanes and require no inlet or outlet valves. Rotary compressors have no clearance volume and so have very high volumetric efficiency but they are limited in application by the stresses set up by the thrust on the tips of the blades.

5.2.2.2 Screw compressor

The working principle of screw compressor is similar to that of a gear pump. The usual form of screw compressor is "twin meshing rotors on parallel shaft". Figure 5.7 shows the male rotor drives the female rotor. As these turn inside the closely fitting casing, the space between two grooves comes opposite the inlet port and the gas enters. On further rotation, this pocket of gas becomes sealed from the inlet port and moves down the barrels. A meshing lobe of male rotor then reduces the pocket volume compressing the gas, which is finally released at the opposite end, where the exhaust port is uncovered by the movement of the rotors.

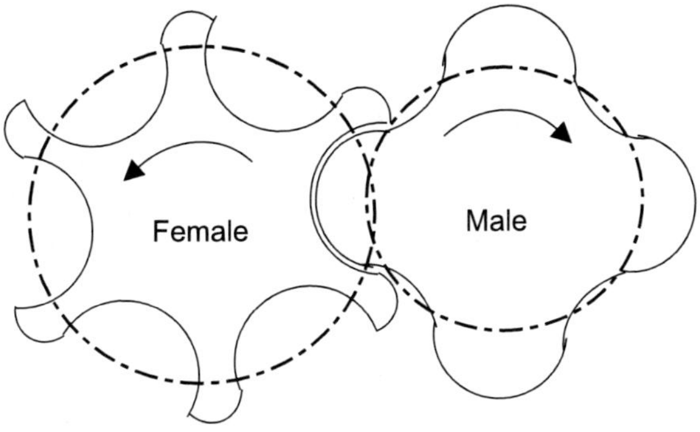

Figure 5.7: Schematic of working principle of rotary (screw type) compressor

Scroll compressor consists of two inter-fitting spiral shaped scroll members. One scroll remains fixed while the other one moves in orbit (not rotates) inside it. Gas is sucked to the pockets formed between the fixed and moving scroll walls at the outer periphery. Compression is achieved as the moving scroll starts orbiting with respect to fixed one, reducing the size of the pocket. Gradually, it moves inward, i.e., towards the centre and is discharged through a port.

However, clearance volumetric efficiency of rotary compressors is considered to be nearly 100% (but less than 100%) as it has negligible clearance volume.

5.2.3 Dynamic Compressors

In dynamic compressors energy is imparted to the gas by high speed rotation of the rotor blades, mostly, generated due to centrifugal force. Initially, this increases its velocity there by the kinetic energy. Then the gas is passed through an increasing cross-sectional area, a diffuser, where the velocity of gas decreases, converting the kinetic energy into pressure energy. The most common type of dynamic compressor is the centrifugal compressor. Suction gas enters axially the eye of a rotor and is thrown out tangentially to the blade circumference due to centrifugal action. The kinetic energy gained by the gas depends upon the density of the gas and the velocity of the tip of the blade.

5.3 EXPANSION DEVICES

Expansion devices are used in the refrigeration system for two reasons:

(i) to expand the liquid refrigerant coming out of the condenser from condenser pressure to evaporator pressure; and

(ii) to regulate the refrigerant flow rate in the refrigeration system as the need of the system.

Expansion devices are broadly classified into two categories: (i) fixed opening type and (ii) variable opening type.

Under, the first type is capillary tube which is a fixed diameter and length tube. So the opening is fixed. The second type valves are: (i) turbine, (ii) thermostatic (superheat controlled) expansion valve, (iii) automatic expansion valve, and (iv) float valve. The opening through which the refrigerant flow varies as it is actuated by spring force.

5.3.1 Capillary Tube

This is a small diameter (internal diameter varies from 0.5 mm to 2.5 mm) tube of certain length (nearly 0.6 m to 6.0 m) (Fig. 5.8a). This is a misnomer as the tube diameter is not so small to initiate capillary action for the flow through the tube. This connects condenser directly to the evaporator. The pressure drop in the capillary tube is due to two factors.

(i) Frictional pressure drop: As the viscous liquid refrigerant flows through the tube pressure drops due to friction.

(ii) As the pressure drops the liquid refrigerant flashes into vapour and the resultant density decreases. The mass flow rate and the flow area being constant, the velocity of refrigerant flow through the tube increases since $m = \rho VA$, hence, pressure decreases. This is known as *momentum pressure drop* due to acceleration of the flow in the tube as the liquid refrigerant flashes into vapour.

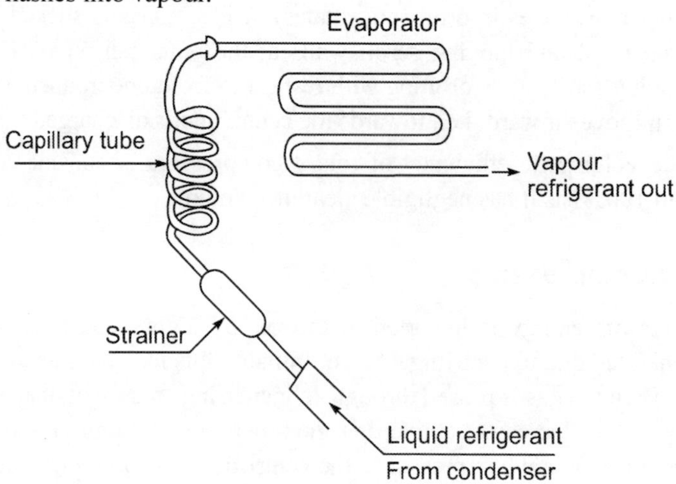

Figure 5.8(a): Schematic of a capillary tube and its relative position

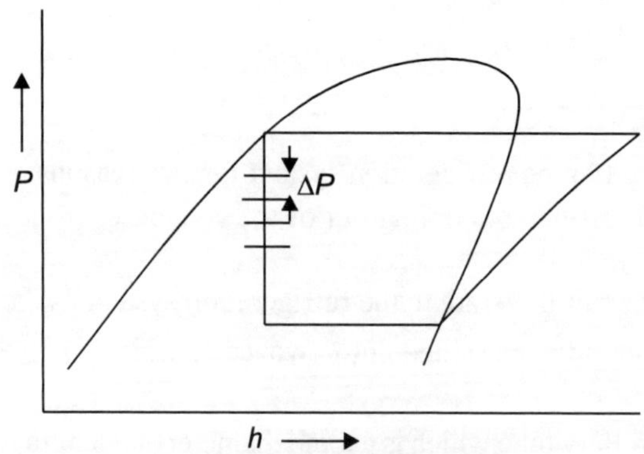

Figure 5.8(b): Incremental pressure drop in capillary tube

So, once a capillary tube of certain dimension is installed in a system, the mass flow rate cannot adjust itself with the refrigeration load variation. But the advantages are its simplicity, low cost and absence of any moving part. It equalizes the pressure in the system when the compressor stops. So the compressor can restart in no-load condition. Hence it is used for small cooling devices like domestic refrigerator, window A/C, split A/C, etc.

5.3.1.1 Selection of a capillary tube

For any new system the diameter and length of the capillary tube are to be selected to ensure the designed pressure drop so that balance point between the compressor and the capillary tube can be achieved. There are two methods for the selection. One is analytical method and the other one is graphical method.

Analytical method: As discussed above the pressure drop is due to (i) friction, and (ii) gain in momentum. The well known equation to calculate the pressure drop due to friction is

$$\Delta P_f = f\rho\Delta L\, V^2/2D \qquad [5.8]$$

where, f is the Darcy's friction factor, V is velocity of flow, and D is diameter of tube. For laminar flow $f = 64/\text{Re}$ and turbulent flow (f) is calculated from the correlation $0.32\ \text{Re}^{-0.25}$ where, Re is the Reynolds number and is calculated from $\rho VD/\mu$ or GD/μ.

For flow through tubes, if $\text{Re} \leq 2500$, flow is laminar, otherwise the flow is turbulent.

Let us define mass velocity which is expressed as $G = \rho V$ [5.9]

For incremental length of ΔL (Fig. 5.8b), instead of V and f, V_{mean}, f_{mean} should be taken, i.e., $(V_1 + V_2)/2$ or $\{(f_1 V_1) + (f_2 V_2)\}/2$.

Gain in momentum, $G\Delta V = \rho V.\Delta V = \rho V\,(V_1 - V_2)$ and this is the pressure drop due to gain in momentum. [5.10]

So total pressure drop, $\Delta P = \Delta P_f + G\Delta V$ [5.11]

From the above two equations, $\Delta L = (\Delta P - G\Delta V)/\{(G/2D)(fV)_{mean}\}$ [5.12]

Steps to calculate the length of a capillary tube

1. Find the mass flow rate of the refrigerant from the cycle calculation.
2. Assume, judiciously, the diameter of the capillary tube.
3. Assume an incremental drop in temperature (smaller is the temperature drop, more is the accuracy).
4. Find the required properties of the refrigerant entering to capillary tube and at each incremental drop, like specific enthalpy, specific volume, dynamic viscosity, etc.
5. Find the dryness fraction of the refrigerant after the incremental drop assuming an isenthalpic process if it is in two-phase zone.
6. Then find the other relevant properties and velocity of the flow, as above, at that state point.
7. Calculate the Reynolds number, friction factor, velocity, mass velocity, etc.
8. Then calculate ΔL from Eq. 5.12.
9. Repeat from step 4 to step 8 till ΔL becomes negative. This means the total pressure required for a segment may become less than the pressure drop required for acceleration alone. The velocity of flow reaches the velocity of sound sqrt (γRT). This will be a choked flow condition.

10. Then add up all ΔL's to get the total length of the capillary tube. However, the exact length is finally determined from cut and trial method.

Example 5.4

Calculate the length of a capillary tube to be used in a 1 TR, R134a refrigeration system working between temperature limit of 46°C and 7°C.

Ans: Assuming S-S-S cycle the refrigerant flow rate, m, will be

$$= 1 \times 3.5167/(402.63 - 265.47) = 0.02564 \text{ kg/s}$$

Normally in many refrigeration systems two numbers of capillary tubes are used to avoid the high turbulence in a small diameter tube. So each tube will carry flow rate of 0.0128 kg/s.

Select the bore diameter 2.3 mm and an incremental drop of 5°C.

Mass velocity, $G = 4 \times 0.0128/\Pi(0.0023)^2 = 3079.42$ kg/m^2-s

Different properties of saturated refrigerant at 46°C, 41°C, 36°C, 31°C, 26°C, 21°C, 16°C, 11°C, 7°C are as follows.

Temp (°C)	46	41	36	31	26	21	16	11	7
P_{sat}(kPa)	1190.3	1034.4	911.85	792.3	685.43	589.73	504.25	429.4	374.8
h_f(kJ/kg)	265.47	257.91	250.48	242.61	235.97	228.9	221.87	215.0	209.5
h_g(kJ/kg)	421.92	419.98	417.65	415.31	412.84	410.0	407.61	404.05	402.63
$v_f \times 10^3$ (m^3/kg)	0.8924	0.8753	0.8595	0.8450	0.8313	0.8185	0.8066	0.7952	0.7866
v_g(m^3/kg)	0.01687	0.01932	0.02238	0.02588	0.03	0.03493	0.04078	0.0479	0.0539
$\mu_f \times 10^6$ (Pa-s)	174.71	185.85	197.73	210.43	224.00	238.54	254.14	270.91	285.26
$\mu_g \times 10^6$ (Pa-s)	13.052	12.861	12.67	12.48	12.286	12.09	11.90	11.71	11.55

The saturated liquid refrigerant at 46°C enters the capillary tube. As it flows down the tube, it will be in two phase.

Now the length of tube required to drop the temperature from 46°C to 41°C is calculated as follows.

Dryness fraction, $x = (h_{f,46} - h_{f,41})/h_{fg,41} = (265.47 - 257.91)/(419.98 - 257.91) = 0.0466$

Dynamic viscosity, $\mu_{46} = 174.67 \times 10^{-6}$ Pa-s

$\mu_{41} = \mu_f + x\mu_g = 186.45 \times 10^{-6}$ Pa-s.

Similarly, specific volume, $v_{46} = 0.8924 \times 10^{-3}$ m^3/kg

$v_{41} = v_f + xv_g = 1.776 \times 10^{-3}$ m^3/kg,

Reynolds number, $\text{Re}_{46} = GD/\mu_{46} = 40548.84$, $\text{Re}_{41} = GD/\mu_{41} = 37986.95$

So friction factor, $f_{46} = 0.32/\text{Re}^{0.25} = 0.0226$, $f_{41} = 0.32/\text{Re}^{0.25} = 0.0229$

Velocity of refrigerant in the tube at 46°C, $Gv_{46} = 3079.42 \times 0.8924 \times 10^{-3} = 2.748$ m/s

Velocity of refrigerant in the tube at 41°C, $V_{41} = Gv = 3079.42 \times 1.334 \times 10^{-3} = 5.469$ m/s

So $\Delta V = 2.721$ m/s

$(fV)_{46} = 0.0226 \times 2.748 = 0.0621$, $(fV)_{41} = 0.0229 \times 5.469 = 0.1252$,

$(fV)_{mean} = (0.0621 + 0.1252)/2 = 0.09365$

$\Delta L = (\Delta P - G\Delta V)/\{(G/2D)(fV)_{mean}\}$

$= \{(1190.3 - 1034.4) \times 1000 - 3079.42 \times 2.721\}/\{(3079.42/2 \times 0.0023)(0.09365)\} = 2.353$ m.

Similarly, the incremental lengths from 41°C to 36°C, 36°C to 31°C,31°C to 26°C,26°C to 21°C, 21°C to 16°C, 16°C to 11°C,11°C to 7°C have been calculated and presented below.

Temp (°C)	46	41	36	31	26	21	16	11	7
X	0.0	0.0466	0.0897	0.132	0.1668	0.202	0.235	0.267	0.290
$\mu_{avg} \times 10^6$ (Pa-s)	174.67	186.45	198.866	212.077	226.05	240.98	256.94	274.036	288.61
Re	40548.84	37986.95	35615.27	33396.67	31332.3	29391.09	27565.45	25845.75	24540.61
F	0.0226	0.0229	0.0233	0.0237	0.0241	0.0244	0.0248	0.0252	0.0256
$v \times 10^3$ (m^3/kg)	0.8924	1.776	2.867	4.261	5.835	7.874	10.39	13.58	16.42
V (m/s)	2.748	5.469	8.83	13.121	17.97	24.25	31.995	41.82	50.56
(fV)	0.0621	0.1252	0.2057	0.311	0.433	0.5917	0.7935	1.054	1.294

Table

Temp (°C)	46-41	41-36	36-31	31-26	26-21	21-16	16-11	11-7
ΔP (kPa)	155.9	122.55	119.55	106.87	95.7	85.48	74.85	54.60
ΔV (m/s)	2.721	3.36	4.291	4.849	6.28	7.745	9.825	8.74
$(fV)_{mean}$	0.09365	0.16545	0.25835	0.372	0.5124	0.6926	0.92375	1.174
ΔL (m)	2.353	1.013	0.6148	0.3691	0.2226	0.1329	0.0721	0.0352
							Total	4.8127

In real fluid flow, adiabatic, constant area tube with friction, the specific enthalpy does not remain constant. Because the flow velocity increases to such a high value (as shown in above tables) that the change in kinetic energy cannot be neglected. The specific enthalpy of the refrigerant goes on decreasing. The line representing the locus of points with the same mass velocity and stagnation enthalpy is called a 'Fanno line'. And the flow is often called as fanno line flow. Hence, the specific enthalpy for different points is calculated from energy conservation equation.

$$h = h_1 + V^2/2 \text{ or } h_1 = h - V^2/2 \quad [5.13]$$

From new h_1 the new properties of refrigerant is calculated and steps 5 to 10 are repeated.

5.3.2 Float Type Expansion Valves

Float type expansion valves are normally used with flooded evaporators in large capacity refrigeration systems. A float type valve opens or closes depending upon the liquid level as sensed by a buoyant member, called float. The float could take the form of a hollow metal or plastic ball, a hollow cylinder or a pan. Thus the float valve always maintains a constant liquid level (i.e., the liquid refrigerant is in saturated state at that temperature and pressure) in a chamber called float chamber. Depending upon the location of the float chamber, a float type expansion valve can be either a low-side float valve or a high-side float valve. Float type expansion valves, both low side and high side, have components: (a) a float and float chamber, and (b) a needle valve with valve seat.

Low-side float valves: (Fig. 5.9) A low-side float valve maintains a constant liquid level in a flooded evaporator or a float chamber attached to the evaporator. When the load on the system increases, more amount of refrigerant evaporates from the evaporator. As a result, the refrigerant liquid level in the evaporator or the low-side float chamber drops momentarily. The float then moves in such a way that the valve (needle valve) opening is increased and more amount of refrigerant flows into the evaporator to take care of the increased load and the liquid level is restored. The reverse process occurs when the load falls, i.e., the float reduces the opening of the valve and less amount of refrigerant flows into the evaporator to match the reduced load. As mentioned, these valves are normally used in large capacity systems and normally a by-pass line with a hand-operated expansion is installed to ensure system operation in the event of float failure.

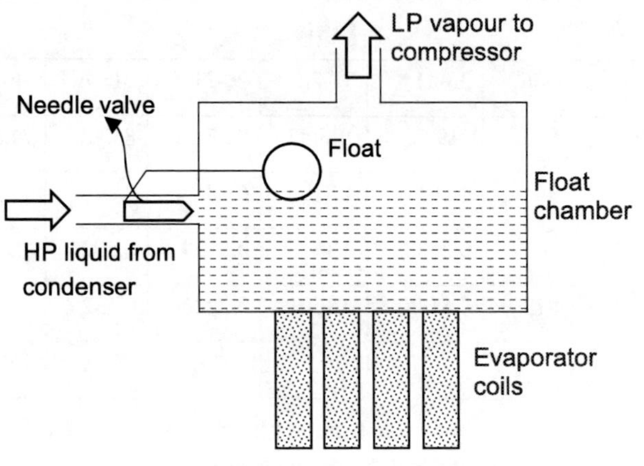

Figure 5.9: Schematic of a low-side float valves

High-side float valves: Figure 5.10 shows the schematic of a high-side float valve. As shown in the figure, a high-side float valve maintains the liquid level constant in a float chamber that is connected to the condenser on the high pressure side. When the load increases, more amount of refrigerant evaporates and condenses. As a result, the liquid level in the float chamber rises momentarily. The float then opens the valve more to allow a higher amount of refrigerant flow to cater to the increased load, as a result the liquid level drops back to the original level. The reverse happens when the load drops. Since a high-side float valve allows only a fixed amount of refrigerant on the high pressure side, the bulk of the refrigerant is stored in the low-pressure side (evaporator). Hence there is a possibility of flooding of evaporator followed by compressor slugging. However, unlike low-side float valves, a high-side float valve can be used with both flooded as well as direct expansion type evaporators.

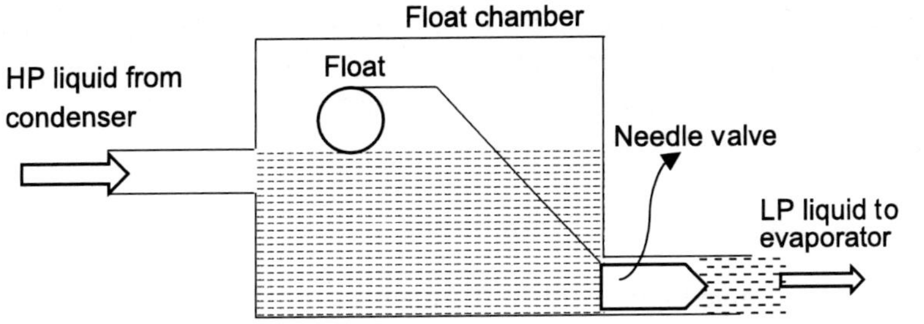

Figure 5.10: Schematic of a high-side float valve

5.3.3 Electronic Type Expansion Valve

The schematic diagram of an electronic type expansion valve is shown in Fig. 5.11 (a) is given in the last page of this chapter). As shown in the figure, an electronic expansion valve consists of an orifice and a needle in front of it. The needle moves up and down in response to magnitude of current in the heating element. A small resistance allows more current to flow through the heater of the expansion valve, as a result the valve opens wider. A small negative coefficient thermistor is used if superheat control is desired. The thermistor is placed in series with the heater of the expansion valve. The heater current depends upon the thermistor resistance that depends upon the refrigerant condition. Exposure of thermistor to superheated vapour permits thermistor to self-heating, thereby lowering its resistance and increasing the heater current. This opens the valve wider and increases the mass flow rate of refrigerant. This process continues until the vapour becomes saturated and some liquid refrigerant droplets appear. The liquid refrigerant will cool the thermistor and increase its resistance. Hence, in the presence of liquid droplets the thermistor offers a large resistance, which allows a small current to flow through the heater making the valve opening narrower. The control of this valve is independent of refrigerant and refrigerant pressure; hence it works in reverse flow direction

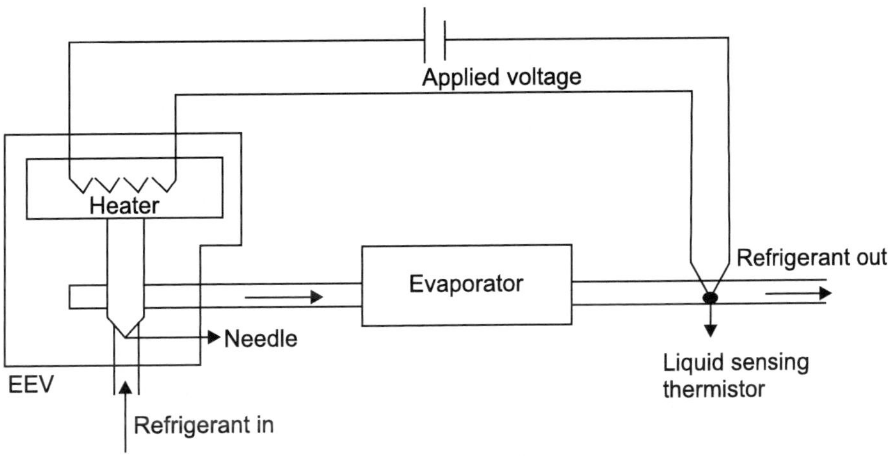

Figure 5.11(a): Schematic of an electronic expansion valve

also. It is convenient to use it in year-round-air-conditioning systems, which serve as heat pumps in winter with reverse flow. In another version of it the heater is replaced by stepper motor, which opens and closes the valve with a great precision giving a proportional control in response to temperature sensed by an element.

5.3.4 Thermostatic Expansion Valve (TEV)

Thermostatic expansion valve Figure 5.11(b) is the most versatile expansion valve and is most commonly used in refrigeration systems. A thermostatic expansion valve maintains a constant degree of superheat at the exit of evaporator; hence it is most effective for dry evaporators in preventing the slugging of the compressors since it does not allow the liquid refrigerant to enter the compressor. The schematic diagram of the valve is given in Fig. 5.11(b). This consists of a feeler bulb that is attached to the evaporator exit tube so that it senses the temperature at the exit of evaporator. The feeler bulb is connected to the top of the bellows by a capillary tube. The feeler bulb and the narrow tube contain some fluid that is called *power fluid*. The power fluid may be the same as the refrigerant in the refrigeration system, or it may be different. In case it is different from the refrigerant, then the TEV is called *TEV with cross charge*. The pressure of the power fluid (P_e) is the saturation pressure corresponding to the temperature at the evaporator exit. If the evaporator temperature is T_e and the corresponding saturation evaporator pressure is P_e then the purpose of TEV is to maintain a temperature $T_e + \Delta T$ at the evaporator exit, where ΔT is the degree of superheat required from the TEV. The power fluid senses this temperature $T_e + \Delta T$ by the feeler bulb and its pressure (P_P) is the saturation pressure at the temperature,

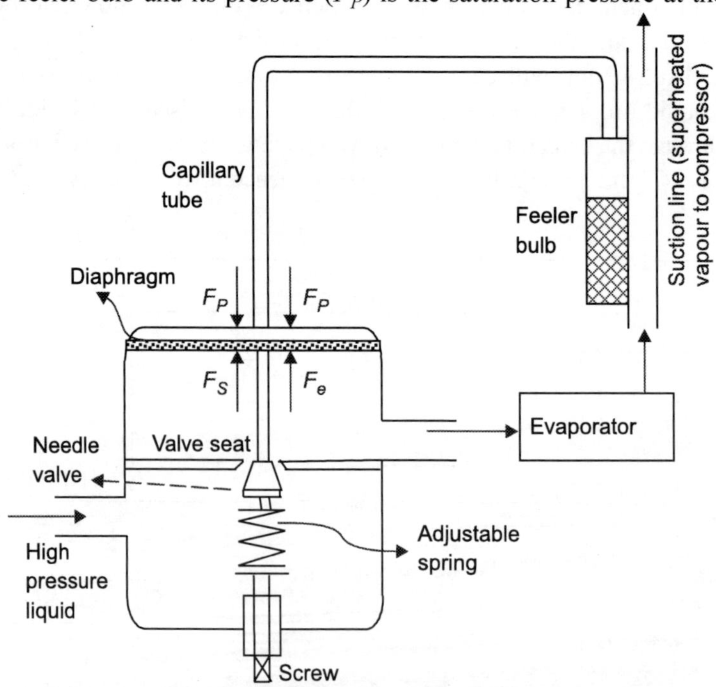

Figure 5.11(b): Schematic of a thermostatic expansion valve

$T_e + \Delta T$. The force (F_P) exerted on top of bellows of area A_b due to this pressure is given by:

$$F_P = A_b P_P \qquad [5.14]$$

The force (F_e) exerted due to this pressure (P_e) on the bottom of the bellows is given by

$$F_e = A_b P_e \qquad [5.15]$$

The difference of the two forces (F_P and F_e) is exerted on top of the needle stand. There is an adjustment spring below the needle stand that exerts an upward spring force (F_S) on the needle stand. In steady state there will be a force balance on the needle stand, that is, $F_P = F_e + F_S$ [5.16]

During off-cycle, the evaporator temperature is same as room temperature throughout, that is, degree of superheat ΔT is zero. If the power fluid is the same as the refrigerant, then $P_P = P_e$ and $F_P = F_e$. Therefore, any small spring force (F_S) acting upwards will push the needle stand against the orifice and keep the TEV closed. If it is TEV with cross charge or if there is a little degree of superheat during off-cycle then for TEV to remain closed during off-cycle, F_S should be slightly greater than $F_P - F_e$.

Thus, for a given setting force (F_S) of the spring, TEV maintains the difference between F_P and F_e or the degree of superheat ΔT constant.

$$\Delta T \propto (F_P - F_e) \qquad [5.17]$$

$$\propto F_S$$

This is irrespective of the level of P_e, that is, evaporator pressure or temperature, although degree of superheat may be slightly different at different evaporator temperatures for same spring force, F_S. It will be an ideal case if the degree of superheat is same at all evaporator temperatures for a given spring force.

Effect of load variation

If the load on the plant increases, the evaporation rate of liquid refrigerant increases, the area available for superheating the vapour increases. As the degree of superheat increases, pressure of power fluid (P_p) increases, the needle stand is pushed down and the mass flow rate of refrigerant increases. This is the ideal case. The evaporation rate of refrigerant is proportional to the load and the mass flow rate supplied through the expansion valve is also proportional to the load.

On the other hand, if the load on the plant decreases, the evaporation rate of refrigerant decreases, as a result the degree of superheat decreases. The thermostatic expansion valve reacts in such a way so as to reduce the mass flow rate through it. The flow rate of refrigerant in this valve is proportional to the evaporation rate of refrigerant in the evaporator. Hence, this valve always establishes balanced flow condition of flow between compressor and itself.

Advantages, disadvantages and applications of TEV

The advantages of TEV compared to other types of expansion devices are:

1. It provides excellent control of refrigeration capacity as the supply of refrigerant to the evaporator matches the demand. It utilizes the whole surface of the evaporator.
2. It ensures that the evaporator operates efficiently by preventing starving under high load conditions.
3. It protects the compressor from slugging by ensuring a minimum degree of superheat under all conditions of load, if properly selected.

However, compared to capillary tubes and AEVs, a TEV is more expensive and proper precautions should be taken at the installation. For example, the feeler bulb must always be in good thermal contact with the refrigerant tube. The feeler bulb should preferably be insulated to reduce the influence of the ambient air. The bulb should be mounted such that the liquid is always in contact with the refrigerant tubing for proper control.

The use of TEV depends upon the degree of superheat. Hence, in applications where a close approach between the fluid to be cooled and evaporator temperature is desired, TEV cannot be used since very small extent of superheating is available for operation. A counter flow arrangement can be used to achieve the desired superheat in such a case. Alternately, a subcooling HEx may be used and the feeler bulb mounted on the vapour exit line of the HEx. The valves with bellows have longer stroke of the needle, which gives extra sensitivity compared to diaphragm type of valve. Valves with bellows are more expensive.

Thermostatic expansion valves are normally selected from manufacturers' catalogues. The selection is based on the refrigeration capacity, type of the working fluid, operating temperature range, etc. In practice, the design is different to suit different requirements such as single evaporators, multi-evaporators, etc.

5.4 CONDENSERS

Condenser, simply a heat exchanger, is the part of the refrigeration system, which is placed after the compressor and exchanges heat with the environment. The high pressure and high temperature (superheated) vapour of refrigerant from compressor enters the condenser and rejects heat to the cooling medium. First it is desuperheated from 2 to 2′ rejecting the sensible heat. Then it condenses from 2′ to 3 rejecting the latent heat of vaporisation. Then it rejects heat from 3 to 3′ the sensible heat of liquid if there is any sub-cooling. Thus, the total heat rejected by the condenser,

$$Q_C = m_{\text{ref}}\,\{C_{PV\,\text{avg}}\,(t_2 - t_2') + h_{fg} + C_{PL}(t_3 - t_3')\} \qquad [5.18]$$

Form conservation of energy of the refrigeration system $Q_C = Q_E + W_C$, where Q_E is the heat absorbed by the evaporator and W_C is the work input to the compressor.

Sometimes condenser capacity is determined from 'heat rejection ratio (R)′ which is expressed as the ratio of heat rejected by the condenser to the heat absorbed by the evaporator.

$$R = Q_C / Q_E = (Q_E + W_C)/Q_E = 1 + 1/\text{COP} \qquad [5.19]$$

For fixed evaporator temperature, as condenser temperature increases, COP decreases. So heat rejection ratio increases. For fixed condenser temperature if evaporator temperature decreases COP decreases. So heat rejection ratio increases. Increase in R indicates either more condenser area or more heat transfer coefficient is required to reject more amount of heat to the cooling medium.

Generally, the condensers are classified into three categories.

(a) Air cooled condensers
(b) Water cooled condensers
(c) A combination of air and water cooled condenser, known as evaporative condensers

5.4.1 Air Cooled Condensers

Air cooled condensers are usually finned tubes. They are cooled by naturally ventilated air (by natural convection, e.g., household refrigerators) or forced ventilated air (by forced convection, e.g., window airconditioners). Fins may be of wire type or plate type (Fig. 5.12 a). Heat capacity of air is less because its specific heat is less (1.005 kJ/kg). Hence, this type of condensers are used for small refrigeration systems (< 5 TR cooling capacity) like household refrigerators, window airconditioners, etc. Nowadays, instead of providing wire fins for enhancing the heat transfer rate, condenser tubes are mounted on plates otherwise known as *plate fins*. Sometimes air cooled condensers are used for high capacity refrigeration systems where freshwater availability is scarce as in desert areas.

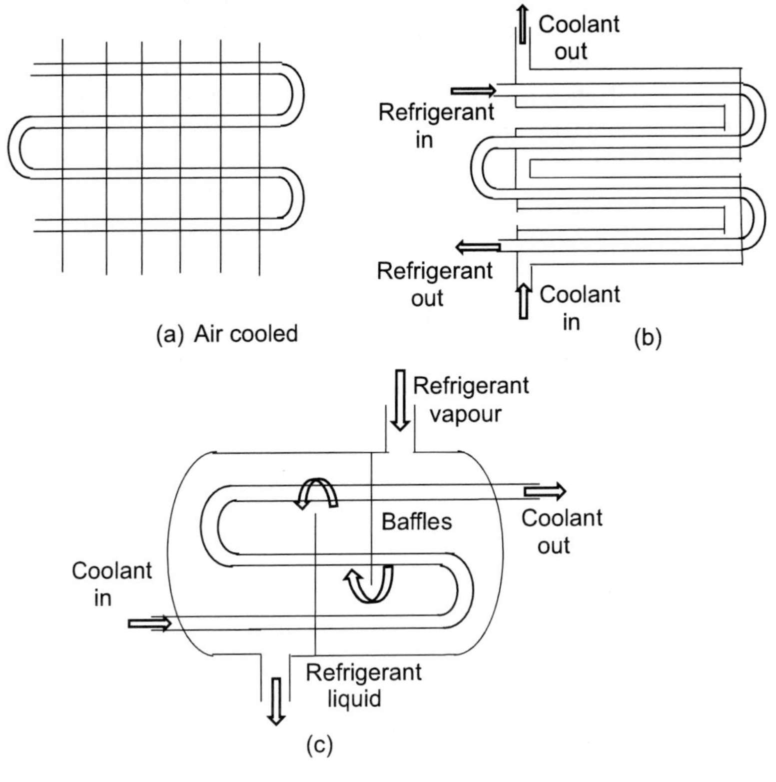

Figure 5.12: (a) Wire and tube condenser,
(b) double tube condenser, and (c) shell and tube condenser

5.4.2 Water Cooled Condensers

Water cooled condensers are used for big installations like central air conditioners of cooling capacities more than 100 TR because water heat capacity is high due to high specific heat of water (4.19 kJ/kg). Water cooled condensers are of three types.

(a) Double tube type (Fig. 5.12 b): Cooling water flows through the inner tube and the refrigerant flows through the annular space in a counter current direction.

(b) Shell and coil type: Cooling water flows through the coil counter to the flow of refrigerant through the shell.

(c) Shell and tube type (Fig. 5.12 c): Cooling water flows through the tubes counter to the flow of refrigerant in the shell. There will be baffles inside the shell so that refrigerant will have a zigzag flow in the shell and the contact area and contact time between the coolant and the refrigerant increase.

The cooling water from the condenser gets heated up by taking the condenser heat. Hence, this hot water is sent to cooling towers where it rejects heat to the environment and this cold water can be reused.

Cooling towers (Fig. 5.13) are seen in the outskirts of the premises where central air conditioning systems or processing plants are in operation. Hot water from the condenser comes to the cooling tower and is sprayed through nozzles from the top of it for atomization of water. Fills (normally made up of wood) are provided on the wall of the cooling tower so that sprayed water is more atomized when it hits the fills. Atmospheric air is drawn from the bottom of the cooling tower by forced draft or induced draft. This causes a direct contact heat transfer between hot water particles and cold air (direct contact type heat exchanger). The air gets heated up by receiving the heat from the hot water and goes out to the atmosphere. The water gets cooled by rejecting heat to the air and goes back to the air conditioning plant to be reused as the coolant in the condensers.

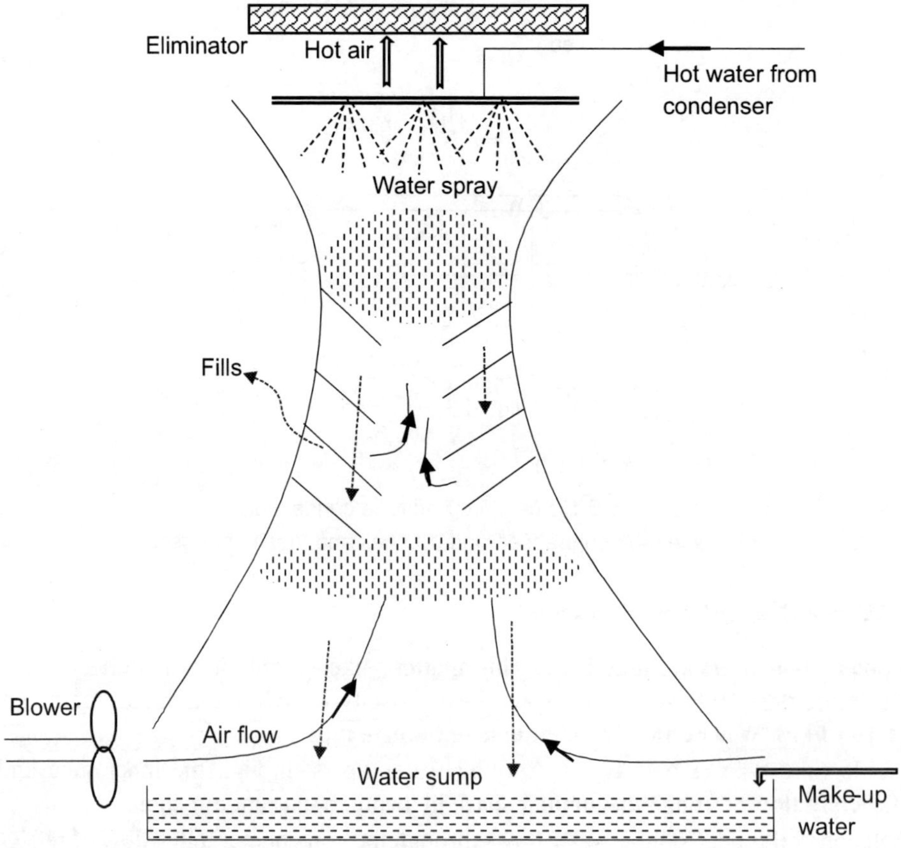

Figure 5.13: Schematic of a cooling tower

5.4.3 Evaporative Condenser

This type of condenser utilizes the advantages of both the air cooled and water cooled condensers (Fig. 5.14). Water is sprinkled through the nozzles over the bank of tubes of condenser. Thus, water forms a thin film over the tubes. Heat is transferred to the water layer first. Air is drawn over the coil by an induced fan which enhances the evaporation rate of water from the condenser coil surface. Evaporation of water cools the condenser tubes and the air along with the evaporated water is discharged to the atmosphere. This type of condenser is more efficient in comparison to water cooled condenser and can be used indoors as well as outdoors.

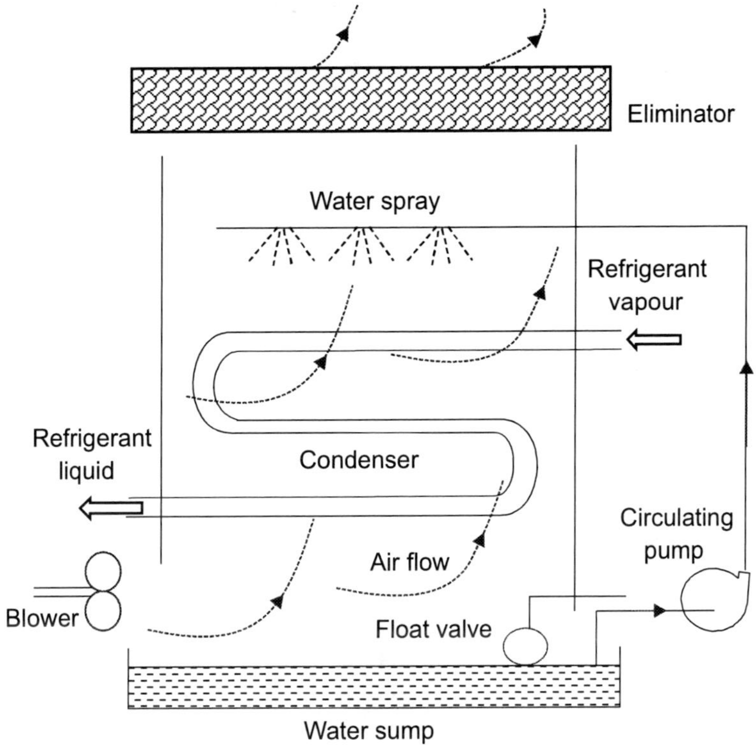

Figure 5.14: Schematic of an evaporative condenser

5.5 EVAPORATORS

This is also a heat exchanger and a part of the refrigeration system where the actual cooling effect is obtained by boiling off of refrigerant. The evaporator is a misnomer because, in true sense, there occurs no evaporation. Cooling effect is obtained as the refrigerant absorbs the latent heat from the refrigerated space and boils off at the desired temperature at corresponding saturation pressure.

Classification of evaporators may be done basing upon heat transfer process or refrigerant flow or condition of the heat transfer surface.

But according to operating conditions evaporators are broadly classified into two categories. One is direct expansion type and the other one is flooded type.

5.5.1 Direct Expansion Type (DX Coil)

This type evaporator (also called dry expansion type) receives pure liquid or a mixture of liquid and vapour refrigerant from the flow control valve. The length of the evaporator coil is so designed that all the refrigerant coming out of the evaporator is in dry, slightly superheated state. To ensure this, refrigerant flows to the evaporator through a thermostatic expansion valve which is set for certain degree of superheat (Fig. 5.15). Thus, in this type of evaporators some part of the tube surface is not wetted. Hence, it is a dry type of evaporator. These evaporators are simple to operate. This DX type of evaporator exists in two types. One is DX cooling coil where air is cooled and another is DX cooling chiller where water is cooled.

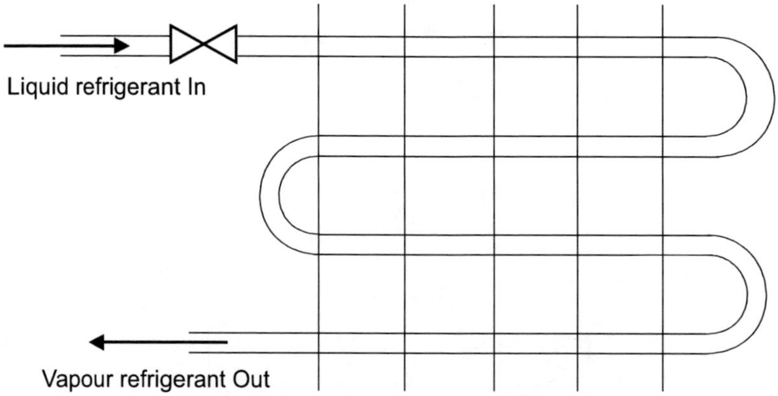

Figure 5.15: Schematic of a DX coil

5.5.2 Flooded Type

An evaporator is said to be flooded type if the liquid refrigerant covers the entire heat transfer surface. This type of evaporator uses a float type expansion valve. (Fig. 5.16) The two-phase refrigerant from a float type expansion valve (dryness fraction x_1) enters a surge tank where the vapour and liquid refrigerant get separated. The flash vapour is directly drawn by the compressor. The remaining liquid refrigerant along with some from surge tank enters from the bottom side of the evaporator. It starts vaporising absorbing the heat and rises up. At the end of the evaporator the vapour and liquid part (dryness fraction x_2) of the refrigerant enters the surge drum from top. This vapour along with the vapour from expansion device is drawn by the compressor. This liquid along with the liquid from expansion device again flows to the evaporator. Thus, all the time the heat transfer area of the evaporator tube remains covered with liquid refrigerant. Hence, the heat transfer rate remains much higher in comparison to DX type evaporator.

Mass balance for the system is as follows. Liquid mass (m), when throttled, x_{1m} amount of vapour gets separated and rest $(1 - x)\, m$ amount of liquid and m_S amount of liquid from surge tank enters evaporator. As this liquid flows through the evaporator, $x_2\, \{(1 - x)m + m_S\}$ gets vaporised absorbing the cooling load and the rest liquid goes back to surge tank.

So $x_2\, \{(1 - x)m + m_S\} + x_1 m = m$

Let us define **recirculation factor** as the ratio of mass through evaporator to mass through expansion valve, i.e., $\{(1 - x)m + m_S\} / m = f$.

Substituting f in the above equation we get $f = (1 - x_1)/x_2$

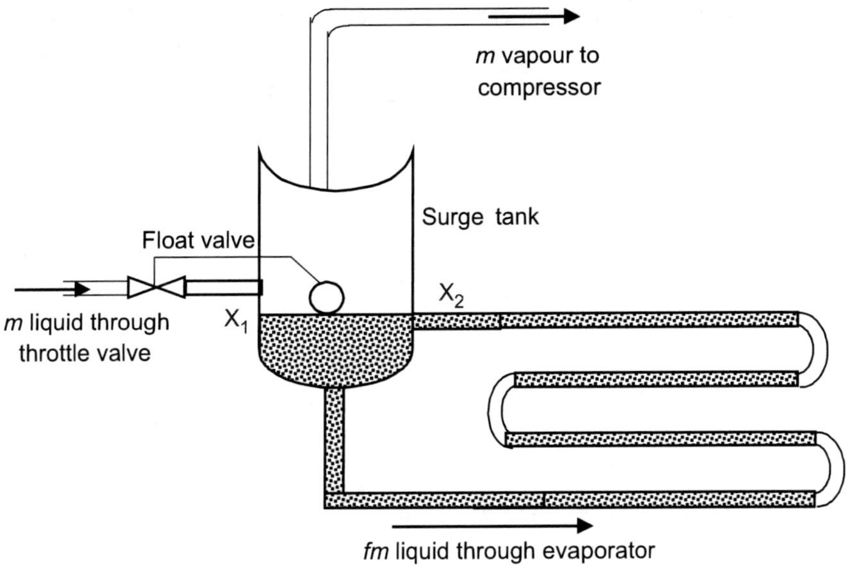

Figure 5.16: Schematic of a flooded evaporator

5.5.3 Classification on the Basis of Constructional Features

Small refrigeration system like the domestic refrigerators or air conditioners have finned (normally plate type) coils through which the refrigerant flows. The fluid (air or water) to be cooled is passed over the coil may be by natural convection (as in domestic refrigerators) or by forced convection (as in room air conditioners). Big refrigeration systems, like central air conditioners, industrial refrigeration systems, use shell and tube type evaporators (heat exchangers). In flooded type shell and tube evaporator refrigerant flows through the shell and the liquid, to be chilled, flows through the tubes. A float valve is provided to maintain the level of the refrigerant. Reverse is the case for dry type shell and tube evaporator and a thermostatic expansion valve is provided to maintain the degree of superheat [Fig. 5.11(b)]. Similarly, there are shell and coil type chillers, double tube type chillers. The refrigerant and water flow in the same pattern as described for the case of condenser.

5.6 STEPS TO BE FOLLOWED IN DESIGNING A CONDENSER/EVAPORATOR

Condenser and evaporator of a refrigeration system are basically heat exchangers. The design of these two components can be done by any of the two methods. (1) LMTD method, (2) ε – NTU method. LMTD method is followed if the temperatures of inlet and outlet of hot fluid and cold fluid are known. ε – NTU method is followed if one of the four temperatures is not known. In the design, the required heat transfer area for the condenser/evaporator to reject/absorb given quantity of heat is estimated.

Q = UA(LMTD); where, Q is the amount of heat to be rejected/absorbed

$$Q = m_C C_{PC} \Delta t_C = m_h C_{Ph} \Delta t_H$$

where, C and h refer to cold and hot fluid respectively.

LMTD is the log mean temperature difference between the hot fluid and the cold fluid and is expressed as

$$\text{LMTD} = (\Delta t_o - \Delta t_i)/\ln(\Delta t_o/\Delta t_i) \qquad [5.20]$$

where, Δt_O is $(t_{hi} - t_{CO})$ and Δt_i is $(t_{hO} - t_{Ci})$. U is the overall heat transfer coefficient and expressed as

$$1/U = A_O/(A_i h_i) + t/(kA_{LM}) + 1/h_O + R_f \qquad [5.21]$$

where, U = overall heat transfer coefficient, A_O = outer surface area, A_i = inner surface area of the tube, t = thickness of the tube, h_O = outside heat transfer coefficient, h_i = inside heat transfer coefficient, R_f is the fouling factor, and A_{LM} = Log mean area = $(A_O - A_i)/\ln(A_O/A_i)$.

In most of the cases, tube material resistance to heat flow is neglected because of high conductivity of the tube material and negligible thickness of the tube wall (this means $A_O \approx A_i$). So overall heat transfer coefficient can also be expressed, without much error, $1/U = 1/hi + 1/h_O + R_f$.

It is a bit tedious to calculate the values of h_i and h_O. For condenser, calculations are done for three zones separately. First is single phase superheated vapour (desuperheating zone). Here Dittus-Boelter equation can be used to calculate h_{sv} for the desuperheating. (However, limitations for the use of Dittus-Boelter equation must be followed.) Second is condensation zone. Many more reliable correlations are available to calculate the Nusselt number depending upon the type of refrigerants, type of flow, etc. Suitable correlation is to be selected for one's own system. From Nusselt number (h_{Co}) can be calculated. Then third is sub-cooling zone (where sub-cooling is done). Here also Nusselt number for the liquid phase can be calculated from Dittus-Boelter equation. For calculating the heat transfer coefficient for coolant side (may be air cooled or water cooled), useful correlations can be used for different flow patterns. (Flow through bank of tubes, flow through fins, cross flow or counter flow, etc.) After calculating overall heat transfer coefficient for each zone, the required heat transfer area for each zone can also be calculated. The sum of all these three required heat transfer areas will give the total heat transfer area of the condenser.

In ε – NTU method, ε is the effectiveness of the heat exchanger and defined as the ratio of actual rate of heat transfer $\{C_h (T_{hi} - T_{ho})\}$ or $\{C_c (T_{c,o} - T_{c,i})\}$ to the maximum possible rate of heat transfer $\{C_{min} X (T_{hi} - T_{c,i})\}$. NTU is number of transfer units.

$$\text{NTU} = \text{AU}/\text{C}_{\text{min}} \qquad [5.22]$$

where A is the required heat transfer area, U is the overall heat transfer coefficient, C_{min} is the minimum of the heat capacities (mass × $C_{P\,fluid}$) of the two fluids. (Its details are given in Heat Exchanger chapter of any heat transfer textbook).

When the tube is finned on one side to enhance heat transfer, the total heat transfer surface area on the finned side will be the summation of bare portion of the tube and the surface area of fins.

$$A_{total} = A_{unfinned} + A_{fin}$$

Total surface area of condenser is also calculated with much accuracy by assuming the average heat transfer coefficient of above three zones as that during condensation only. The reason is that during desuperheating and sub-cooling convective heats are less but the temperature difference between the refrigerant and the coolant are more, whereas during condensation convective heat transfer coefficient is high but the temperature difference is low. Hence, the heat transfer rate per unit surface area is approximately same in each of these three phases.

Design of evaporator

This is also a heat exchanger, the design procedure of evaporator is same as that of the condenser and other heat exchangers. The only difference in designing is that in condenser refrigerant side heat transfer coefficient is calculated from available correlations for condensation and in evaporator refrigerant side heat transfer coefficient is calculated from available correlation for boiling heat transfer. Most frequently used correlation to estimate the boiling heat transfer coefficient is expressed as

$$C_f\Delta t/h_{fg} = C_{sf}\{(Q/A)/(\mu_f h_{fg})[\sigma/g(\rho_f - \rho_g)]^{0.5}\}^r(C\mu/k)_f^s \qquad [5.23]$$

where, σ is the vapour-liquid surface tension of the fluid and other fluid properties in the correlations are evaluated at the saturation temperature corresponding to the evaporation pressure. C_{sf} is a function of refrigerant and heating surface combination. For example, it is 0.013 for R22 and also other CFCs and copper combination. Exponents r and s have the values 0.33 and 1.7 for CFCs, respectively but 1.0 for water. This correlation is suggested by Rohsenow. There are also many more correlations to estimate the boiling heat transfer coefficient depending upon the flow pattern, type of refrigerants, etc.

Example 5.5

Design of condenser: A water cooled condenser (double tube heat exchanger type) of a 5 TR capacity R22 refrigeration system working between temperature limit of 40°C and –5°C is to be designed. The tube dimensions are as follows. Refrigerant flowing through the inner tube whose d_i = 8 mm and d_O = 9.5 mm. The coolant water at the rate of 1.0 kg/s flowing through the annulus having D_i = 16.5 mm and D_O = 19 mm. Water inlet temperature is 28°C. Calculate the length of the condenser tube.

Ans: Assume the refrigeration system works on S-S-S cycle. (Refer Fig. 4.4)

Mass of refrigerant flow rate

$$= 5 \times 3.5167/(h_{g((-5°C)} - h_{f(40°C)})$$
$$= 17.5835/(249.29 - 95.40)$$
$$= 0.1143 \text{ kg/s}$$

Specific entropy, $s_{g((-5°C)}$ = 0.9393 kJ/kg-K

From interpolation in superheat table at 40°C, t_2 = 64°C and h_2 = 283.2 kJ/kg

Hence, the total heat to be rejected from the condenser = 0.1143 (283.2 – 95.4) = 21.47 kW.

As a first hand design, bare tubing is considered for the design of condenser, i.e., condenser without fins. As explained, the properties of refrigerant and coolant at the average temperature are to be determined. The properties of refrigerant at 40°C: k_L = 0.08 W/m-K, $\mu_L = 1.39 \times 10^{-4}$ kg/m-s, $\rho_L = 1.13 \times 10^3$ kg/m^3, ρ_V = 65.8 kg/m^3, C_{PL} = 1.34 kJ/kg-K. Water properties at 30°C: k_W = 0.623 W/m-K, $\mu_W = 7.65 \times 10^{-4}$ kg/m-s, C_P = 4.19 kJ/kg-K, So Prandtl number = 5.15.

To calculate the refrigerant side heat transfer coefficient, instead of considering the desuperheating zone and condensation zone separately, the heat transfer coefficient during condensation is only considered (as discussed earlier). The reliable correlation for condensation inside tube, suggested by Chato (P.K. Nag, HT, pp. 484), is taken.

The correlation: $h_C = 0.555\ [\{k_L{}^3\rho_L(\rho_L - \rho_V)gh_{fg}\}/(\mu_L \Delta t D)]^{1/4}$

where, K – thermal conductivity, ρ – density, μ – dynamic viscosity, Δt – temperature difference between wall and the saturation temperature of refrigerant, D – diameter of the tube, g – acceleration due to gravity, subscript L and V stand for liquid and vapour, respectively and h'_{fg} is $h_{fg} + 3(C_{PL}\Delta t)/8$.

Now substituting the respective values, we get $h'_{fg} = 168.553$ kJ/kg and $h_C = 2030$ W/m^2-K.

Temperature rise of cooling water is calculated from $Q = m_W C_{PW} \Delta t_W$. So $\Delta t_W = 21.47/(1 \times 4.19) = 5.12°C$. Water is flowing in counter direction in the annulus. The hydraulic diameter is calculated from $D_h = 4A_C/P_{Wet}$; where A_C is the cross-sectional area of annulus $(\pi/4)(D_i^2 - d_O^2)$ and P_{Wet} is the wetted perimeter $\pi(D_i + d_O)$. So $D_h = (D_i - d_O) = (0.0165 - 0.0095) = 0.007$ m

$Re_W = (m_W/A_C)(D_h/\mu_W)$; substituting the respective values $Re_W \approx 64000$.

So Nusselt number, $Nu_W = 0.023Re^{0.8}Pr^{0.4} = 310$. So h_O = htc of water side $= 310 \times k/D_h = 27590$

So, overall heat transfer coefficient, $1/U_O = 1/h_O + (A_O / A_i)R_{foulig} + (A_O / A_i)1/h_i$,

where, $R_{fouling}$ is the 'fouling factor'. For copper (tube material) and water and refrigerant R22, its value is 0.00009. $(A_O / A_i) = d_O / d_i = 0.0095/0.008 = 1.1875$. The resistance to heat flow through the tube material, i.e., (t/k) is neglected.

Now substituting the respective values, we get $U_O = 1373.44$ W/m^2-K.

$LMTD = (\Delta t_O - \Delta t_i)/\ln(\Delta t_O/\Delta t_i) = (6.88 - 12)/\ln(6.88/12) = 9.2°C$

From the expression $Q = U_O A(LMTD)$ and substituting the respective values

Required heat transfer area, $A = 21.47 \times 10^3/(1373.44 \times 9.2) = 1.7$ m^2

So the length of the tube required, $L = 1.7/\pi d_O = 1.7/(\pi \times 0.0095) = 57$ m

For such high length we have to have multi-pass condenser so that the face length may be limited to small value (1 m or so).

Solution by ε - NTU method

Effectiveness of the heat exchanger, ε = Actual heat transferred $m_W C_{PW}(t_{CO} - t_{Ci})$ / Maximum possible heat transferred $m_W C_{PW}(t_{hi} - t_{Ci}) = 5.12/12 = 0.427$.

Effectiveness of a heat exchanger during condensation is also expressed as $(1 - e^{-NTU})$; so NTU = 0.557

NTU is expressed as $NTU = AU_O/C_{minimum}$. The condensing fluid has infinite heat capacity (very large). Hence, $C_{minimum}$ of water will be $= 1 \times 4.19 = 4.19$ kJ/K

Therefore, $U_O = 1373.44$ W/m^2-°C. And the required heat transfer area is calculated to be 1.7 m^2.

5.7 EFFECT OF PRESENCE OF AIR OR NON-CONDENSABLE GAS

The reasons of entering air or non-condensable gases into refrigeration system are: (a) maintaining the evaporator pressure below atmospheric due to high NBP of the refrigerants (like R11, R113, R114, etc) used; (b) after evacuation of the system for charging refrigerant, some amount of air is still left behind.

These gases along with refrigerant vapour go to the condenser and cause the following problems.

(a) These gases or air do not condense along with refrigerant vapour. Rather they remain as gas in the condenser tube and cover the inner surface of the tube. The thermal conductivity of these gases, being very low, the heat transfer coefficient decreases drastically. Hence, proper condensation cannot take place. That reduces the cooling effect and hence COP.

(b) The condenser pressure is constant at P_c. Since refrigerant vapour and air/non-condensable gas are in mixed state in condenser tube, total condenser pressure, $P_c = P_R + P_{air}$. P_R and P_{air} are the partial pressures of refrigerant and air, respectively. Thus, the refrigerant vapour is less than the condenser pressure, which is the saturated pressure of the refrigerant. So the refrigerant cannot be condensed fully.

5.8 DEFROSTING

This is an important aspect of all the refrigeration systems whose evaporators work below 0°C. Frost is accumulated around the cooling coil when the water vapour in air comes in contact with it. We mostly come across such phenomena in household refrigerators. This frost should be removed periodically because it gives resistance to the heat flow to the cooling coil and the required temperature cannot be achieved. In an environment where the outdoor temperature is more than 10°C, the compressor should be stopped for a small period of time so that the frost will be melted off. This is possible in case of small systems like household refrigerators whose service can be interrupted for that period. In large systems the following methods of defrosting are followed.

1. Electric resistance heaters may be placed inside the cooling coil or directly under it. When there will be accumulation of certain thickness of frost, electric heater will be on and there will be defrosting.
2. Reverse cycle method: If the refrigeration cycle is reversed by using a two-way valve, the evaporator will act as condenser and vice versa. So the hot gas from the compressor will flow through the evaporator coil and defrost it. This is a commonly used method of defrosting.
3. Hot gas bypass method: Hot gas from the compressor discharge is made to pass over the cooling coil through a branch pipe and arrangement should be made to compensate that much of gas from an auxiliary evaporator so that there will be no interruption in the cooling process. Sometimes, hot air/water, which is used as the coolant for the condenser and comes out of it, can be used for the purpose.

After defrosting, the water must be collected and drained out properly by using a **condensate pump**.

5.9 BALANCING OF FOUR COMPONENTS

Performance of one component depends upon that of the other components (compressor, condenser, expansion device and evaporator) of the refrigeration system and the operating conditions. When the operating conditions of a refrigeration system do not match with that of the designed conditions, each component has to adjust itself with the other. Suppose there is Q_e cooling load on evaporator and required amount of refrigerant is vaporised in the evaporator to meet Q_e. Same amount of vapour must be sucked and compressed by the compressor to the condenser pressure. Condenser with the cooling equipment should have the capacity to condense all the vapour to liquid state (saturated or sub-cooled). Then the liquid refrigerant passes through the expansion device where the refrigerant must have the required pressure drop from condenser to evaporator and allow that much of refrigerant to flow to the evaporator so that all of the liquid refrigerant will be vaporised to meet Q_e amount of cooling load. Thus, there will be no flooding or starvation of refrigerant in any of the components of the refrigeration system. Any change in any of the parameters that affect the performance of the system component will lead to unbalanced condition. Even the change in coolant temperature to the condenser has a major effect on the condensation process and consequently the performance of the system is seriously affected. A brief discussion on balancing of one component with the other and consequently balancing of the whole system is given. This discussion is done for graphical method with reference to the evaporator temperature as this is the most desired parameter.

5.9.1 Performance of Compressor

(Figure 5.17) As evaporator temperature increases (keeping condenser temperature constant) the compression ratio decreases, and volumetric efficiency increases. So mass flow rate increases and the cooling capacity (Q) of compressor increases. For constant evaporator temperature, if the condenser temperature increases, pressure ratio increases and this decreases the cooling capacity.

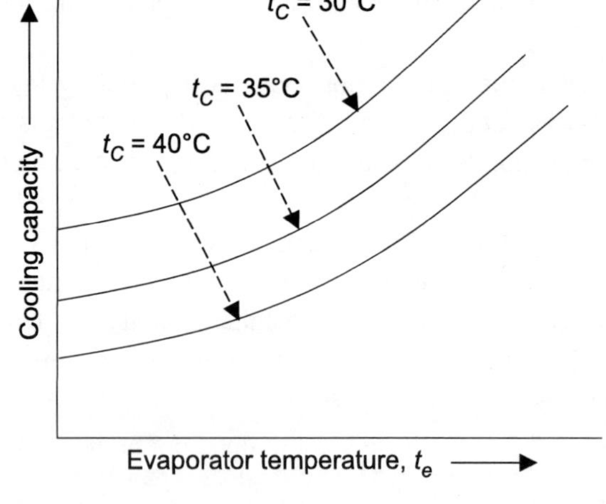

Figure 5.17: Variation of Q of compressor with t_e for fixed t_C

5.9.2 Performance of Condenser

(Figure 5.18) As evaporator temperature increases, compression ratio decreases for constant temperature of condenser. Hence, the discharge temperature decreases. So condensation rate increases and this increases the cooling capacity of condenser. When condenser temperature increases, condenser cooling capacity increases because of increase in LMTD value and thus the heat transfer rate from the condenser.

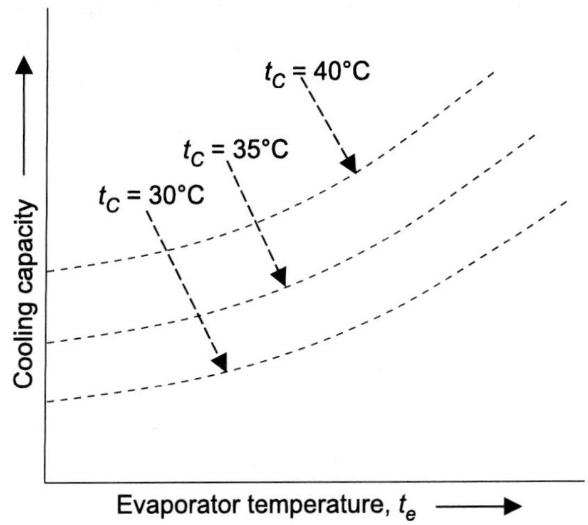

Figure 5.18: Variation of Q of condenser with t_e for fixed t_C

Condensing Unit Characteristic

Combining the performance of these two components, otherwise called 'condensing unit' performance is shown in (Fig. 5.19).

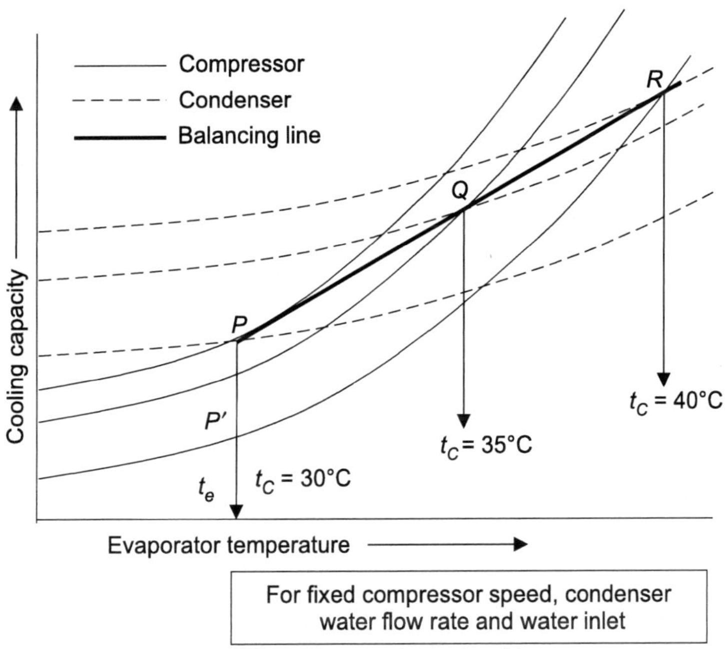

Figure 5.19: Variation of cooling capacity with evaporator temperature

It is observed that as the evaporator temperature decreases, the condensing temperature for the combination also decreases. At lower evaporator temperatures, the volumetric efficiency and the mass flow rate through the compressor decreases. This decreases the load on condenser. A large condenser heat transfer area is available for small mass flow rate, hence condensation can occur at lower condenser temperature. At condenser temperature of 40°C and evaporator temperature t_e, the compressor cooling capacity corresponds to point *P*', while the condensing unit has the capacity corresponding to point *P*. That is, one gets a larger cooling capacity because of the reduced condenser temperature. The balance points are shown as *P*, *Q* and *R*.

5.9.3 Balancing of Compressor and Capillary Tube

Refer to Fig. 5.20. It shows the mass flow rate through the compressor and the capillary tube. As the evaporator pressure increases pressure ratio (P_2/P_1) decreases. This results in increase of mass flow rate through compressor (keeping condenser temperature constant) due to increase in volumetric efficiency of compressor (Eq. 5.3) and decrease in specific volume of refrigerant vapour. But there will be decrease in mass flow rate through capillary tube. If the condenser temperature increases (keeping evaporator temperature and pressure constant) *m* through compressor decreases because volumetric efficiency decreases due to increase in pressure ratio. But *m* through capillary tube increases as the pressure ratio increases. Thus, the figure shows the balance points, (i.e., whatever mass of refrigerant sucked through compressor from the evaporator would be delivered by the capillary tube to the evaporator) between the compressor and the capillary tube for different condenser temperatures and evaporator pressures.

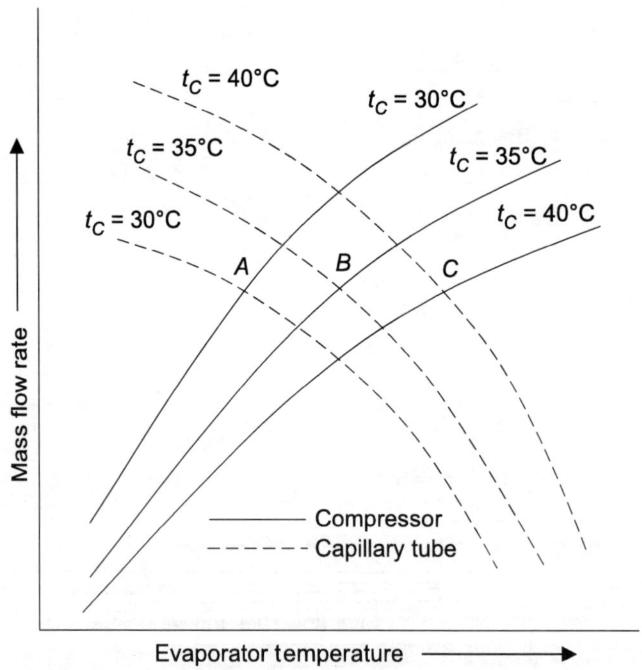

Figure 5.20: Balancing of mass flow rate between compressor and capillary tube

Now let us consider a particular case (Fig. 5.21). The refrigeration system is designed for evaporator pressure, P_A. The corresponding balance point is '*A*' for a particular condenser temperature. Now the load on the evaporator increases. It increases the evaporation rate and the pressure of the evaporator increases to P_B. So mass flow rate through the compressor will increase but at the same time refrigerant flow rate through the capillary tube decreases as shown in figure. Thus, *m* through the evaporator decreases and this condition called the '*starving of evaporator*'. Now the system takes the corrective measure. Excess liquid refrigerant flows back to the condenser. It decreases the heat transfer area of the condenser which increases the condenser temperature and thereby the condenser pressure. Increase in condenser pressure decreases *m* through compressor and increases *m* through capillary tube. Thus the system adjusts itself to a new balance point.

Let us consider the case of decrease of load on evaporator. Less amount of refrigerant will be boiled off in the evaporator. So the evaporator pressure will decrease to P_C. Compressor will suck less amount of refrigerant vapour from the evaporator but at the same time capillary tube will feed more amount of refrigerant to the evaporator. Thus, there will be an overflow of refrigerant and the condition is called 'flooding of evaporator'. In this case liquid refrigerant may enter the compressor. So in capillary tube refrigeration system the amount of refrigerant charged is carefully calculated so that in eventuality, if all refrigerant flows down to evaporator, it will just fill to the brim of the evaporator. There will be no overflow of liquid refrigerant to enter the compressor. This is called '*critical charge*'. Now the corrective action is taken by the system. Due to less amount of vapour from compressor, there will be excess cooling of the refrigerant vapour which causes a decrease in condenser temperature and hence its pressure. So compressor will now pump more amount of refrigerant and mass flow rate will be reduced through the capillary tube. Thus, both the components will be adjusted to a new balance point.

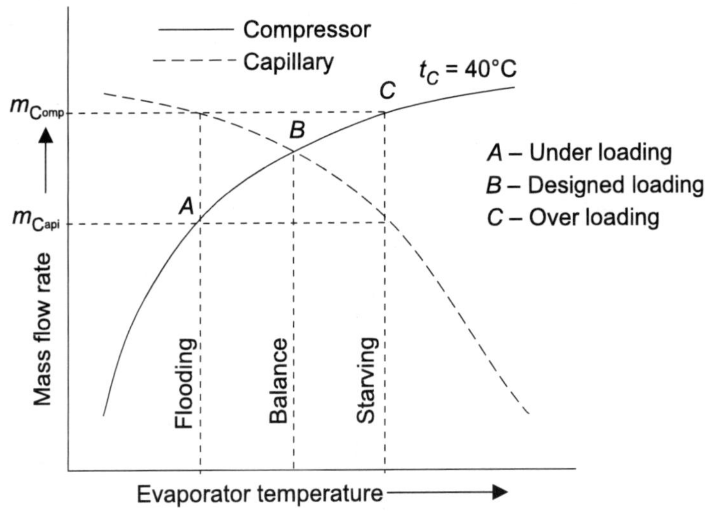

Figure 5.21: Balancing of mass flow rate between compressor and capillary tube

However, the compressor and capillary tube cannot fix the balance point as it is described above. The evaporator heat transfer characteristics must also be satisfied so that the required amount of refrigerant is boiled off from the evaporator to reach the balance point.

5.9.4 Inference

It has been observed from the above analysis that the compressor capacity has the dominating effect on system's capacity and the evaporator is next in importance. An increase in compressor capacity by 10% has the effect of 6.3% increase in system capacity. A 10% increase in evaporator capacity gives 2.1% increase while 10% increase in condenser capacity gives 1.3% increase in system capacity. These observations can be used for system optimization of the initial cost of the system.

5.10 CONTROLS AND ACCESSORIES OF A REFRIGERATION SYSTEM

Besides compressor, condenser, expansion device and evaporator, there are other controls and accessories required for a vapour compression refrigeration system for safe running, control and better performance of the system.

For compressor performance and durability, the devices mounted are high pressure and low pressure cut-outs, oil separators and oil pumps, pressure gauges, receivers and accumulators, filter-drier, sight glass, solenoid valve, fans and blowers.

Suction line accumulator: It is provided in the refrigeration circuit in between compressor and evaporator to separate liquid from vapour coming out of the evaporator (if there is no complete evaporation) so that liquid refrigerant cannot pass to the compressor. This liquid carries lubricating oil with it. Therefore, in the accumulator a small bleed hole, at the bottom, is provided to suck the oil out and return to the compressor.

Receiver: A refrigerant liquid receiver is placed in between the condenser and expansion device. It is required to temporarily store the refrigerant charge within the system or to accommodate the excess refrigerant arising from changing operating conditions. The total refrigerant charge required in a circuit will vary with different operating loads, and must be sufficient at all times so that only the liquid can enter the expansion valve.

Oil separator: During the compression stroke, the high temperature refrigerant vapour carries away some of the lubricating oil along with it. For small self-contained systems it quickly finds its way back to the compressor. For large compressors an oil separator is incorporated just after the discharge of the compressor. The hot refrigerant vapour is made to impinge on a spiral to lose much of the oil on the surface by centrifugal force. This is how 95% of the entrained oil is separated from the refrigerant vapour and can be returned to the crank case. The oil returned line is controlled by the float valve or it may have a bleed orifice.

Filter-drier: Sometimes atmospheric air which contains water vapours is leaked into the refrigeration system. Presence of water vapours in the system reduce the partial pressure of refrigerant and also reduce the heat transfer coefficient. Specifically, for halocarbons, it is essential to reduce the water content in the refrigeration system. So, a solid desiccant such as activated alumina, zeolite or silica gel in the form of capsules is used in the liquid line just ahead of the expansion valve. These capsules must have strainers to prevent loss of drying agent into the circuit and so form an effective filter-drier to also protect the valve orifice from damage by fine debris. In some cases the dirt, scales, debris, etc., is filtered by using a strainer separately. Strainer is placed very close to the automatic valve to protect it from blockage.

Sight glass: This is provided in the liquid line of the refrigeration circuit to see whether there is any gas bubble entrapped in the liquid line. Presence of gas bubbles indicates the shortage of refrigerant or poor functioning of filter drier. It is normally placed very near to the expansion device to ensure that only liquid refrigerant enters the expansion device.

Solenoid valve: This is an electrically operated shut-off valve used in refrigeration circuit as a flow control valve. Solenoid valve opens when energized and closes when de-energized. This valve consists of a coil of insulated copper wire and an armature (or sometimes called plunger), which is drawn into the centre of the coil magnetic field when the coil is energized. By attaching a valve stem and pin to a coil armature, a valve port can be opened or closed as the coil is energized or de-energized, respectively. As it is a flow control valve, its opening amount should be determined from the maximum amount of fluid flow through it. It is normally fitted in liquid line. A simple example of its use in multi-cylinder compressor is as follows. Suppose cooling load is decreased. Then the system adjusts itself to the reduced load condition by unloading (cutting off) one cylinder. When the control system receives signal of under loading, the coil is de-energized (electric connection is cut off). Thus the valve is closed. The refrigerant vapour from compressor retraces its path to the suction valve instead of being circulated in the refrigeration circuit. When the load increases, the coil is energized by receiving the signal from the control unit (thermostat) and the valve is opened. That is how this automatically operated valve is used in place of a hand shut-off valve.

Charging connection: The refrigeration system is to be charged initially with refrigerant or to be charged with more refrigerant whenever required. This requires a charging connection. The usual place to introduce refrigerant is in the liquid line and in between condenser and expansion device. The expansion device controls the flow and prevents the liquid from entering the compressor. The position of the charging point is in a branch of the liquid line and it is fitted with a shut-off valve and a suitable connector with a sealing cap or flange.

Thermostat: A refrigeration or air conditioning system is normally used to maintain a particular temperature of the refrigerated space. During the cooling (refrigerator)/heating (heat pump) process, if the temperature goes beyond the pre-set value the thermostat senses it and stops the equipment or reduce the capacity till the required condition is reached. Different types of thermostats are used: (1) movement of a bimetallic element, (2) expansion of a fluid, (3) vapour pressure of a volatile fluid, (4) electric resistance, and (5) electronic type.

Humidistat: Sometimes the system needs to maintain a predetermined level of humidity. There a humidistat, may or may not be with a thermostat, is used to maintain the humidity level. Mechanical humidistats employ materials which change dimension with humidity, such as animal hair, plastics, cellulosic, etc. Electronic humidistat generally depends upon the properties of a hygroscopic salt.

Pressure switches: These are 'high pressure cut-outs' and 'low pressure cut-outs'. The compressor of a refrigeration system works between a lower and a higher pressures limit. Sometimes non-condensable gas ingresses to the system. This gas increases the condenser pressure because it does not condense, as refrigerant does, in the condenser. This causes high compressor discharge pressure. If the environment temperature becomes high, heat rejection from the condenser is reduced and the system may resort to higher condensing temperature and corresponding higher saturation pressure. Due to malfunctioning of other components or incorrect closure of a valve system pressure may go high. Due to high pressure, compressor temperature also increases and compressor may get damaged. The system performance is also adversely affected if the system pressures go beyond the limit. In this case high pressure cut-out disconnect the compressor from electric supply and the compressor stops. The compressor discharge pressure is brought to one side of a bellows or diaphragm and balanced by an adjustable spring. A scale on the control indicates the pressure setting. If the spring pressure is overcome, the switch will open and stop the compressor.

If a plant is shut down, pressure of each component becomes equal. As the plant restarts, the suction pressure will pull down below the normal until the liquid refrigerant has begun to circulate. Sometimes evaporator pressure becomes abnormally low when the refrigerating load on the plant decreases. Under such circumstances, low pressure cut-out operates and it stops the compressor. Abnormally, low pressure may not be an unsafe condition but abnormally low suction pressure will lead to high discharge temperature because of high compression ratio and possible malfunctioning of other components.

A low pressure switch can also be used in conjunction with a thermostat and a solenoid valve in the pump down circuit. In this method of control, the thermostat does not stop the compressor but de-energize the liquid line solenoid valve to stop the supply of refrigerant to the evaporator. The compressor continues to run and pumps down the evaporator until stopped by the low pressure switch. When the thermostat again calls for cooling, it opens the solenoid valve, liquid enters the evaporator and the low pressure switch will close again to restart the compressor.

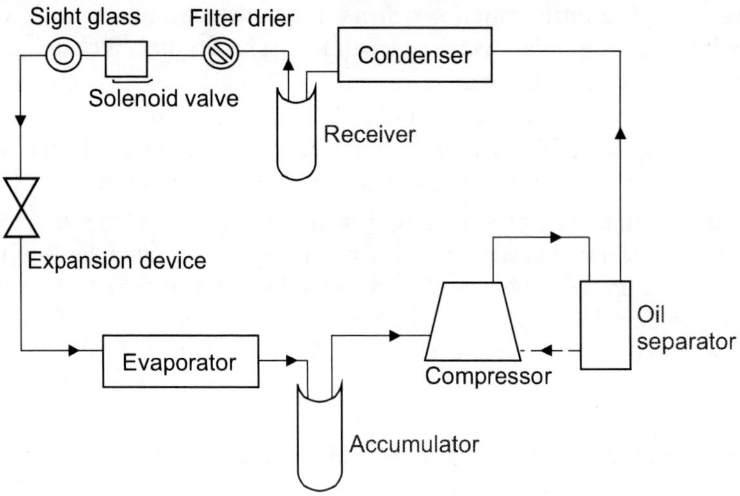

Figure 5.22: Schematic of a refrigeration system with some accessories

Besides the above discussed accessories and controls there are some other accessories also. Oil pressure switches are provided with a cut-out so that in case of low oil pressure (i.e., starving of lubricating oil), this switch will stop the compressor. For direct indication of operating condition, pressure gauges are mounted at suction, discharge of the compressor and oil delivery. Similarly, provisions for temperature indicators (electronic probes) at these points should be given. Manual shut-off valves are required throughout the circuit to permit isolation during partial operation, service or maintenance. Piping circuits usually contain a small quantity of dirt, scale and swarf, in spite of best care being taken to avoid it. A strainer is fitted in the compressor suction to trap such particles before they can enter the machine.

5.11 INSULATION

Insulation of a refrigerating system is done to reduce the heat transfer from the warmer outdoor to the cooler indoor space. The selection of it depends upon its thermal conductivity and fire hazardness. Various insulating materials are discussed here.

Foamed polyurethane (PUF): This chemical, in liquid state mixed with foaming agent swells into a low-density foam. Then it is turned into a rigid mass by polymerization process. Since this is a swelling material (the shape can easily be adjusted), it can be fitted into the core of sandwich panels. Its thermal conductivity is 0.025 – 0.03 W/m-K but resistance to fire is not so good.

Mineral fibre: This is a bonded Rockwool with vertical fibre structure. This has great thermal insulation property with excellent fire and acoustic resistance property. This is water repellant and most suitable for internal temperature above 0°C. This is free of CFCs. The thermal conductivity is 0.03 – 0.04 W/m-K.

Polystyrene: This had been a most popular insulating material. Because of high fire risk this material is now being rejected by the refrigeration industry. The thermal conductivity is 0.035 – 0.045 W/m-K.

Polyisocyanurate (PIR): This is an efficient insulating material having thermal conductivity 0.02 W/m-K and very low combustibility. This is also a low-density rigid foam and free of CFCs.

5.12 CHARGING OF REFRIGERANT: (HOW IT IS DONE AND HOW MUCH IS TO BE CHARGED)

Charging procedure: Figure 5.23 shows the schematic of a HVAC kit. It has three connecting points. Point A is connected to the low pressure side of the refrigeration system (evaporator exit) through a blue hose pipe and controlled by a valve, V_L. Similarly, point C is connected to the high pressure side of the refrigeration system (condenser/receiver exit) through a red hose pipe and controlled by a valve, V_H. Point B is connected to the charging/refrigerant cylinder.

Refrigerant charging may be done by two methods. One is by charging the system with refrigerant vapour, and other one by charging it with liquid (Refer Fig. 5.24).

Charging with vapour: Small systems are generally charged with vapour. First the system is evacuated so that air along with moisture is completely removed. (If the system is to be recharged, then it is evacuated to a recovery bottle.) This is done in the following manner. Point B is connected to a vacuum pump through a hose pipe and the valves (V_L and V_H) are kept open. The vacuum pump is started. It is run till the vacuum gauge reads the desired pressure (around 4.0 kPa). The final vacuum pressure should be lower than the water vapour pressure at the lowest part of the system. Hence, the vacuum to be maintained depends upon the temperature at the coldest point. Now the pump is stopped and the two valves are closed for nearly 20 minutes

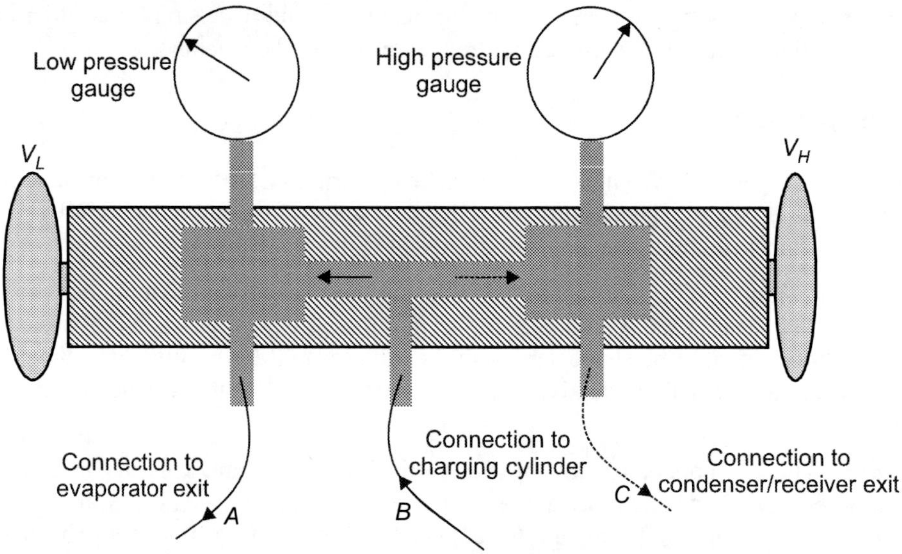

Figure 5.23: HVAC kit

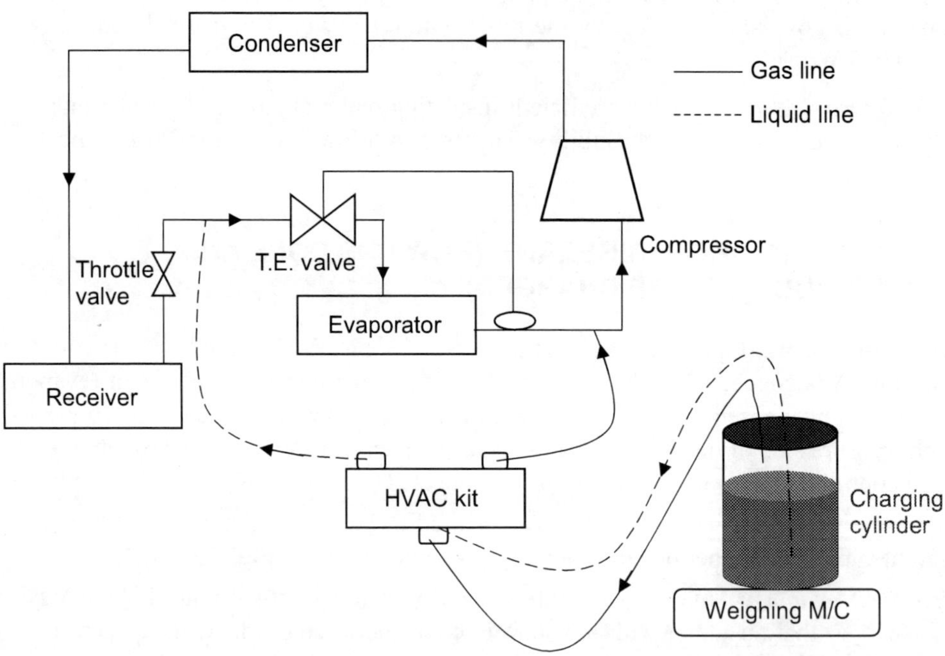

Figure 5.24: Refrigerant charging method

to see any increase in pressure which indicates any leakage in the system. Then the hose pipe is removed and connected to the charging cylinder. The blue pipe (vapour) of the HVAC kit is connected to the connector placed at the exit of the evaporator. Valve (V_L) and charging cylinder valve are opened. Initially, the hose pipes are purged. This is done by slightly loosening the evaporator connector and then quickly and repeatedly doing off and on of the cylinder valve, so that all the air along with some amount of charge goes out at the connecting points and the hoses will be free of air and moisture. The connector is tightened. Now the system is charged with refrigerant vapour till the gauge pressure shows zero. Then the compressor is started. Wait for 40–50 seconds and see whether the suction gauge pressure reaches a higher value. This process is repeated till the suction pressure reaches a predetermined value, or by weighing the charging cylinder, the amount of refrigerant charge can be known. It should match with the amount recommended by the manufacturer. Once the refrigerant is charged the charging cylinder and the valve (V_L) are closed and the pipes are disconnected.

Charging with liquid: Large systems are charged with liquid refrigerant. The same procedures are adopted to evacuate the system. Then the red pipe is connected to the connector at a point at the exit of the condenser/receiver (i.e., high-pressure side).

(The vapour pipe is placed just above the refrigerant level in the charging cylinder and the liquid pipe is placed at the bottom of the charging cylinder. Hence, the liquid refrigerant carrying pipe also called 'dip tube'.)

EXERCISE

1. In which thermodynamic process work of compression is minimum?
2. How the hermetic compressor gets cooled?
3. What is meant by *DX* coil and how it is different from flooded evaporator?
4. Saturated ammonia at 0.236 MPa enters a 15 cm × 14 cm twin cylinder, single acting compressor. Its volumetric efficiency is 80% and speed is 250 rpm. The head pressure is 1.17 MPa. Liquid ammonia at 23°C enters expansion valve: Find (a) ammonia circulation rate, (b) refrigeration capacity, and (c) COP.
5. A R134a refrigeration system operating between temperature limit of –15°C and 45°C. It has a four-cylinder compressor, each with bore of 10 cm and stroke of 11.5 cm and runs at 750 rpm. Assuming a clearance volume ratio of 0.045, find: (a) clearance volumetric efficiency, (b) swept volume, (c) mass flow rate, (d) refrigerating effect, and (e) COP.
6. A four-cylinder R22 refrigerant compressor operates between a temperature range of 7°C and 52°C. It gives 100 kW of refrigeration at 1740 rpm and consumes power 32 kW. The vapour entering the compressor is superheated by 10°C and liquid refrigerant leaving the condenser is sub-cooled to 5°C. The bore and stroke are 67 mm and 57 mm, respectively. Calculate: (a) clearance volumetric efficiency, (b) actual volumetric efficiency, and (c) compression efficiency. Assume clearance volume ratio 4.5%.
7. Classify the compressors used in refrigeration system.
8. What is the difference between compression ratio and pressure ratio?
9. What is volumetric efficiency for a rotary compressor? Justify your answer.
10. How does the pressure decrease in throttling process?
11. Why power fluid is provided in TEV?

MULTIPLE–CHOICE QUESTIONS

1. Minimum work is done on the system in —— compression process.

(a) adiabatic (b) ientropic

(c) isothermal (d) isobaric

2. The motors of the hermetic compressors are cooled by

(a) refrigerant vapour from evaporator (b) atmospheric air

(c) water jacket (d) none

3. Volumetric efficiency of a compressor is defined as the ratio of

(a) actual mass discharged to mass occupied in the swept volume

(b) actual volume to swept volume

(c) clearance volume to swept volume

(d) clearance volume to actual volume

4. The clearance volumetric efficiency of a rotary compressor is almost

(a) 100% (b) 90%

(c) 85% (d) 75%

5. Expansion device is used to

(a) flow regulation and pressure drop (b) absorb heat

(c) reject heat (d) increase pressure

6. In capillary tube, mass flow rate —— as per the load variation.

(a) can be adjusted (b) cannot be adjusted

(c) need not vary (d) stops

7. —— type of expansion device is suitable for flooded evaporators.

(a) Capillary tube (b) Thermostatic expansion valve

(c) Float valve (d) Turbine

8. —— type of expansion device is suitable for dry type evaporators.

(a) Capillary tube (b) Thermostatic expansion valve

(c) Float valve (d) Turbine

9. The fluid inside the feeler bulb of the TEV is called

(a) refrigerant (b) alternative fluid

(c) secondary fluid (d) power fluid

10. TEV is more importantly used where there is

(a) more flow variation (b) maintain degree of superheat

(c) less space is available (d) for low cost

11. In TEV degree of superheat is maintained by

(a) fixing the adjustable spring (b) changing the power fluid

(c) changing the bellow setting (d) power fluid variation

12. Heat rejection ratio to estimate the condenser capacity is expressed by

(a) 1/COP (b) (1/COP) – 1

(c) COP + 1 (d) 1 + (1/COP)

13. In evaporative condenser, the working fluid is cooled by evaporation of

(a) working fluid itself (b) coolant sprayed over it

(c) water in cooling tower (d) none

14. In Direct expansion (DX) coil type evaporator, refrigerant vapour comes out as

(a) saturated (b) wet

(c) liquid (d) two-phase

15. In Flooded type evaporator, refrigerant vapour comes out as

(a) saturated (b) wet

(c) liquid (d) superheated

16. In an evaporator of a refrigeration system, liquid refrigerant

(a) evaporates at low pressure (b) boils at low pressure

(c) boils at high temperature (d) evaporates at high temperature

17. Starving of evaporator condition occurs when

(a) capillary tube flow decreases (b) load on the system increases

(c) (a) and (b) correct (d) none

18. Flooding of evaporator condition occurs when

(a) capillary tube flow increases (b) refrigeration load increases

(c) (a) and (b) correct (d) none

19. Increase in which component capacity will have more effect on refrigeration system capacity?

(a) Compressor (b) Condenser

(c) Evaporator (d) Expansion device

20. For a two-stage reciprocating air compressor, the suction pressure is 1.5 bar and the delivery pressure is 54 bar. What is the value of the ideal intercooler pressure?

(a) 6 bar (b) 9 bar

(c) 27.75 bar (d) 19 bar

21. Which statements are correct:

1. Reciprocating compressors are best suited for high pressure and low volume capacity.
2. The effect of clearance volume on power consumption is negligible for the same volume of discharge.
3. While the compressor is idling, the delivery valve is kept open by the control circuit.
4. Intercooling of air between the stages of compression helps to minimize losses.

(a) 1 and 2 (b) 1 and 3
(c) 1 and 4 (d) 3 alone

22. The special features of a hermetically sealed compressor of a refrigerator:

1. The compressor may be reciprocating to rotary type
2. No shaft seal is necessary
3. More silent in operation
4. COP is more than that of open compressor

(a) 2 and 4 (b) 1, 2 and 3
(c) 1, 3 and 4 (d) 2, 3 and 4

23. Find optimum pressure in the last intercooler if four-stage compressor with perfect inter-cooling between stages compresses air from 1 bar to 16 bar.

(a) 6 bar (b) 8 bar
(c) 10 bar (d) 12 bar

24. Volumetric efficiency of a reciprocating air compressor increases with

1. increase in clearance ratio
2. decrease in delivery pressure
3. multistaging

(a) Only 1 and 2 (b) Only 2 and 3
(c) Only 3 (d) 1, 2 and 3

25. A rotary compressor is employed in a refrigerating system having

(a) low specific volume and high pressure difference
(b) low specific volume and low pressure difference
(c) large specific volume and high pressure difference
(d) large specific volume and low pressure difference

26. The flash chamber in a single-stage simple VCR cycle

(a) increases the refrigerating effect (b) decreases the refrigerating effect
(c) increases the work of compression (d) has no effect on refrigerating effect

27. Which type of condenser is used for low capacity refrigeration systems?

(a) Shell and cube type
(b) Shell and coil type
(c) Double tube type
(d) Air cooled type

28. Which expansion valve is employed for a refrigeration plant operating at constant load?

(a) Thermostatic expansion valve
(b) Automatic expansion valve
(c) Capillary tube
(d) None of the above

29. An automatic expansion maintains constant

(a) pressure in the evaporator
(b) temperature in the freezer
(c) pressure in the liquid line
(d) temperature in the condenser

30. A thermostatic expansion valve

(a) ensures the evaporator is completely filled with refrigerant of the load
(b) is suitable only for constant load system
(c) maintains different temperatures in evaporator in proportion to load
(d) none of the above.

31. A capillary tube controls the flow of refrigerant from the

(a) expansion valve to the evaporator
(b) evaporator to the thermostat
(c) condenser to the expansion valve
(d) condenser to the evaporator

32. The throttling device used in the domestic refrigerator is

(a) internally equalized thermostatic expansion valve
(b) externally equalized thermostatic expansion valve
(c) automatic expansion valve
(d) capillary tube

33. The pressure in a capillary tube of a refrigerator decreases because

1. tube wall offers frictional resistance.
2. refrigerant accelerates in the tube
3. tube transfer the heat
4. Potential energy decreases

Of these statements

(a) 1 and 2 are correct
(b) 1, 2 and 3 are correct
(c) 2 and 4 are correct
(d) 3 and 4 are correct

34. Which expansion device is used in Window AC

(a) capillary tube (b) thermostatic expansion valve

(c) automatic expansion valve (d) float valve

ANSWER

1. (c)	**2.** (a)	**3.** (a)	**4.** (a)	**5.** (a)	**6.** (b)	**7.** (c)	**8.** (b)	**9.** (d)	**10.** (b)
11. (a)	**12.** (d)	**13.** (b)	**14.** (a)	**15.** (b)	**16.** (b)	**17.** (c)	**18.** (a)	**19.** (a)	**20.** (b)
21. (c)	**22.** (b)	**23.** (b)	**24.** (b)	**25.** (d)	**26.** (a)	**27.** (d)	**28.** (c)	**29.** (a)	**30.** (a)
31. (d)	**32.** (d)	**33** (a)	**34.** (a)						

6

Vapour Absorption Refrigeration System

6.1 INTRODUCTION

A vapour compression system is well established in the market due to its efficient performance in comparison to any other refrigeration system. But the use of CFCs and HCFCs as working fluids and need for high grade energy, such as electricity, to produce low grade effects like heating and cooling have shifted the attention of the scientists to alternate systems. The immediate alternative to the vapour compression system is 'sorption type' refrigeration system. This system is again divided into two categories. One is vapour absorption type (sometimes called wet type) and the other one is adsorption type (sometimes called dry type). The later type is discussed in Chapter 7.

The idea of producing cooling effect by vapour absorption came while Michael Faraday was conducting experiments to liquefy certain gases. The observation was as follows.

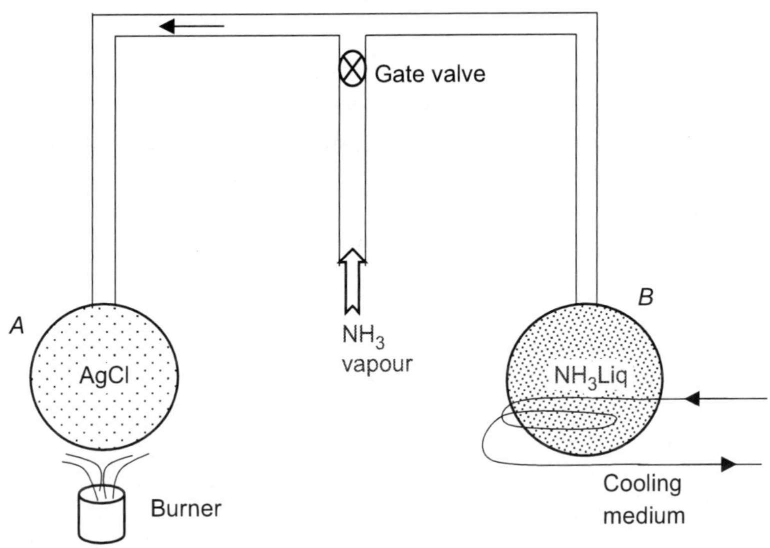

Figure 6.1: Schematic of the experiment conducted by Faraday

Ammonia vapour was supplied through a Gate valve and that was readily absorbed by the powdered silver chloride (AgCl) kept in the vessel *A* till AgCl became saturated. This happened so because Ammonia has a great affinity for silver chloride at normal temperature and pressure. Then the supply of ammonia vapour was stopped and the vessel was heated. The absorbed ammonia vapour gets liberated and came to the vessel *B* which was continuously cooled by circulating water. By this liquid ammonia was obtained in vessel *B*. Then the heating source is removed. It is observed that the liquid ammonia in vessel *B* started boiling and the ammonia vapour was reabsorbed by the silver chloride. When touched, vessel *B* was felt very cool. This was very much an intermittent cooling process. But the inference was 'the strong affinity for absorption is the driving force for drawing of refrigerant vapour from the evaporator'. The same action, i.e., the drawing of vapour from the evaporator in vapour compression system, is done by the compressor.

6.2 WORKING PRINCIPLE OF VAPOUR ABSORPTION REFRIGERATION SYSTEM (VARS)

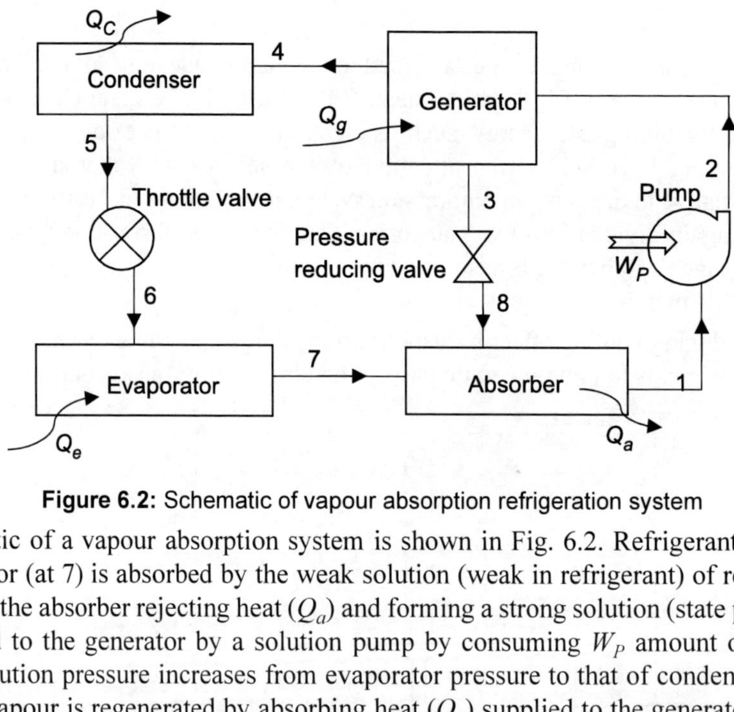

Figure 6.2: Schematic of vapour absorption refrigeration system

The schematic of a vapour absorption system is shown in Fig. 6.2. Refrigerant vapour from the evaporator (at 7) is absorbed by the weak solution (weak in refrigerant) of refrigerant and absorbent in the absorber rejecting heat (Q_a) and forming a strong solution (state point 1). Then it is pumped to the generator by a solution pump by consuming W_P amount of work (state point 2). Solution pressure increases from evaporator pressure to that of condenser. There the refrigerant vapour is regenerated by absorbing heat (Q_g) supplied to the generator (state point 4). The refrigerant vapour flows to the condenser where it is condensed to saturated liquid (state point 5) rejecting heat (Q_C) to the sink. Then it enters the evaporator after being throttled to the evaporator pressure (to state point 6). There it gets vaporised by absorbing heat (Q_e) from the source (refrigerated space). The remaining weak solution in generator (state point 3), after the refrigerant vapour goes out at state point 4, is throttled back to absorber (state point 8) where it absorbs the refrigerant vapour coming from the evaporator (state point 7) and the strong solution of refrigerant and absorbent is formed. Now this strong solution is pumped to the generator. Thus, the absorption refrigeration cycle goes on.

The COP of the above system can be calculated as follows.

Desired effect is refrigerating effect = Q_e, and Energy input to the system = $Q_g + W_P$

So COP = $Q_e / (Q_g + W_P)$. But W_P is very small. It is obvious from the following example.

$W_P = \int(vdP)$ where v is the specific volume. For liquid v (v_f) is less than that of the vapour (v_g) in the order of 10^3. (Refer to the property table of any refrigerant and compare the value of v_f and v_g). So W_P for liquid is less than that required for vapour for same compression ratio in the order of 10^3. (If P kW of power is needed to compress m kg of vapour by compression ratio 'r', around ($P \times 10^{-3}$) kW power is needed to compress m kg of liquid by same compression ratio.)

6.3 COMPARISON WITH VAPOUR COMPRESSION SYSTEM

From the above discussion it is seen that the compressor of the vapour compression refrigeration system is replaced by absorber, pump and generator. Absorber draws the low pressure and low temperature vapour from the evaporator and the pressure is increased by the pump. Then temperature is increased in the generator. Thus, high pressure and high temperature refrigerant vapour comes out of the generator. However, a vapour absorption refrigeration system can be called a 'heat operated refrigerating machine'. Schematically this is represented in the Fig. 6.3. Heat is drawn from the source T_g, rejected heat to the sink at T_a while producing W amount of work. This work is being utilized to drive the heat from low temperature source at T_e to high temperature sink at T_C. Thus, the heat engine working between the temperature limit of T_g and T_a runs the refrigerator working between temperature limit T_e and T_C. So the COP of the system will be the efficiency of the heat engine $[(T_g - T_a)/T_g]$ multiplied by the COP of the refrigerator, $T_e/(T_C - T_e)$. The maximum COP will be attained when the heat engine will work on Carnot cycle and the refrigerator will work on reverse Carnot cycle. In other words, there will be no irreversibility; i.e., $\Delta s = 0$. Mathematically, this can be expressed as

$$(\text{COP})_{\text{Max}} = (T_g - T_a)/T_g \times T_e/(T_C - T_e) \qquad [6.1]$$

As per the working principle, both the condenser and absorber reject heat to the environment. So T_c may be considered to be equal to T_a and

$(\text{COP})_{\text{Max}}$ can be expressed as $(T_g - T_a)/T_g \times T_e/(T_a - T_e)$ [6.2]

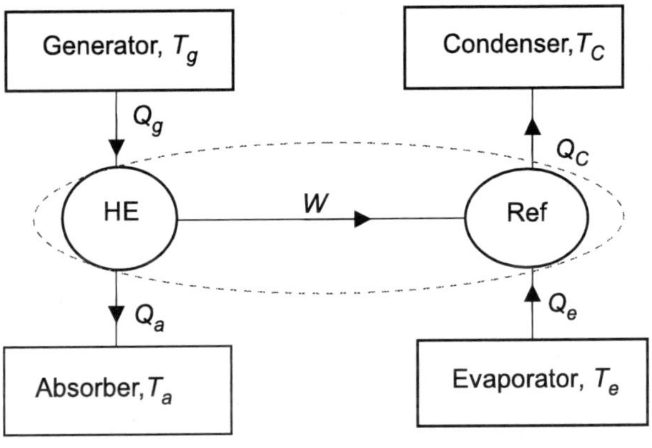

Figure 6.3: Schematic of a refrigeration system driven by a heat engine

Or it can be expressed in terms of heat flow work output from the heat engine as $(Q_g - Q_C)$ which is used to produce refrigerating effect of Q_e.

Efficiency of the heat engine is $(Q_g - Q_a)/Q_g$ [6.3]

and the COP of refrigerator is $Q_e/(Q_C - Q_e)$ [6.4]

So the maximum COP of the absorption system is $(Q_g - Q_a)/Q_g \times Q_e / (Q_C - Q_e)$ [6.5]

From energy balance: Taking the C.V. around HE and Ref combined (Reffer Fig. 6.3),

Energy In $(Q_g + Q_e)$ = Energy Out $(Q_C + Q_a)$ or $Q_g - Q_a = Q_C - Q_e$.

So maximum COP = Q_e / Q_g [6.6]

The advantages of this system are as follows:

1. It can use low grade energy for the generator like solar energy, or the waste heat from the cogeneration plants.
2. It does not require a compressor and the pump needs very less power to raise the fluid pressure.
3. Pressure ratio is not a limiting factor for high temperature lift.
4. The working fluids used are environment friendly.

Due to the above advantages the absorption system finds applications for space cooling and heating and in places where electricity is not available.

The disadvantages of the system are as follows:

1. The system COP is low. However, this may not be a serious problem if the system runs on solar energy or waste heat.
2. It requires more space.
3. As the number of heat exchangers is more (at least 4), the initial cost is high. Their reversibilities are high due to heat transfer with finite temperature difference and this results in low COP.

Hence, the absorption system can be a good substitute to the conventional vapour compression system if its COP can be increased. The different methods to increase the COP are as follows.

1. Increasing the generator temperature
2. Having internal heat exchange like generator-absorber heat exchange (GAX)
3. Using suitable binary/ternary mixture as working fluids
4. Decreasing the circulation ratio (defined as the ratio of the total mass of the mixture to the mass of the refrigerant vapour).

The adaptability of the existing absorption refrigeration system to the above means for increasing COP is achieved by making some modifications or by adding more components/accessories to the system. To name some of these advanced absorption systems, as suggested by different authors, are two-stage triple effect, three-stage triple effect, generator-absorber heat exchange, double-lift series flow, and double-lift vapour exchange cycles. These systems require more number of heat exchangers which increases the complexity and cost of the system. In practice it is also difficult to obtain higher COP as the greater complexity generally causes greater sensitiveness to the approach temperature.

6.4 WORKING FLUID

As mentioned, the working fluid is the mixture of a refrigerant and absorbent for absorption refrigeration system. The important requirements of a refrigerant and absorbent pair are as follows:

(a) The refrigerant should have more than Raoult's law (negative) solubility in the absorbent.

(b) There should be large temperature difference in the normal boiling points of the two components (>200°C).

The commonly used refrigerant and absorbent pairs are NH_3-H_2O and H_2O-LiBr. Some of the organic absorbents like DMETEG (also known as E-181), DMF, NMP, etc., are also considered suitable to form the working fluid pair with refrigerants like R22, R134a, TFE, etc. These organic absorbents also have good lubricating properties. The physical properties of some of the organic absorbents are given in Appendix 8. Some of the important properties of various refrigerants are given in Table 2.2.

6.4.1 NH_3–H_2O as Working Fluid

In NH_3-H_2O VARS, NH_3 is used as the refrigerant and water is used as the absorbent. The difference between boiling points of NH_3 and H_2O being less (i.e., 138°C) some amount of water vapour is carried away along with ammonia vapour while heated in the generator section. Hence, this system is provided with a 'rectification coloumn' where most of the water vapour is separated to get more than 99.9% of pure ammonia.

Figure 6.4 shows a schematic of a rectification coloumn. The lower part where the strong solution enters the rectification coloumn is called *exhausting coloumn* or *analyser*. The upper part is called *rectification coloumn* and the top part of it is called dephlegmator. The purification of ammonia vapour occurs as described.The low temperature strong solution from the pump enters the rectification coloumn and trickles down the analyser as the high temperature NH_3-H_2O vapour mixture moves up from the generator. There occurs a direct contact heat transfer from hot fluid to cold fluid. Thus, some amount of water vapour from NH_3-H_2O vapour mixture is condensed. Also the low temperature strong solution gets preheated. Now more concentrated in NH_3 and high temperature NH_3-H_2O vapour mixture enters the rectification coloumn. As it moves up it comes in direct contact with the condensed water droplets, which is at lower temperature, from the dephlegmator. Like it happens in analyser, some amount of water vapour is again condensed making NH_3-H_2O vapour mixture more concentrated in NH_3. Thus purified NH_3-H_2O vapour mixture enters the dephlegmator where it is again cooled by the circulating coolant water. Now the mixture becomes more concentrated in NH_3 by condensing some more water vapour which trickles down the rectification coloumn and draws heat from the NH_3-H_2O mixture as described above. In this process the NH_3-H_2O vapour mixture becomes enriched to the maximum possible (even up to 99.9%) and heat is recovered by the liquid mixture entering the generator from the pump. That is how less amount of cooling is required in the condenser chapter for condensation of ammonia vapour and less amount of heat required in the generator by this internal heat exchange process.

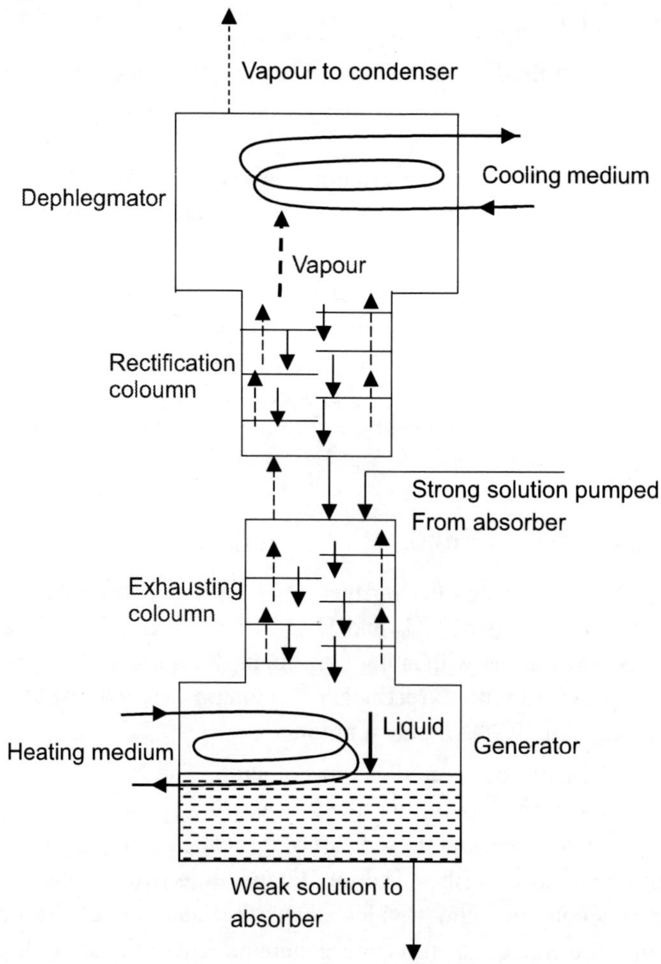

Figure 6.4: Schematic of the analyser of NH_3-H_2O VAR system

6.4.2 H_2O-LiBr as Working Fluid

In case of H_2O-LiBr system, H_2O is used as refrigerant and LiBr is used as absorbent. Lithium bromide is a salt and it does not evaporate when heated in the generator. So a rectification column is not required for this system. But the disadvantage is that as water solidifies at 0°C, this system cannot be used for temperature for 0°C or less (i.e., for refrigeration purpose) but most suitable for comfort air conditioning where the evaporator temperature is maintained at around 7°C. Also the system cannot use higher concentration of LiBr because the strong solution of LiBr is prone to crystallization of LiBr. Hence, a proper system design is needed to prevent this.

6.5 PERFORMANCE ANALYSIS

The properties of pure fluid vary from that of the mixture in many ways. Let us take a mixture of fluid *A* and *B* of concentration x_A and x_B. It is necessary to have minimum three properties to fix a point on a thermodynamic plane for a mixture of two fluids as stated by phase rule. The sub-cooled liquid is now heated. At a certain temperature (T) the first bubble is found. That temperature is the saturation temperature (boiling point) of the mixture at that pressure and concentration. If the concentration changes the boiling point of the mixture also changes. When these boiling points of different concentrations (from 0% to 100%) are joined a line is generated and is called '*bubble point curve*'. As the heat supply continues, more number of bubbles are formed. Concentration remaining same all liquids are vaporised at another temperature. That temperature is the saturation temperature of the mixture for that concentration and pressure. In other way if superheated vapour mixture is cooled down at certain temperature the first liquid droplet will appear. That temperature is saturation temperature (dew point) of the mixture for that concentration and pressure. As the concentration varies keeping the pressure constant the dew point also varies. When these dew points for different concentrations are joined a line is generated and is called as '*dew point curve*'. If the pressure is varied different set of bubble point and dew point curves are obtained as shown in Fig. (6.5).

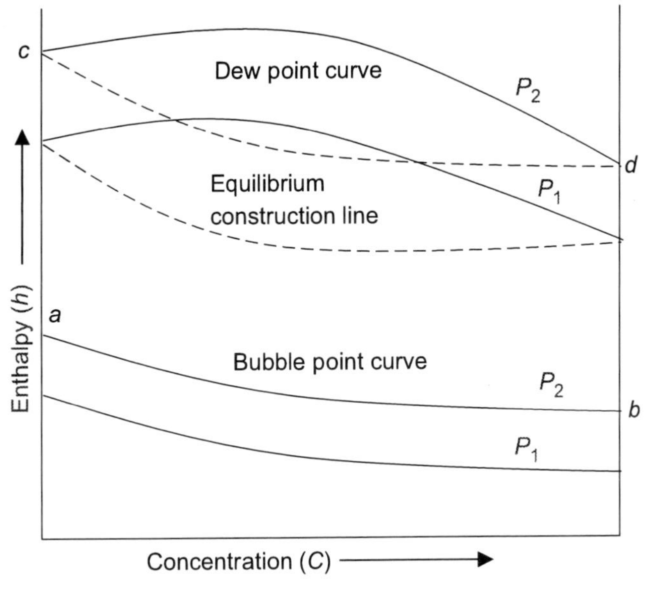

Figure 6.5: Equilibrium lines of a mixture on *h-C* plane

Similarly, specific enthalpy vs concentration diagram for any of the refrigerant absorbent pair is obtained by the above method. In the diagram curve *ab* is the bubble point curve and *cd* is the dew point curve for pressure P_1. Left side of the diagram is meant for the absorbent, i.e., high boiling point fluid. Thus, (for NH_3-H_2O) point '*a*' signifies there is 100% water and the corresponding enthalpy is that of the saturated water at pressure, P_1. Point '*c*' signifies 100% water vapour and the corresponding enthalpy is that of the saturated water vapour at pressure, P_1. Vertical length '*ac*' gives the enthalpy of vaporisation. Similarly, right side of the diagram is meant for the refrigerant, i.e., low boiling point fluid. Thus point '*b*' signifies there is 100% NH_3 and the corresponding enthalpy is that of the saturated NH_3 at pressure P_1. Point '*d*' signifies 100% NH_3 vapour and the corresponding enthalpy is that of the saturated NH_3 vapour at pressure, P_1. Vertical length '*bd*' gives the enthalpy of vaporisation. The mid region between these two extreme points gives the enthalpy of the mixture of NH_3-H_2O for different concentrations. In the same procedure bubble point and dew point curves for different pressures can be drawn. In this diagram the dotted lines are the isotherms which are only drawn in pure liquid and pure vapour zone. In the two-phase **zone the isotherms cannot be reliably predicted**. However, for getting the state points of the liquid and vapour mixture region, an approximate isotherm can be drawn as discussed below.

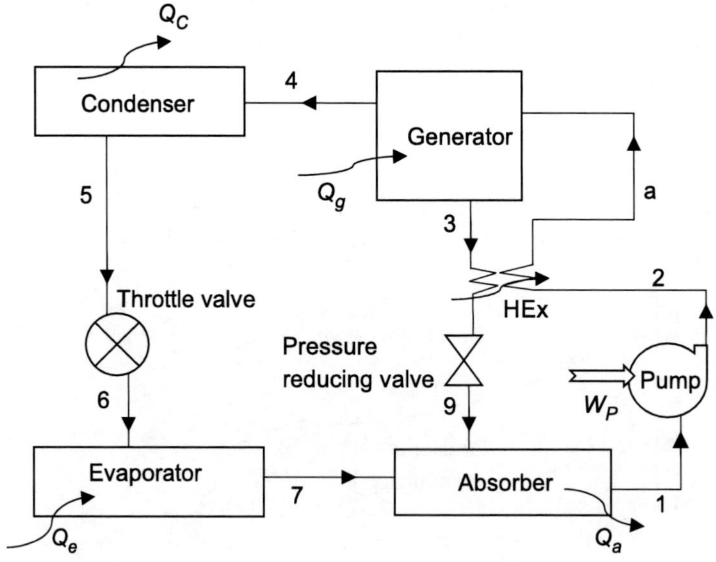

Figure 6.6: Schematic of the NH_3-H_2O VAR system with HEx

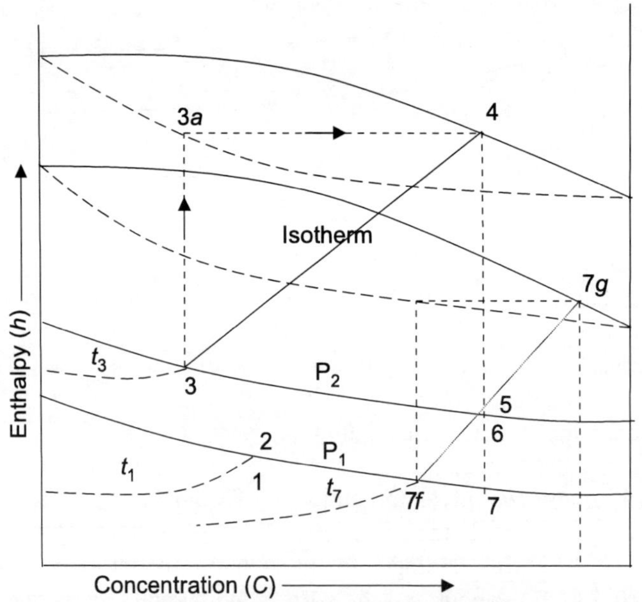

Figure 6.7: Performance of NH_3-H_2O VAR system on *h-C* plane

6.6 FIXING THE DIFFERENT STATE POINTS OF A SIMPLE VARS ON ENTHALPY-CONCENTRATION CHART OF NH_3-H_2O

Refer to Fig. (6.7). Strong solution of NH_3-H_2O mixture at lower pressure P_1 comes out of the absorber at 1. The solution is assumed to be saturated and fixed at point 1 on bubble point curve P_1 for the NH_3 concentration of x_A. Then the solution is pumped to a higher pressure P_2 concentration of the solution remaining unchanged. If the work done or the enthalpy added by the pump is neglected (as it is small), point 2 can be shown at the same place of point 1. But point 2 will be in reference to bubble point curve, P_2. Thus, that will be a sub-cooled liquid. Then the high pressure strong solution passes through the heat exchanger and becomes saturated by receiving heat from the hot solution coming from the generator. So point 'a' is shown bubble point curve P_2 corresponding to the same concentration. Then the solution is heated in the generator. First point 3 which is the intersection point of generator temperature and P_2 is fixed. It means that the weak solution coming out of generator is assumed to be saturated at generator temperature and pressure. To get point 4 an equilibrium constant line *OQP* is drawn. (The procedure of drawing this line is beyond the scope). A vertical line is drawn from point 3 which intersects *OQP* at point 3′. Then a line parallel to the abscissa is drawn which intersects the dew point curve at point 4. This is state of the vapour which comes out of the generator. **Line 3-4 is called the isotherm** in the two-phase zone. Point 4 gives the percentage of ammonia vapour present in the NH_3-H_2O vapour mixture. It is condensed by rejecting heat to saturated liquid state at pressure, P_2. Thus, point 5 is shown on the bubble point curve of P_2. Then it is throttled which is a constant enthalpy process but the pressure is dropped from P_2 to P_1. So the point 6 will be same as point 5 but with reference to bubble point curve of P_1. Hence, it is clear that the mixture coming out of the throttle valve is in two-phase zone. After throttling the liquid vapour mixture enters the evaporator. Point 7 can be fixed at the intersection of evaporator temperature isotherm and P_1. Similar to the procedure of fixing point 4, point 7*g* is fixed as shown in the diagram. That shows the state of the vapour coming out of the evaporator. After fixing the state points enthalpy can be calculated from the *h-c* chart. Thus, refrigerating effect, energy requirement and finally COP of the system can be calculated.

6.7 FIXING THE STATE POINTS ON ENTHALPY-CONCENTRATION CHART FOR H_2O-LiBr SYSTEM

For H_2O-LiBr mixture two charts are to be referred to get the state points. One is pressure-temperature-concentration (p-t-c) diagram and another is enthalpy-temperature-concentration (h-t-c) diagram. After the generator section pure water vapour is obtained as LiBr does not evaporate with water. So water vapour properties can be obtained from the steam tables. In absorber and generator, where H_2O-LiBr mixture is there, the enthalpy of the mixture is obtained by knowing the concentration of the mixture from the p-t-c diagram. Generator and condenser, being at the same pressure, the concentration of the mixture in generator is obtained from solution temperature in the generator and the refrigerant temperature in the condenser. Similarly, absorber and evaporator, being at the same pressure, the concentration of the mixture in absorber is obtained from the mixture temperature in the absorber and the refrigerant temperature in the evaporator. After knowing the concentrations of the solutions in the generator and absorber, the respective enthalpies can be obtained from the enthalpy-concentration diagram. However, it is observed that there is a maximum amount of LiBr which can dissolve in water for a particular temperature and pressure. When this amount exceeds, crystallization of LiBr occurs. This is shown (Fig. 6.8) as a dotted line in p-t-c diagram of H_2O-LiBr.

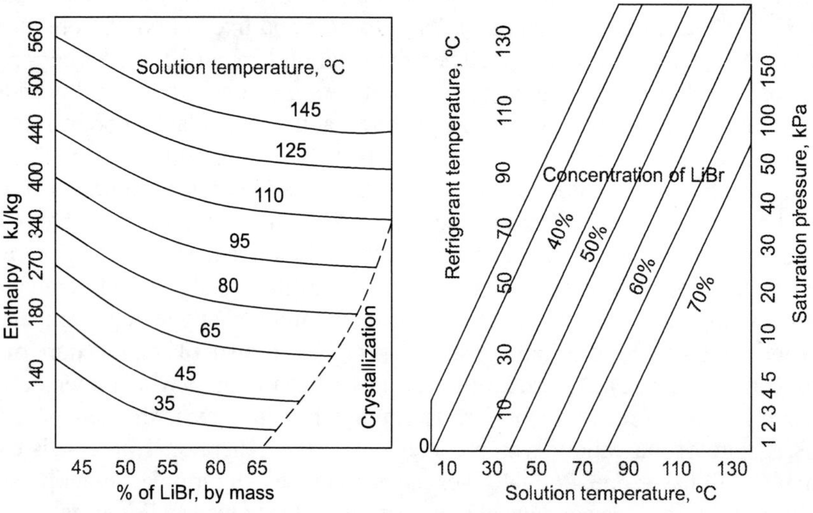

(a) (b)

Figure 6.8: (a) Equilibrium lines of H_2O-LiBr solution on h-t-c plane, (b) p-t-c chart of H_2O-LiBr solution

Example 6.1

The following data refer to a simple vapour absorption refrigeration system. Generator temperature 120°C, condenser temperature 40°C, evaporator temperature 0°C and absorber temperature 35°C. Total heat rejected to the outdoor is 10 kW. Calculate the heat required in the generator and the refrigerating effect.

Ans: From the above data $(COP)_{Max} = (T_g - T_a)/T_g \times T_e/(T_C - T_e)$

$= (120 - 35)/(273 + 120) \times (273 + 0)/(40 - 0) = 1.476$

From energy conservation $Q_g + Q_e = Q_C + Q_a = 10$ kW (as condenser and absorber reject heat to the environment).

Again $(COP)_{Max}$ can be expressed as Q_e/Q_g. Adding 1 on both the sides $(Q_g + Q_e)/Q_g = (COP)_{Max} + 1$

This gives 10 kW/Q_g = 1.476 + 1.

So Q_g = 4.039 kW and Q_e = 4.039 × 1.476 = 5.96 kW (1.695 TR)

Therefore, the heat required in the generator is 4.039 kW and the refrigerating effect is 5.96 kW.

Example 6.2

A reversible refrigeration system is driven by a reversible heat engine. Heat is supplied from condensation of saturated steam at 200°C. Cooling capacity of the refrigerator is 22.0 kW at –10°C. Environment temperature is 25°C. Calculate: (a) the steam condensation rate, and (b) power input to the refrigerator.

Ans: $(COP)_{Max} = \{(T_g - T_a)/T_g\} \{T_e/(T_C - T_e)\}$; $T_a = T_C = 25°C$

Substituting the respective values $(COP)_{Max} = (175/473) \times (263/35) = 2.78$

$2.78 = 22.0$ kW/Q_g, So $Q_g = 22/2.78 = 7.914$ kW

So condensation of steam at 200°C is $7.914/h_{fg}$ (at 200) = 7.914/1939.8 = 0.00408 kg/s

COP of refrigerator = 263/35 = 7.51

So work input to refrigerator = 22/7.51 = 2.93 kW

Example 6.3

A simple vapour absorption refrigeration system of 2 TR capacity uses ammonia-water as the refrigerant-absorbent pair and works within a pressure limit of 14.0 bar and 1.8 bar pressure. Temperatures for absorber, condenser, generator and evaporator are 30°C, 30°C, 120°C, and 0°C, respectively. Calculate heat transfer from/to absorber, generator and condenser. Calculate the COP of the system.

Assume equilibrium state/saturated state for exit of absorber, generator, condenser and evaporator. Neglect the pressure drop in the system.

Ans: (Refer Fig. 6.7 and the chart)

State 1: At 1.8 bar and 30°C, from the chart, $x_1 = 0.364$ and $h_1 = 20$ kJ/kg.

State 3: This is the saturated liquid and located at the intersection point of 120°C and 14 bar. From the chart $x_2 = 0.28$ and $h_2 = 435$ kJ/kg.

State 4 is located by drawing a vertical line from state point 3, that intersects the auxiliary or equilibrium construction line of 14 bar at point p. Then draw a horizontal line from p that intersects the 14 bar dew point curve at 4. From the chart at state point 4, $x_4 = 0.87$ and $h_4 = 1980$ kJ/kg.

State point 5, that is, the condenser outlet is located by dropping a vertical line from point 4 to intersect the 14 bar bubble point curve at 5. From this point it is read that the condenser temperature is 42°C and $h_5 = 410$ kJ/kg. Thus, it is seen that the condenser temperature is set automatically from system condenser pressure and concentration of the fluid.

After throttling state point 6 will be the same as state point 4, but with reference to 1.8 bar pressure. So state point 6 will be in the two-phase zone and if an isotherm is drawn through 6, the temperature is seen to be around –15°C.

NH_3 vapour at the outlet of the evaporator is assumed to be saturated. Actually, at 0°C, the saturation pressure of NH_3 is nearly 4.3 bar. But the evaporator pressure is 1.8 bar. This is the demerit with the absorption system that the evaporator pressure is to be maintained lower than the saturation pressure to attain the required evaporator temperature because it is not a pure refrigerant. Secondly, all the refrigerant will not be evaporated and part of refrigerant liquid goes to the absorber without producing any cooling effect, thereby reducing the COP. However by drawing the isotherm for 0°C and 1.8 bar pressure the liquid and vapour state for the evaporator is fixed as 6_f and 6_g as it is done for the case of generator. From the chart, for the state point 7_f, fraction of liquid NH_3 in liquid mixture = 0.575 and $h_f = -49$ kJ/kg and for 7_g, fraction of vapour NH_3 in vapour mixture, $x = 0.995$ and $h_g = 1655$ kJ/kg

Assuming 1 kg/s flow of mixture at the evaporator outlet, the mass balance of NH_3 gives

$$0.87\,(1) = w_6(0.575) + (1 - w_6)0.995; \text{ So } w_6 = 0.2976 \text{ kg/kg of mixture.}$$

Here w_6 is the liquid mixture.

This gives $h_6 = w_6 h_{6f} + (1 - w_6)h_{6g} = 0.2976(-49) + (1 - 0.2976)1655 = 1147.89$ kJ/kg

For 2 TR cooling capacity, $m_6 = (2 \times 3.5167) / (h_6 - h_4) = (2 \times 3.5167)/(1147.89 - 410)$
$= 9.532 \times 10^{-3}$ kg/s

Overall mass conservation gives $m_2 + m_6 = m_1$

NH_3 mass conservation gives $m_2x_2 + m_6x_6 = m_1x_1$. Substituting the respective values, we get $m_2 = m_2h_2 + m_6h_6 - m_1h_1 = 57.42 \times 10^{-3}$ kg/s, and $m_1 = 66.95$ kg/s

Energy balance gives

For absorber, $Q_a = m_2h_2 + m_6h_6 - m_1h_1 = 57.42 \times 10^{-3} \times 435 + 9.532 \times 10^{-3} \times 1147.89 - 66.95 \times 10^{-3} \times 20 = 34.58$ kW

For generator, $Q_g = m_2h_2 + m_3h_3 - m_1h_1 = 57.42 \times 10^{-3} \times 435 + 9.532 \times 10^{-3} \times 1980 - 66.95 \times 10^{-3} \times 20 = 42.51$ kW

For condenser, $Q_C = m_3(h_3 - h_4) = 9.532 \times 10^{-3} (1980 - 410) = 14.965$ kW

Overall energy balance, for check, yields

$Q_C + Q_a = 14.965 + 34.58 = 49.545 \approx Q_g + Q_e = 42.51 + 7.0334 = 49.543$ (neglecting the pump work)

$COP = Q_e / Q_g = (2 \times 3.5167)/42.51 = 0.165$.

Circulation factor (f) is defined as the ratio of the mass flow rate of the weak solution to the mass flow rate of vapour out of the generator.

Circulation factor, $f = m_1/m_6 = 66.95 \times 10^{-3}/9.532 \times 10^{-3} = 7.024$

Example 6.4

A H_2O-LiBr absorption refrigeration system with a heat exchanger as shown in Fig. 6.9 has the following data. Absorber, generator, condenser and evaporator temperatures are 36°C, 110°C, 40°C and 5°C, respectively. Mass flow rate through the pump is 0.7 kg/s. The solution enters the generator at 65°C. Calculate the heat transfer from/to absorber, generator, condenser and evaporator. Calculate the COP of the system.

Ans: For these systems we use two charts for calculation. One is pressure-temperature-concentration chart (p-t-c chart). Another is enthalpy-concentration-temperature (h-t-c) chart.

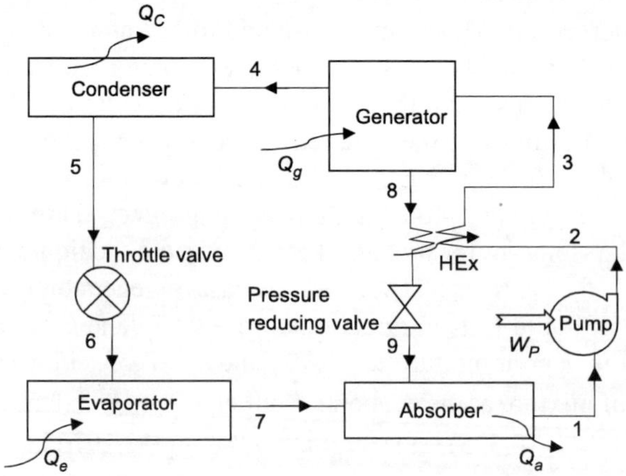

Figure 6.9: Schematic of the H_2O-LiBr VAR system with HEx

Two pressure limits, i.e., higher pressure is the saturation pressure of water at condenser temperature and lower pressure is the saturation pressure of water at evaporator temperature. Equilibrium states of the solution are assumed at generator exit and absorber exit.

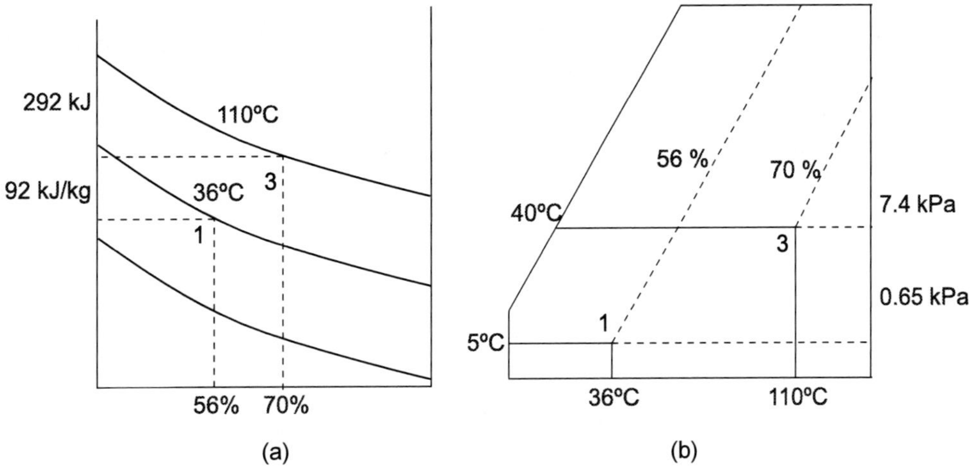

Figure 6.10: Fixing of state points of the problem on (a) h-t-c plane and (b) on p-t-c plane

Mass flow rate m_1 and $m_2 = 0.7$ kg/s

From mass conservation, $m_2 = m_3 + m_5 = m_1$

For 36°C absorber temperature and 5°C evaporator temperature (from p-t-c) chart equilibrium pressure is 0.65 kPa, and $x_2 = 0.56$ (x is the amount of LiBr per kg of solution) (Fig. 6.10).

For 110°C generator temperature and 40°C condenser temperature, equilibrium pressure is 7.4 kPa, and $x_3 = 0.70$

Mass balance for LiBr gives $m_2x_2 = m_3x_3$; $0.7 \times 0.56 = m_3 \times 0.7$; So $m_3 = 0.56$

Also $m_5 = m_6 = m_7 = 0.7 - 0.56 = 0.14$ kg/s

From enthalpy concentration chart $h_1 = 92$ kJ/kg, $h_3 = 292$ J/kg and $h_2 = 148$ kJ/kg (for $x_2 = 0.56$ and 65°C)

From steam tables h_5 = water vapour at 110°C = h_g = 2691.5 kJ/kg; $h_6 = h_f$ at 40°C = 167.57 kJ/kg

Again $h_7 = h_g$ at 5°C = 2510.6 kJ/kg

Heat transfer from condenser, $Q_C = m_5(h_5 - h_6) = 0.14(2691.5 - 167.57) = 353.35$ kW

Heat transfer from evaporator, $Q_e = m_7(h_7 - h_6) = 0.14(2510.6 - 167.57) = 328.02$

For heat exchanger, $m_1(h_2 - h_1) = 0.7(148 - 92) = 40.32 = m_3(h_3 - h_4) = 0.57(292 - h_4)$

So $h_4 = 221.26$ kJ/kg

Heat transfer to generator, $Q_g = m_5h_5 + m_3h_3 - m_2h_2 = 0.14(2691.5) + 0.56(292) - 0.7(148)$
$= 436.73$ kW

Heat transfer from absorber $Q_a = m_7h_7 + m_4h_4 - m_8h_1$
$= 0.14(2510.6) + 0.56(221.26) - 0.7(92) = 411.0$ kW

(As $h_8 = h_1$, pump work is being neglected)

COP = 328.02/436.73 = 0.75

For check: Energy balance yields: $Q_g + Q_e = 764.75$ kW $\approx Q_C + Q_a = 764.35$ kW

Example 6.5

Following data was recorded for a 1 TR NH_3-H_2O vapour absorption refrigeration system operating between pressure limits of 15 bar and 2 bar.

Condenser temperature = Absorber temperature = 45°C

Generator temperature = 140°C and evaporator temperature = 5°C

Saturated conditions of the working fluid are assumed at absorber exit, at generator exit, and at condenser exit. Calculate the pump work and the COP of the refrigeration system. Compare this value with the maximum possible COP.

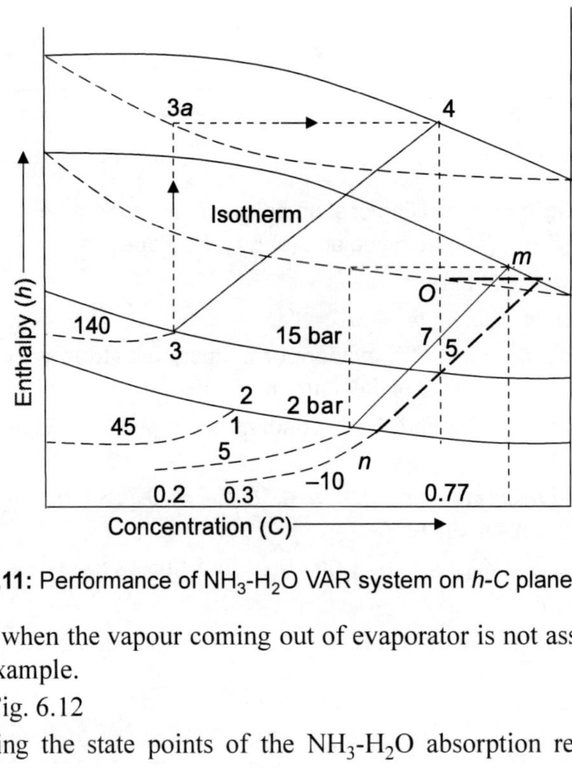

Figure 6.11: Performance of NH_3-H_2O VAR system on *h-C* plane for Ex. 6.5

Ans: This is a case when the vapour coming out of evaporator is not assumed to be saturated as in the previous example.

Refer Fig. 6.2 and Fig. 6.12

The process of fixing the state points of the NH_3-H_2O absorption refrigeration system is discussed below.

State point 1 is the low pressure saturated strong solution from absorber exit, corresponds to 2.0 bar and 45°C. The corresponding concentration (x_1) and enthalpy (h_1), as determined from the chart, are 0.3 and 80.0 kJ/kg, respectively.

State point 2 is the strong solution coming out of the pump. It is just above point 1 by an amount equal to pump work and it corresponds to 15 bar.

Specific volume (v_f) of solution at entry to the pump

$$= 0.3 \times v_{f(\text{Ammonia})} + 0.7 \times v_{f(\text{water})}$$

$$= 1.15 \text{ m}^3/\text{kg}$$

So pump work is calculated from $v_f \times \Delta P = (1.15 \times 10^{-3})\,(15 - 2) \times 10^5 / 10^3 = 1.5$ kJ/kg

So $h_2 = h_1 + 1.5 = 80.0 + 1.5 = 81.5$ kJ/kg and the concentration, $x_2 = 0.3$.

State pont 3 is the saturated weak solution coming out of generator, corresponds to 15 bar and 140°C. The property values, from chart, are enthalpy $h_3 = 540$ kJ/kg and concentration $x_3 = 0.2$

State point 4 is fixed by drawing the isotherm as discussed in Example 6.3 and Section 6.6. The property values, from chart, are enthalpy $h_4 = 2120$ kJ/kg and concentration $x_4 = 0.77$.

State point 5 is the intersection point of the vertical line drawn from point 4 (i.e., concentration is 0.77) and the pressure of 15 bar and isotherm of 45°C. This point is not exactly on the saturation curve but slightly sub-cooled. The property value, from chart, enthalpy, $h_5 = 320$ kJ/kg.

State point 6, after throttling, is the same as point 5 as the enthalpy and concentration remain same. But it corresponds to pressure of 2.0 bar. That is how it is in two-phase zone and its temperature, from chart, is –10°C. This is found out by drawing the isotherm (*m-n*), by trial and error method, through 6 and corresponding to 2 bar.

The state of solution coming out of the evaporator, point 7, lies on the intersection point of the isotherm of 2.0 bar and 5°C and concentration of 0.77. The property value at this point, from chart, enthalpy, $h_7 = 780$ kJ/kg.

The mass flow rate through different components are calculated as follows.

Mass flow rate through evaporator $(m_7) = 1\ \text{TR}/(h_7 - h_6) = 3.5167/\,(780 - 320)$

$= 7.645 \times 10^{-3}$ kg/s

Considering ammonia conservation around absorber $m_3 x_3 + m_7 x_7 = m_1 x_1$

From mass conservation $m_3 + m_7 = m_1$

Substituting the numerical values $(m_1 - 0.007645) \times 0.2 + 0.007645 \times 0.77 = m_1 \times 0.3$

So mass flow at exit of absorber $m_1 = 0.04358$ kg/s = mass flow to generator

Mass flow rate from generator (weak solution) = 0.04358 – 0.007645 = 0.036 kg/s

So the work done by pump = 0.04358 × 1.5 = 0.0654 kW

Heat input to generator = $Q_g = 0.007645 \times 2120 + 0.036 \times 540 - 0.04358 \times 81.5 = 32.1$ kW

So COP = 3.5167/ (32.1 + 0.0654) = 0.11

The minimum temperature could be obtained as shown above is –10°C.

So maximum COP = {263/(318 – 263)} × {(413 – 318)/ 413} = 1.1

6.8 THREE FLUID REFRIGERATION SYSTEMS

The schematic of the system is shown in Figure (6.11). The system is so named because three working fluids are used in this system. NH_3 and H_2O are used as the refrigerant and absorbent pair and the inert hydrogen gas is the third fluid used as carrier gas. The working principle is as described below. Refer Fig. 6.11.

The liquid NH_3-H_2O mixture is heated in the generator section. The vapour mixture moves up. The entrained liquid particles are separated in separator and moves down to the absorber through a liquid seal. The NH_3vapour moves to condenser where it condenses to liquid. After condenser there is a liquid seal which only allows liquid NH_3 to enter to the evaporator. In the evaporator H_2 gas is added. H_2, being inert, it does not react with NH_3 but the partial pressure of NH_3 reduces. It goes like this.

The total pressure, $P_{\text{total}} = P_{\text{ammonia}} + P_{\text{hydrogen}}$

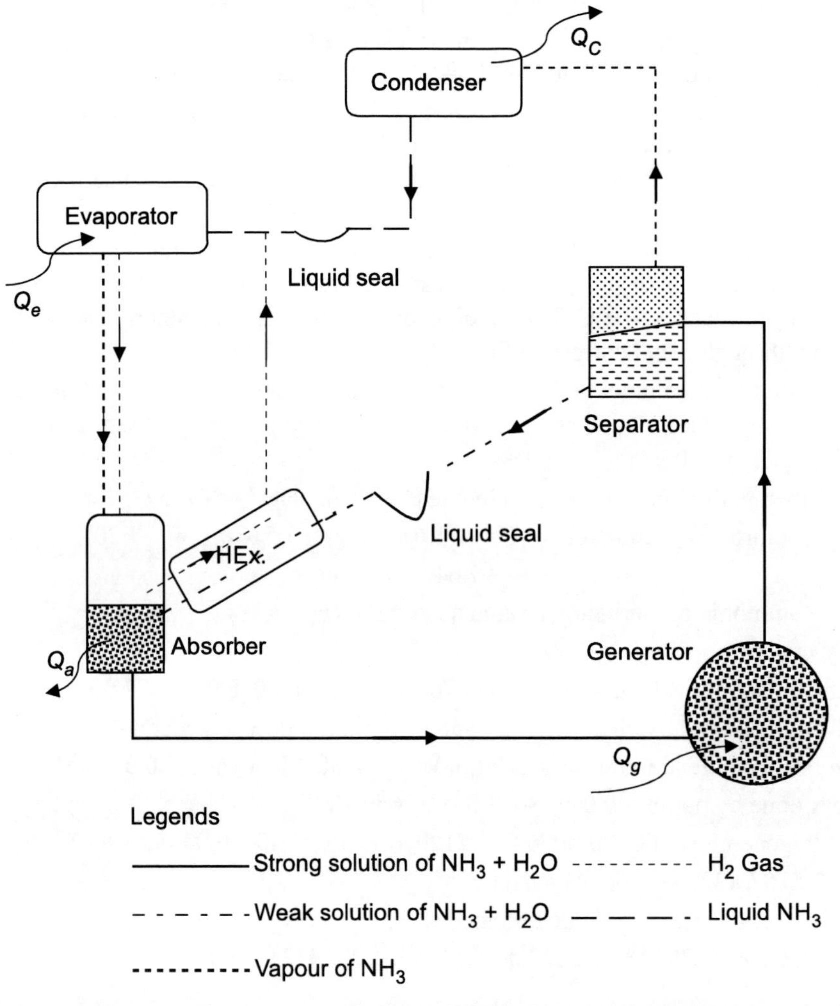

Figure 6.12: Schematic of a three-fluid absorption refrigeration system

Thus the (partial) pressure of NH_3 in the evaporator remains less than the pressure in the condenser which is maintained at P_{total}. So NH_3 evaporates at a lower temperature than it condenses in the condenser. Thus, by adding hydrogen gas in the evaporator pressure differential is maintained between the condenser (high pressure) and evaporator (low pressure).

The NH_3vapour along with H_2 gas comes out of the evaporator and enters the absorber. In the absorber the NH_3vapour is absorbed by the weak solution from the separator to form the strong solution and the H_2 gas is separated. The strong solution goes to generator and the H_2 gas passes through the hot weak solution. Thereby it gets heated up by exchanging heat and moves up to evaporator. Thus, it completes the cycle.

6.8.1 Advantages and Disadvantages

The advantage is that there is no moving part and the flow of the working fluid (motive force) is totally based upon the gravity. So the component of the system should be positioned relative to each other in such a manner that the flow should always be gravity assisted (may be due to variation in density or due to relative vertical height). The total pressure of the system at any section is same. Hence, pressure of the whole system is well balanced. The main disadvantage in the system is that the working fluid may not return to the generator or it cannot complete the cycle. Gravity force is not sufficient to overcome the opposing force to flow generated due to improper separation of liquid and vapour, frictional losses in the pipes during the flow. So the continuity of flow is hampered and the system becomes unreliable. Though the system is simple, less costly and environment friendly, it has not been accepted on commercial basis. Commercially, this system was first fabricated by Electrolux company in commercial basis. That is why this system sometimes is referred to as Electrolux Refrigeration System.

EXERCISE

1. What should be NBP difference of refrigerant and absorbent in vapour absorption system? Justify the answer.
2. Which components of vapour absorption refrigeration system replace the compressor of vapour compression system?
3. State the reasons of very low COP of absorption refrigeration system.
4. Why H_2O-LiBr system does not require a rectification column like NH_3-H_2O system?
5. In a three-fluid refrigeration system, what is the function of the third fluid, i.e., inert gas (H_2)?
6. Differentiate between VCR and VAR systems.
7. In a VAR system, heating, cooling and refrigeration take place at the temperatures of 100°C, 20°C and –5°C, respectively, Find the maximum COP of the system.
8. In a H_2O-LiBr absorption refrigeration system following temperatures are given.

   Generator temperature 105°C, condenser temperature 40°C, absorber temperature 35°C,

   evaporator temperature 7°C. Pump delivers 0.55 kg/s of solution. Calculate the refrigerant flow rate, refrigerating capacity and COP. Ans: 0.105 kg/s, 246.4 kW, 0.7.

MULTIPLE–CHOICE QUESTIONS

1. COP of an ideal vapour absorption refrigeration system is

   (a) $\eta_{HE} \times COP_{Ref}$ (b) COP_{Ref}/η_{HE}

   (c) η_{HE} / COP_{Ref} (d) $\eta_{HE} + COP_{Ref}$
2. Compressor of a VCR system is replaced by —, —, and — in VAR.

   (a) pump, condenser, TEV (b) absorber, pump, TEV

   (c) absorber, pump, generator (d) pump, generator, condenser

3. Work done by the pump in VAR is reduced by —— in comparison to that done by compressor in VCR.

(a) 1/10 (b) 1/100

(c) 1/1000 (d) 1/10000

4. In VAR work done is less by 10^3 in comparison to VCR because

(a) specific volume of liquid is less by 10^3 than that of vapour

(b) density of liquid is less by 10^3

(c) compression ratio requirement is high

(d) compression ratio requirement is low.

5. The refrigerant and the absorbent pair used for VAR should follow the —— Raoult' law.

(a) positive deviation (b) negative deviation

(c) no deviation (d) Dalton's law

6. The difference of NBPs of refrigerant and absorbent should be, preferably, more than

(a) 25°C (b) 50°C

(c) 100°C (d) 200°C

7. Rectification coloumn is required for —— VAR system.

(a) H_2O-LiBr (b) R22-R134a

(c) H_2O-Glycol (d) NH_3-H_2O

8. H_2O-LiBr VAR system is not used for refrigeration purpose because

(a) LiBr does not vaporise (b) H_2O vaporises

(c) LiBr freezes below 0°C (d) H_2O freezes at 0°C

9. Three-fluid (Electrolux) refrigeration system works on

(a) electrical energy (b) mechanical Energy

(c) gravitational Energy (d) chemical Energy

10. Inert gas (like H_2) is used in three fluid refrigeration system to

(a) increase the cooling capacity

(b) maintain differential pressure between condenser and evaporator

(c) decrease the heating requirement

(d) maintain a constant pressure in the system

11. Following statements in respect of VAR and VCR systems are as follows:

1. The former runs on low grade energy.
2. The pumping work in the former is negligible since specific volume of strong liquid solution is small.
3. The latter uses an absorber while former uses a generator.
4. The liquid pump alone replaces compressor of the latter.

Which of these statements are correct?

(a) 1 and 2 (b) 1 and 3

(c) 1 and 4 (d) 2 and 4

12. Solar energy can be directly used in

(a) vapour compression refrigeration system

(b) vapour absorption refrigeration system

(c) air refrigeration system

(d) jet refrigeration system

13. Which refrigeration system uses waste heat

(a) Vapour compression cycle (b) Vapour absorption cycle

(c) Air refrigeration cycle (d) Vortex refrigeration system

ANSWERS

1. (a) **2.** (c) **3.** (c) **4.** (a) **5.** (b) **6.** (d) **7.** (d) **8.** (d) **9.** (c) **10.** (b) **11.** (a) **12.** (b) **13.** (b)

7

Thermoelectric and Other Refrigeration Systems

7.1 INTRODUCTION

Vapour compression refrigeration system is widely used due to its high COP and reliability. Next to it is the vapour absorption refrigeration systems which uses low grade energy, but gives very low COP (less than 1). Besides these two refrigeration systems there are other refrigeration systems, sometimes termed non-conventional refrigeration systems. They are used for specific application, though they give very low COP in comparison to vapour compression refrigeration system.

7.2 PRINCIPLE OF THERMOELECTRIC REFRIGERATION

When two dissimilar metals [for example bismuth and tellurium (Bi_2Te_3)] are joined together and electric current is passed through the junctions, following thermoelectric effects are observed.

1. *Seebeck effect (1822):* (Fig. 7.1) When two dissimilar metals are joined together and both the ends are maintained at different temperatures, there is generation of emf which is directly proportional to the difference in temperature of both the ends. Mathematically expressed as

$$\text{Emf} = \alpha \Delta T \qquad [7.1]$$

where, α is the proportionality constant, called Seebeck coefficient and depends upon the type of materials used.

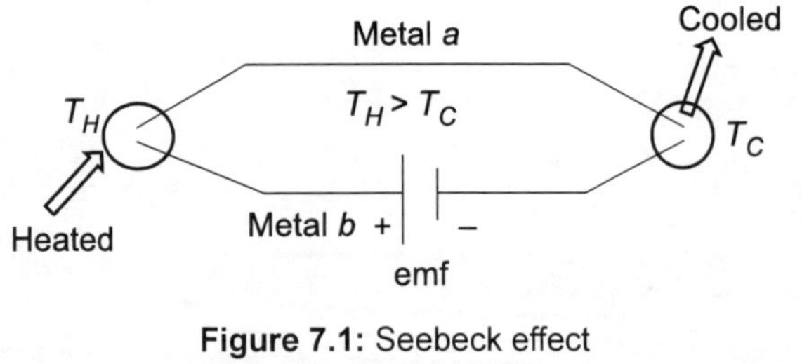

Figure 7.1: Seebeck effect

2. *Peltier effect (1834):* (Fig. 7.2) When two dissimilar metals are joined together and a current is passed through the two junctions, one end becomes hot and the other end becomes cool.

The heat transfer from either end of the junctions is proportional to the amount of current passing through it. Mathematically expressed as

$$q = \phi I \qquad [7.2]$$

where, ϕ is the Peltier coefficient and I is the amount of current passing through it.

Thermodynamic analysis gives the relationship between Seebeck coefficient and Peltier coefficient as

$$\varphi_{ab} = \alpha_{ab} T \qquad [7.3]$$

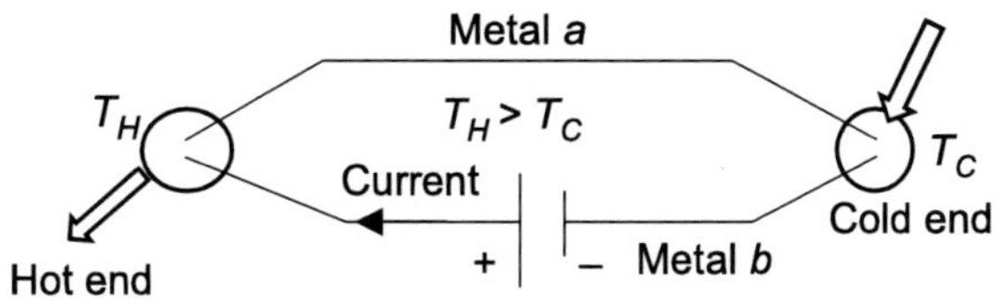

Figure 7.2: Peltier effect

3. *Joulean effect:* As the current flows through a conductor heat (q) is generated amounting to I^2R where I is the amount of current and R is the electrical resistance of the material.
4. *Thompson effect:* When a current is passed through a single conductor where there already exists a uniform temperature gradient, heat is absorbed or rejected in excess of the Joulean effect and mathematically expressed as

$$q = \tau I(dT/dx) \qquad [7.4]$$

   where, τ is the Thompson coefficient, I is the current, dT/dx is the temperature gradient across the conductor.
5. *Fourier effect of conduction:* When the two ends are maintained at different temperatures heat is conducted from higher temperature to lower temperature as per the Fourier's law of heat conduction and mathematically expressed as $q = U\Delta T$

   where, U is the overall thermal conductance and ΔT is the temperature difference.

From the above stated five effects, it is the Peltier effect that gives the refrigerating effect when current is passed through the dissimilar metals junctions as shown in Fig. 7.2. Conversely Seebeck effect generates electric power when two ends of the junction are subjected to two different temperatures. It has a very broad application, starting from a power supplier in space mission (around 300 W) to that for domestic purpose (several kW). In thermoelectric refrigeration system electrons may be regarded as the working fluid that carries the energy from one part of the system to the other. Applied emf is the external power supplied as in case electric power to run the compressor in a VCS. Hot end and cold end are considered to be condenser and evaporator, respectively. Drop in energy level of the electrons at the junction and electron flow resistances may be regarded as the throttling action.

7.2.1 Thermodynamic Analysis

Thermoelectric effects are more observed with semiconducting materials and are called thermoelectric elements. Mostly used combination is bismuth and bismuth alloys with antimony. To attain below –100°C, bismuth telluride alloys are used. Refer to the Fig. 7.3.

Assumptions made for the thermodynamic analysis are:

(a) Heat transfer takes place through semiconductors at the ends only.

(b) The physical properties like conductivity, resistivity, Seebeck coefficient, etc. are independent of temperature.

Cooling and heating due to thermoelectric effect (Peltier effect) are expressed as

$$Q_{PC} = \alpha_{PN} IT_C, \text{ and } Q_{Ph} = \alpha_{PN} IT_h$$

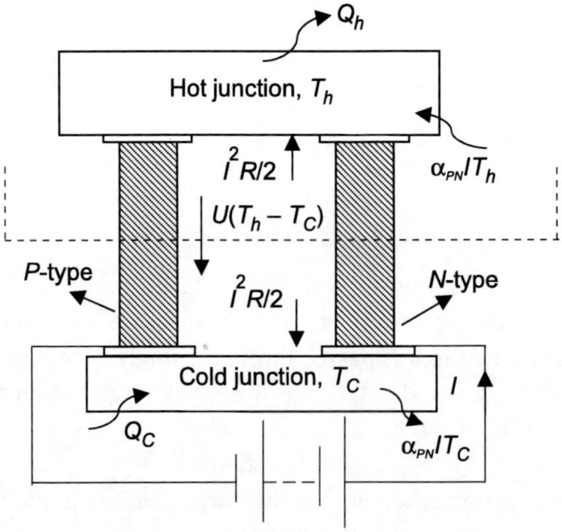

Figure 7.3: Schematic of a thermoelectric refrigeration system

Thompson effect is zero when α is invariant with temperature.

The Joulean effect is equally divided between the hot end and the cold end of the circuit. So the heat generation in either of the ends will be $I^2R/2$ where, R is the total resistance of the two semiconductors and given by

$$R = L_P/A_P + L_N\rho_N/A_N$$

where, ρ is the specific resistance. [7.5]

Fourier's heat conduction from hot end to cold end will be:

$$Q_{\text{cond}} = (T_h - T_C)[k_PA_P/L_P + k_NA_N/L_N] = U(T_h - T_C)$$

where, U is the overall conductance of the two semiconductors.

Now the energy balance for the control volume at the cold end gives:

Cooling effect produced due to Peltier effect = Refrigerating effect (Q_C) + Heat conducted into control volume due to Fourier heat conduction + Joulean heat generated

$$IT_C\alpha_{PN} = Q_C + U(T_h - T_C) + I^2R/2;$$

So $$Q_C = \alpha_{PN} IT_C - U(T_h - T_C) - I^2R/2$$ [7.6]

Now the energy balance for the control volume at the hot end gives:

Heat generated due to Peltier effect + Joulean heat generated = Heat rejected at hot end (Q_h) + Heat leaving the control volume due to Fourier heat conduction

$$\alpha_{PN} IT_h + I^2R/2 = Q_h + U(T_h - T_C);$$

So $$Q_h = -U(T_h - T_C) + I^2R/2 + \alpha_{PN} IT_h \quad [7.7]$$

Net power supplied (from energy balance)

= Heat rejected by the system – Heat absorbed by the system

$$= Q_h - Q_C = I^2R + \alpha_{PN} I(T_h - T_C) \quad [7.8]$$

This power is utilized to overcome the power loss in Joulean effect and to counteract the Seebeck effect. This power is supplied by the battery. The first term is the product of Seebeck effect and the current and the second term is the Joulean effect.

$$\text{COP} = Q_C/(Q_h - Q_C) = [\alpha_{PN} IT_C - U(T_h - T_C) - I^2R/2] / [\alpha_{PN} I(T_h - T_C) + I^2R] \quad [7.9]$$

If the internal irreversibility like Joulean effect and Fourier's heat conduction are neglected

$(\text{COP})_{RC} = T_C/(T_h - T_C)$, which is the reverse Carnot COP.

From the above equation it is clear that if the thermoelectric material is selected, the physical properties remain unchanged. Hence, the COP is a function of current *I*. For maximizing the COP, Eq. 7.9 is differentiated with respect to *I* and equated to zero. We get

$$\alpha_{PN}I^2R\,(T_h + T_C)/2 + \alpha_{PN}\,U\,\Delta T^2 + 2UIR\Delta T = 0; \text{ (where, } \Delta T = T_h - T_C) \quad [7.10]$$

Now introducing a term 'figure of merit' which is expressed as $Z = \boldsymbol{\alpha}_{PN}^{2}/UR$ [7.11]

And substituting $T_{\text{Mean}} = (T_h + T_C)/2$ and

We get $I = (\alpha_{PN}\,\Delta T)/[R\{\sqrt{(1 + ZT_{\text{Mean}})} - 1\}]$ [7.12]

Hence, the COP_{Max} obtained is $T_C\,\{\sqrt{(1 + ZT_{\text{Mean}})} - (T_h/T_C)\}/[(T_h - T_C)\{\sqrt{(1 + ZT_{\text{Mean}})} + 1\}]$

$$= \{T_C/(T_h - T_C)\}\,[\{\sqrt{(1 + ZT_{\text{Mean}})} - (T_h/T_C)\}/\{\sqrt{(1 + ZT_{\text{Mean}})} + 1\}]$$

$$= (\text{COP})_{RC}\,[\{\sqrt{(1 + ZT_{\text{Mean}})} - (T_h/T_C)\}/\{\sqrt{(1 + ZT_{\text{Mean}})} + 1\}] \quad [7.13]$$

In the above expression, it is evident that $(\text{COP})_{\text{Max}}$ is the product of $(\text{COP})_{RC}$ and the effect of actual thermoelectric material behaviour.

Similarly, maximum refrigerating effect can be obtained by differentiating equation for Q_C with respect to *I*,

we get $I = \alpha_{PN}\,T_C/R$ [7.14]

So $(\text{COP})_{\text{Opt}} = (ZT_C^{\,2}/2 - \Delta T)/ZT_hT_C$ and $Q_{C,\text{Max}} = U\,[(Z_{TC}^{\,2}/2) - \Delta T]$ [7.15]

The maximum temperature difference that can be obtained, i.e., $T_h - T_C = Z_{TC}^{\,2}/2$ [7.16]

In all the above expressions it is noticed that *Z* value must be as high as possible so that more refrigeration effect can be obtained. Once material is selected, Seebeck coefficient, electrical resistance and the overall conductance which depend upon the material properties become fixed. So *Z* value can be varied by varying cross-sectional areas and the lengths of the semiconductors. *Z* can be maximized by minimizing *UR* as $Z = \alpha_{PN}^{2}/UR$

The overall conductance for both the conductors *U* is expressed as $A_Pk_P/L_P + A_Nk_N/L_N$;

Total resistance for both the conductors is expressed as

$$L_P\rho_P/A_P + L_N\rho_N/A_N = L_P/(\sigma_P A_P) + L_N/(\sigma_N A_N)$$

as $\sigma = 1/\rho$, where ρ is the specific resistance.

So $UR = (A_Pk_P/L_P + A_Nk_N/L_N)\,(L_P\rho_P/A_P + L_N\rho_N/A_N)$ [7.17]

Now taking aside the geometrical parameters $(A_P L_N)/(L_P A_N)$ as x and differentiating UR with respect to x and equating it to zero to minimize the UR value,

we get $x = \sqrt{(k_N \rho_P)}/\sqrt{(k_P \rho_N)}$ [7.18]

The minimum value of UR is $\{\sqrt{(k_P \rho_P)} + \sqrt{(k_N \rho_N)}\}2$ [7.19]

and $Z_{max} = \alpha_{PN}2 / \{\sqrt{(k_P \rho_P)} + \sqrt{(k_N \rho_N)}\}2$ [7.20]

For special case if $k_P = k_N = k$ and $\rho_P = \rho_N = \rho$;

The value of Z_{Max} will be $\alpha_{PN}^2 / 4k\rho$; OR $Z_{Max} = \sigma\alpha_{PN}^2/4k$ [7.21]

7.2.2 Thermoelectric Materials

The discussion on thermoelectric refrigeration system and the expression for Figure of Merit 'Z' shows that the materials used for the purpose must have the following properties. (1) high Peltier coefficient, (2) low electrical resistivity, (3) low thermal conductivity.

Insulators have small electrical conductivity, i.e., large electrical resistance. So Joulean heating will be very high. Seebeck coefficient is large for insulators. Hence, more refrigerating effect can be obtained. But this is counteracted by large Joulean heating effect. So an insulator cannot be a good thermoelectric material. On the other hand metals have low electrical resistance. They have small Seebeck coefficient but high thermal conductivity. This small refrigerating effect will be counteracted by the high heat conduction to the cold end from the hot end. Hence, the semiconductors are chosen as the good thermoelectric material. $\sigma\alpha_{PN}^2$ value of Eq. (7.21) is very high for semiconductors like alloys of bismuth, tellurium and antimony for p-type element and bismuth, tellurium and selenium for n-type element.

As discussed above, thermoelectric material works on the principle of Seebeck effect (for thermoelectric generation) and/or Peltier effect (for thermoelectric cooling). Its effectiveness is described on the basis of a parameter (Z) known as 'Figure of Merit' and more conveniently, is expressed in a non-dimensional form for performance evaluation as

$$ZT = \frac{\alpha^2 \sigma}{4\kappa} T \qquad [7.22]$$

where the notations have their usual significance. It is because Figure of Merit (Z) of a material is temperature dependent. From this expression it is evident that a good thermoelectric material must have high Seebeck coefficient and electrical conductivity, combinedly called 'power factor' and a low thermal conductivity. But these properties of any thermoelectric material contradict each other as shown in Fig. (7.3a).

[Carrier concentration or charge density means it is the amount of charge carriers per unit Volume of material. For P-type material, since hole is the charge carrier, hole concentration is very high and for n-type material, since electron is the charge carrier, free electron density is very high.]

However, thermoelectric material performance (ZT) and power factor can be enhanced by (i) decreasing the thermal conductivity specifically decreasing the lattice and bipolar conductivity, because thermal conductivity of a material is contributed by electron movement, lattice vibration and bipolar effect. If electron movement is reduced the electrical conductivity will be reduced which is not desirable; (ii) increasing the Seebeck coefficient; and (iii) increasing the electrical conductivity. This is possible only by impregnating the bulk material with suitable doping material or by layering the bulk material using nano-engineering techniques. Increase in carrier concentration decreases the Seebeck coefficient but increases the electrical conductivity, thermal conductivity as well. Maximum value of ZT is obtained at a carrier concentration of around $10^{19}/cm^3$. However, typical carrier concentration is considered to be in a range of 10^{19} to $10^{20}/cm^3$. Experiments have been conducted with Bi_2Te_3 being alloyed with Sb for P-type and with Se for N-type as bulk materials. P-type material is doped with SiC and N-type material is doped with SbI_3 by nano-engineering respectively.

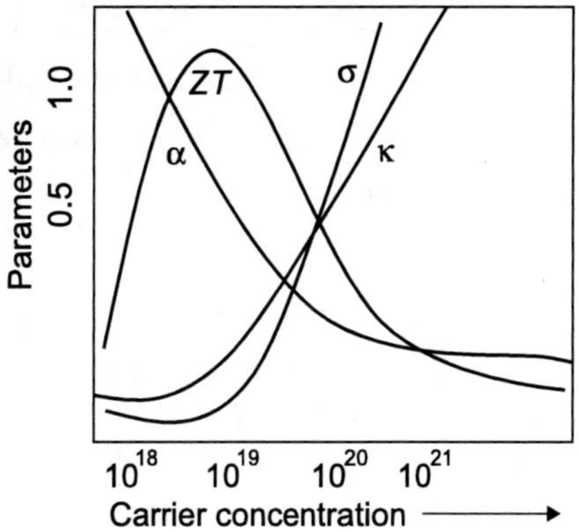

Figure 7.3: (a) Variation of TE material properties with carrier concentration

Results show that inclusion of suitable nanomaterial into the bulk material causes an increase in carrier concentration. This increases both the seebeck coefficient and the electrical conductivity. Simultaneously, it decreases the lattice thermal conductivity though lattice thermal conductivity is independent of carrier concentration. It has been found out that reduction of thermal conductivity by 25% through Na doping and inclusion of nanostructured SrTe again decreases the thermal conductivity by 55%. Thus, ZT increases to 1.7 at 800 K. Similarly, by doping thalium into PbTe gives an enhanced Seebeck effect due to distortion in the density of the electron states. Thus, the thermoelectric materials, in bulk form, are classified into three categories on the basis of their optimal working range of temperatures. Bi_2Te_3 based alloys show ZT_{max} at low temperature (<400 K), PbTe based alloys show the peaks at middle range of temperature (600 K to 900 K) and SiGe based alloys show their peaks at high temperatures (>900 K). From these observations it may be concluded that for refrigeration purpose best suitable bulk material is Bi_2Te_3 whereas PbTe and SiGe based bulk materials are suitable for power generation. It is experimentally verified that thermoelectric properties of copper selenide at higher temperature (at 700 K) this material changes from α-phase to β-phase and a ZT value of 1.6 could be achieved at that temperature.

Example 7.1

For a thermoelectric refrigeration system the current corresponding to COP_{Max} is half to that of the maximum cooling. The system operates between temperature limit of 40°C and –5°C. Determine the Figure of Merit and the ratio of COP_{Max} and COP_{Opt} corresponding to maximum refrigerating effect.

Ans: For COP_{Max}; $I = (\alpha_{PN}\,\Delta T)/[R\{\sqrt{(1 + ZT_{Mean})} - 1\}]$

For $Q_{C},_{Max}$; $I = \alpha_{PN}\,T_C/R$

So $\alpha_{PN}\,T_C/2R = (\alpha_{PN}\,\Delta T)/[R\{\sqrt{(1 + ZT_{Mean})} - 1\}]$

or $T_C/2 = \Delta T/\{\sqrt{(1 + ZT_{Mean})} - 1\}$; Putting the values $\sqrt{(1 + ZT_{Mean})} = 179/134$

$T_{Mean} = 290.5$ K; $Z = 2.7 \times 10^{-3}\,K^{-1}$

$$\text{COP}_{\text{Max}} = \text{COP}_{RC} \text{ X } [\{\sqrt{(1 + ZT_{\text{Mean}})} - (T_h / T_C)\}/\{\sqrt{(1 + ZT_{\text{Mean}})} + 1\}]$$
$$= 5.955556 \times 0.07184 = 0.4278 \text{ (From Eq. 7.13)}$$
$$\text{COP}_{\text{Opt}} = (ZT_C^2/2 - \Delta T) / ZT_hT_C = 0.2294 \text{ (From Eq. 7.15)}$$

The ratio = 1.865

Example 7.2

A thermoelectric refrigeration system works between the temperature limits of 305 K and 275 K. The thermoelectric elements used have total resistance R = 0.0025 ohm, conductance U = 0.02 W/K. α_{PN} = 0.0004 V/K. For cooling capacity of 600 kJ/Hr Calculate (a) Number of couples for maximum COP, (b) COP, (c) power requirement. Find the electric bill per month if the system runs for 10 hours/day and power rate is Re 1 /kWH.

Ans: $T_{\text{Mean}} = (305 + 275)/2 = 290$ K; $Z = \alpha_{PN}^2 / UR = 3.2 \times 10^{-3}$

Corresponding to (COP_{Max}) $I = (\alpha_{PN}\, \Delta T)/[R\{\sqrt{(1 + ZT_{\text{Mean}})} - 1\}] = 12.354$ amp

Refrigerating effect, $Q_C = \alpha_{PN} IT_C - U(T_h - T_C) - I^2R/2 = 0.5681$ W/pair

For refrigerating effect of 600 kJ/Hr = 166.6667 W the number of couple pairs needed

= 166.6667/0.5681 = 293.37, i.e., 294 pairs.

$(\text{COP})_{\text{Max}} = \text{COP}_{\text{RC}} \times [\{\sqrt{(1 + ZT_{\text{Mean}})} - (T_h / T_C)\}/\{\sqrt{(1 + ZT_{\text{Mean}})} + 1\}] = 1.0725$

So power required = 166.6667/1.0725 = 155.4 W

1 kWH = 1000 × 3600 W = 1 unit

For one month total power consumption = (155.4 × 3600 × 10 × 30) W, the total unit

= (155.4 × 3600 × 10 × 30)/ 1000 × 3600

= 46.62 unit, i.e., 46.62 kWh.

Total electricity charges will be Rs. 46.62.

Example 7.3

A thermoelectric refrigeration system (one pair of *p-n* junction) works within a temperature limit of –10°C and 35°C. Diameter and length of both the *p* and *n* materials are 0.01 m. Properties of the material, as obtained from the data book, are as follows:

$\alpha_P = 0.000015$ V/K, $\alpha_N = -0.0002$ V/K, $\sigma_P = 1000\ \text{cm}^{-1}\ \text{ohm}^{-1}$, $\sigma N = 1500\ \text{cm}^{-1}\ \text{ohm}^{-1}$,
$k_P = k_N = 1.2$ W/m-K

Find the maximum COP achievable, maximum cooling capacity and power requirement.

Ans: Area of cross section for both the *p* and *n* materials is $A_P = A_N = 0.00007854\ \text{m}^2$

From Eq. 7.17, $UR = (2 \times 1.2 \times 0.00007854/0.01)\ [\{0.01/(1000 \times 0.00007854)\}$
$+ \{0.01/(1500 \times 0.00007854)\}]/100 = 0.00004$ W-ohm-K.

$\alpha_{PN} = 0.000015 - (-0.0002) = 0.00035$ V/K

So $Z = \alpha_{PN}^2 / UR = 3.0625 \times 10^{-3}$/K

$T_{\text{Mean}} = (263 + 308)/2 = 285.5$ K

So $\sqrt{(1 + ZT_{\text{Mean}})} = 1.369$

For Maximum COP (From Eq. 7.13) 263 [1.369 – 308/263]/[45 {1.369 + 1}] = 0.488

$I_{Max} = (\alpha_{PN}\,\Delta T)/[R\{\sqrt{(1 + ZT_{Mean})} - 1\}] = 0.00035 \times 45/\ \{2.122 \times 10^{-3}\ (1.369 - 1)\} = 20.1145$ amp

$Q_C = \alpha_{PN}\,IT_C - U(T_h - T_C) - I^2R/2 = 0.00035 \times 20.1145 \times 263 - 0.01885 \times 45^2 - \{20.1145^2 \times 2.122 \times 10^{-3}/2\} = 0.574$ W.

Power required = 0.574 / 0.488 = 1.176 W

For the case of maximum cooling:

Maximum COP = $(ZT_C^2/2 - \Delta T)/ZT_h\,T_C = 0.24555$

So $I_{Max} = \alpha_{PN}\,T_C/R = 0.00035 \times 263/(2.122 \times 10^{-3}) = 43.38$ amp

$Q_{C,\,Max} = U\,[(ZT_C2/2) - \Delta T] = 0.01885[3.0625 \times 10^{-3} \times 263^2/2 - 45] = 1.1482$ W

7.3 LIQUEFACTION OF GASES

In vapour compression refrigeration system, the minimum temperature that can be achieved (by cascade system) is nearly –150°C. The science of low temperature, i.e., less than –150°C (though this is not the exact limiting temperature), and the study of different refrigeration processes and behaviour of materials at these low temperatures is called *cryogenics*. Process of liquification of gases comes under cryogenics.

7.3.1 Liquification Principle

The critical points of commonly used liquefied gases, helium, hydrogen and nitrogen are –268°C (2.3 bar), –240°C (13 bar) and –147°C (33.9 bar), respectively. So, none of these gases exist in liquid form at atmospheric condition. The refrigeration processes commonly adopted cannot produce such a low temperature. Hence, a different thermodynamic principle is followed for this purpose. The principle of liquefaction of gases is based upon the inference drawn from the Joule-Thompson porous plug experiment. When a high pressure and low temperature gas is throttled to low pressure, a part of it is obtained in liquid form.

(If a gas that is at high pressure, is made to do some work, its enthalpy decreases resulting in decrease in temperature. This is called *restrained expansion*. Example is the expansion through turbine or expander. There is no work output during throttling process. The pressure drop is due to an irreversible expansion, mainly due to friction. This is called *unrestrained expansion*. Though expanders, by restrained expansion, can give a larger pressure drop, it cannot operate in a two-phase zone where the liquid particles may damage the blades. Also there is high conduction loss in low temperature zone. Hence, unrestrained expansion is preferred for liquefaction of gases.)

The porous plug experiment: Refer to Fig. 7.4.

A gas at certain pressure P_1 and temperature T_1 is made to pass through a valve or a constricted passage (throttled). At the outlet the fluid pressure will be reduced to *Pa* and the temperature will be increased to *Ta*. If this process is continued varying the opening of the valve or constricted passage keeping the inlet pressure constant at P_1, the outlet pressure and temperature of the gas will vary as shown in the Fig. 7.5. As the outlet pressure decreases the temperature increases till it reaches a peak value at '*b*'. After the temperature reaches the peak (maxima) point at (b) the temperature decreases as the pressure decreases. That point (b) is called 'inversion point' for that particular initial pressure and temperature of the gas. As this is a steady flow and adiabatic process the enthalpy of the fluid at inlet and outlet will remain same, i.e., the process is isenthalpic. It is important to note that the state points '*a*' and '*c*' are at same temperature and enthalpy, but are at different pressures. (Explain what is the significance). Now in the second set of reading initial pressure and temperature of the fluid are fixed at P_2 and T_2 respectively. Similar trends of observations are also recorded.

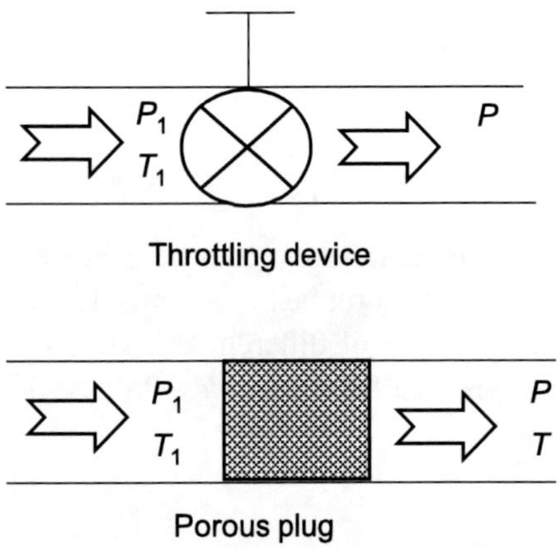

Figure 7.4: Porous plug experiment

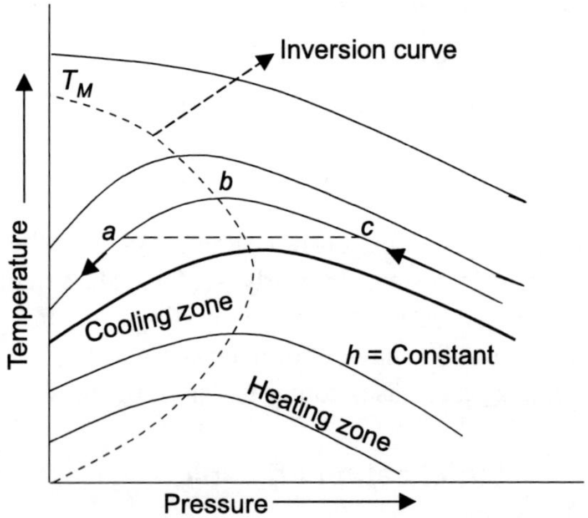

Figure 7.5: Inversion curve

First the temperature of the gas goes on increasing and after reaching the maxima it decreases as the pressure decreases. Thus, sets of experiments are carried out for different initial pressure and temperature of the fluid. For each set one inversion point is obtained. If all the inversion points are joined, a curve is obtained, known as 'inversion curve'. The temperature at the inversion point is called inversion temperature. TM is the maximum inversion temperature. This indicates that a cooling effect cannot be achieved by throttling unless the gas is below its maximum inversion temperature.The left-hand side of the inversion curve is called the cooling zone and the right hand side of it is called as the heating zone. The gradient (μ) of the isenthalpic curve of Fig. 7.5 is expressed by

$$\mu = (\partial T/\partial P)_h \quad [7.23]$$

and μ is called as Joule-Thompson coefficient. The significance of μ is that when $\mu > 0$ the fluid will be cooled after throttling and when $\mu < 0$ the fluid will be heated. But for ideal gas $\mu = 0$, i.e., an ideal gas temperature does not change when it is throttled. The value of μ shows the deviation of a real gas from ideal gas.Let us take the example of air liquefaction.

7.3.2 Air Liquefaction (Linde's) Process

Linde (or may be called Linde-Hampson) Air liquefaction process: Figure 7.6 shows the schematic of the Linde's system and corresponding thermodynamic process is shown in Fig. 7.7. Air is a gas at normal pressure and temperature. As it is stated in the gas liquefaction principle, air is compressed in multi-stage to a very high pressure from state point 1 to 2. During multi-stage compression, there are also multi times inter-cooling. Thus, the process approaches isothermal compression. (Isothermal compression is more economical as minimum work input and theoretically no temperature rise in comparison to more work input and high temperature rise in isentropic compression.) After this isothermal compression, the compressed air is cooled by the cold water from cooling towers or from any external refrigeration system to state point 3. Then it is cooled to a temperature below the inversion temperature to state point 4 in a heat exchanger by the cold air from the separator at state point 7. Then it is throttled to state point 5 which is in two-phase zone. The liquid and vapour are separated in a separator. The liquid part is stored in a cylinder specially made to sustain that high pressure (at 6) and the chilled vapour is passed through the heat exchanger where it cools the high pressure and high temperature air coming from the compressor and reaches state point 8. Then it mixes with the make-up air and reaches the state point 1. Thus, it completes the thermodynamic cycle.

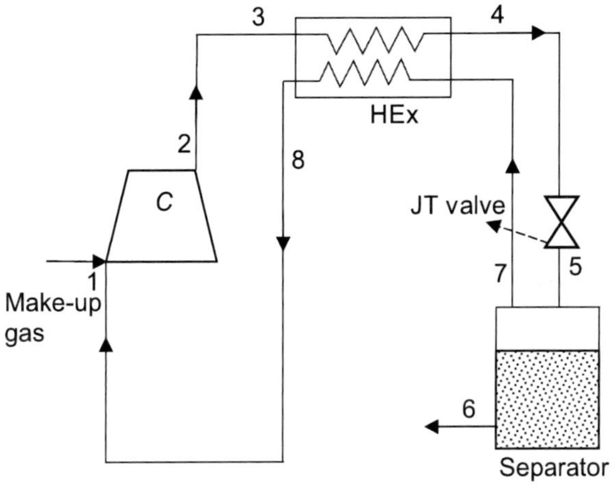

Figure 7.6: Schematic of air liquification by Linde's process

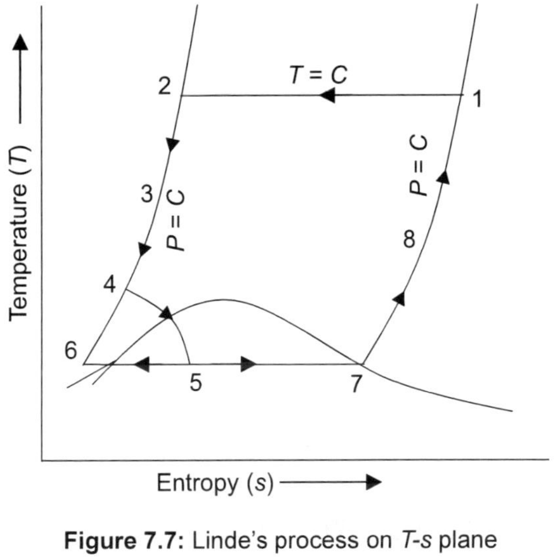

Figure 7.7: Linde's process on *T-s* plane

7.3.2.1 Thermodynamic analysis of Linde process

Thermodynamic analysis of the system is carried out by considering the control volume around the heat exchanger and the separator.

Mass conservation, $m_3 = m_6 + m_8$ [7.24]

Energy conservation, $m_3h_3 = m_6h_6 + m_8h_8$ [7.25]

Substituting the value of m_8, we get $m_3(h_3 - h_8) = m_6 (h_6 - h_8)$ [7.26]

Defining Z_L as the yield in Linde's process, i.e., the ratio of mass of liquefied air to the mass of air compressed

$Z_L = m_6 / m_3 = (h_3 - h_8)/(h_6 - h_8) = (h_8 - h_3)/(h_8 - h_6)$ [7.27]

Considering the isothermal compression

$Tds = dh - vdp$; So $vdp = dh - Tds$;

Integrating from 1 to 2 $W_{2\text{-}1} = (h_2 - h_1) - T_1 (s_2 - s_1)$ [7.28]

Specific work requirement, as defined, kW of power per kg of liquid air produced is expressed as

$W_{S,L} = W_{2\text{-}1}\, m_3 / m_6 = W_{2\text{-}1}/Z_L$ [7.29]

Analysis of the above equation gives that h_8 should always be greater than h_3, though t_8 is smaller than t_3.This is possible because they are different pressures.

7.3.3 Claude's Air Liquefaction Process

It is learnt that at lower temperatures the Joule-Thompson coefficient is large. Therefore, air must be cooled to lowest possible temperature to get the maximum cooling effect by J-T expansion process. Figure 7.8 shows the schematic of the Linde's system and corresponding thermodynamic process is shown in Fig. 7.9. As shown in figure, an expander is used to have the extra cooling of the compressed air before J-T expansion. Compressed air (state point 2), first gets cooled by the water from cooling tower to state point 3. Then it passes through the

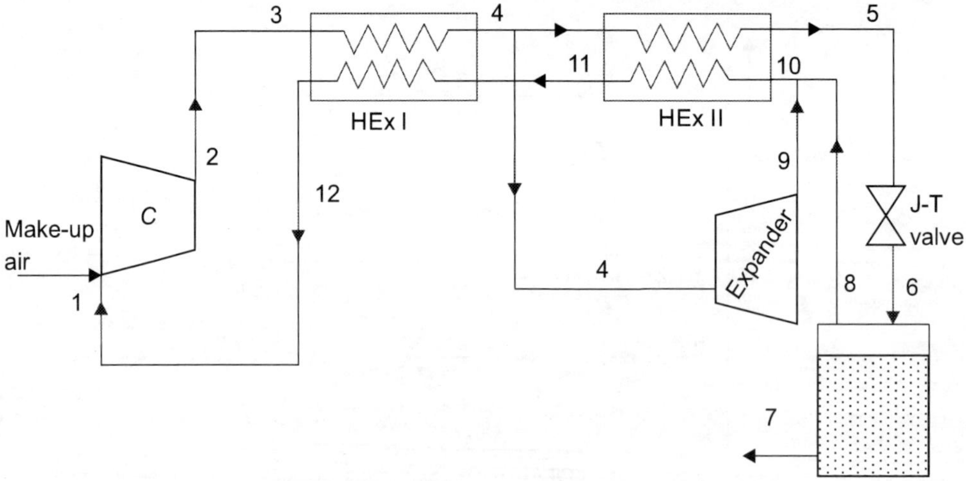

Figure 7.8: Schematic of Claude's air liquification process

first heat exchanger (state point 3), and is divided into two streams. Main stream (state point 4) with nearly 20% of mass passes through the second heat exchanger (state point 5) and auxiliary stream with 80% mass is expanded to low temperature through an expander (state point 9). The air at state point 5 passed through a J-T valve to state point 6 (two-phase zone) into a separator. Here the liquid air is separated at state point 7 and air in gaseous state (state point 8) comes out. It mixes with low temperature air at state point 9 and enters the second heat exchanger at state point 10. It precools the main stream air to below inversion temperature to state point 5. Then it enters the first heat exchanger and comes out at state point 12. It mixes with the make-up air and enters the compressor at state point 1. Thus, the cycle is completed.

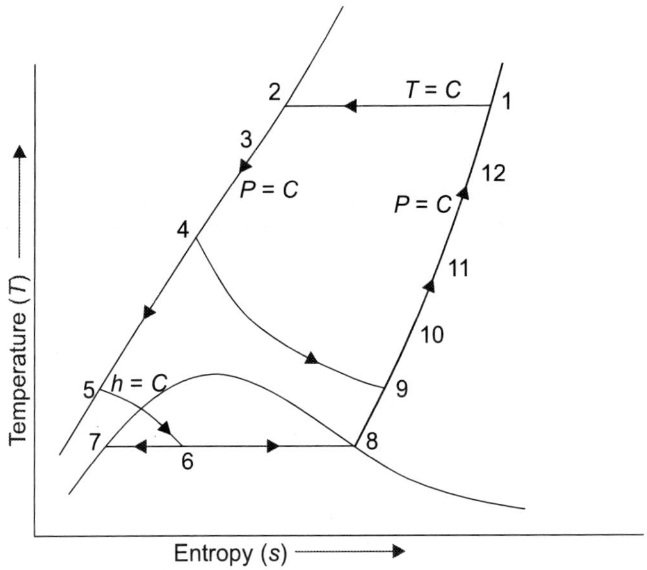

Figure 7.9: Claude's air liquification process on *T-s* plane

7.3.3.1 Thermodynamic analysis of Claude's process

For thermodynamic analysis control volume is considered to be around the two heat exchangers and the separator.

Mass conservation, $m_3 = m_7 + m_{12}$ [7.30]

Energy conservation, $m_3h_3 + m_4h_9 = m_4h_4 + m_7h_7 + m_{12}h_{12}$ [7.31]

Substituting the value of m_{12}, we get $m_3h_3 + m_4h_9$

$$= m_7h_7 + (m_3 - m_7)h_{12} + m_4h_4 m_7 / m_3$$

$$= (h_{12} - h_3)/(h_{12} - h_7) + (m_4/m_3)\{(h_4 - h_9)/(h_{12} - h_7)\}$$

Defining Z_C as m_7/m_3, the yield in Claude's air liquification process

So $Z_C = Z_L + (m_4/m_3)\{(h_4 - h_9)/(h_{12} - h_7)\}$ [7.32]

Work requirement, $W_C = m_3W_{2\text{-}1} - m_4W_{4\text{-}9} = m_3\,(h_2 - h_1) - T_1\,(s_2 - s_1) - m_4\,(h_4 - h_9)$ [7.33]

So specific work requirement, $W_{S,C}/m_7$ [7.34]

Thus it is observed that the yield is more and specific work requirement is less for Claude system in comparison to Linde's system.

7.4 EJECTOR REFRIGERATION SYSTEM

This is a heat engine driven refrigeration system. Refer to the Fig. 7.10. This system consists of two loops, top one is the power loop and the bottom one is the refrigeration loop. In the power loop high pressure and high temperature refrigerant vapour is generated in the boiler by supplying low grade energy from any nonconventional energy source like biogas, solar, etc. This vapour (state point 2) goes to the ejector which has mainly three parts. Nozzle, mixing chamber, and diffuser. As the vapour passes through the nozzle, its velocity increases and the pressure decreases. The decrease in pressure induces vapour from the evaporator (state point 3). Thus, the vapour of state points 2 and 3 get mixed in the mixing chamber. The mixed vapour passes through the diffuser where the vapour velocity decreases and hence the pressure increases, i.e., pressure recovery occurs. The high pressure vapor enters the condenser (state point 4) where it condenses to saturated liquid (state point 5). A part of the saturated liquid refrigerant is pumped to the boiler (state point 1) and thus completes the power loop. Another part of the saturated refrigerant liquid is throttled to evaporator pressure (state point 6) and enters the evaporator. In the evaporator it absorbs the heat from the refrigerated space and comes out as saturated vapour (state point 3). This vapour is induced into the mixing chamber of the ejector. Thus, it completes the refrigeration loop. Figure 7.11 shows the different components in an ejector and the variation of pressure in the ejector is shown in Fig. 7.12.

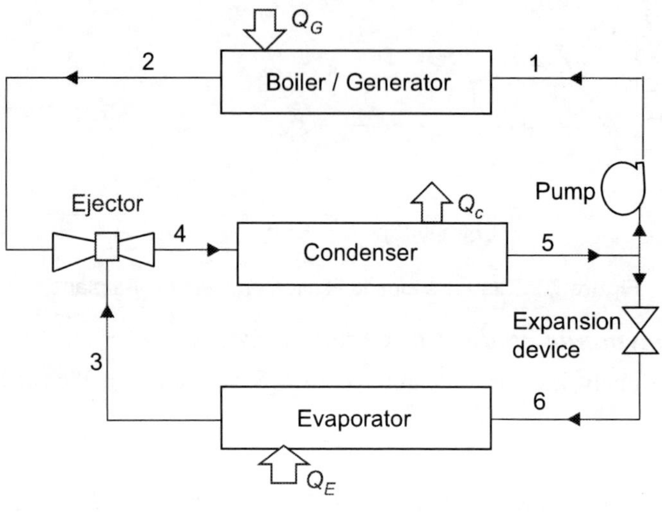

Figure 7.10: Schematic of an ejector refrigeration system.

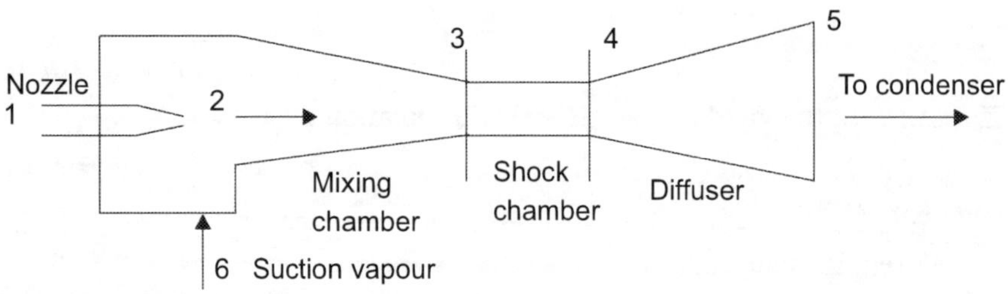

Figure 7.11: Schematic of components of an ejector system

Dry and saturated, high pressure steam, otherwise called *motive steam*, from the boiler expands through the steam nozzle. So it attains a high velocity and its pressure drops from P_1 to P_2. The high velocity steam jet entrains the flash vapour from flash chamber (point 6). They get mixed up at constant pressure in a process 2 to 3. As the velocity is high, i.e., more than sonic velocity, a shock occurs before entering the diffuser section. For the shock to remain stable, a section of constant area of cross section is provided (3 to 4). There the pressure rises from P_3 to P_4 and the velocity comes down. As it enters the diffuser section, its pressure increases from P_4 to P_5. Then this high pressure steam enters the condenser.

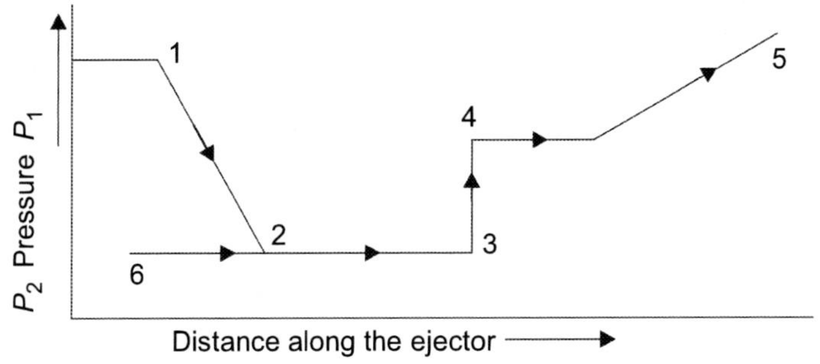

Figure 7.12: Pressure variation along the length of the ejector

This system has the advantages like there is no moving part, it uses waste heat or solar heat and environment-friendly refrigerants. The main disadvantage is that its COP is very low, i.e., of the order of 0.2 or less. Now this system is not commercially available. But some companies have developed the ejector system using steam as the working fluid and so this system can only be used for the temperature above 0°C.

7.5 VORTEX TUBE REFRIGERATION

Kinetic theory of gases says the molecules of a gas move in a random manner with velocities of different magnitudes in different directions. Some molecules have higher velocities whereas others have lower velocities. Temperature is a measure of the root mean square velocity of the molecules. Hence, the faster moving molecules do have higher temperature than the slower moving molecules. This is the physics behind the 'vortex tube refrigeration'. It is required to develop technique to segregate high temperature molecules from low temperature molecules so that hot air and cold air can be obtained from two different outlets.

(Refer Fig. 7.13) Compressed air at pressure, P_C, is supplied through the nozzles which are located tangentially to the internal surface of the vortex tube. These nozzles may be convergent, divergent or convergent-divergent type. Nozzles and nozzle chambers are designed in a manner to impart high velocity, spiral or swirling motion to the compressed air. The high swirling motion of the compressed air causes high velocity molecules (also high temperature) to move to periphery and the low velocity molecules (low temperature) to remain in the core. As it travels along the length of the tube, the air molecules lose some amount of kinetic energy due to friction with the wall of the tube and the generated heat makes the peripheral air warmer. This segregated air reaches the right end where there is a throttle valve. Depending upon the position of the throttle valve (i.e., fraction of opening), some of the warmer air escapes to the hot air collection tank. The rest amount of air stops. Velocity becomes zero and there is rise in pressure, i.e., stagnation pressure is developed. It can also be said that 'there is development of back pressure'. This pressure is higher than the pressure at the nozzle inlet. Hence, there is a reverse flow of air from the right end (hot air end) to the nozzle side in the core region. Rotational motion is in the same direction, only the motion in axial direction is reversed. As this air travels, the centrifugal force ($F = mr\omega^2$, where r is the radius of rotation and ω is the angular velocity) on the air molecules away from the centre is higher than that near to the centre. Thus, there is an increase in pressure from centre to periphery and the air mass, in the periphery, is in a compressive state. At this state there is a momentum and energy transfer from core mass to the peripheral mass. Due to this act the core mass loses energy and becomes cooler and on the other hand the outer mass gains energy and becomes warmer. Then this core mass reaches the left end where there is an orifice whose diameter is nearly half of the tube diameter. The core mass, i.e., cold air passes through this orifice and is collected in the cold chamber. Again the peripheral mass strikes the wall and gets back to mix with the fresh compressed air being delivered at the nozzle. Thus, we get hot air at one end and the cold air at the other.

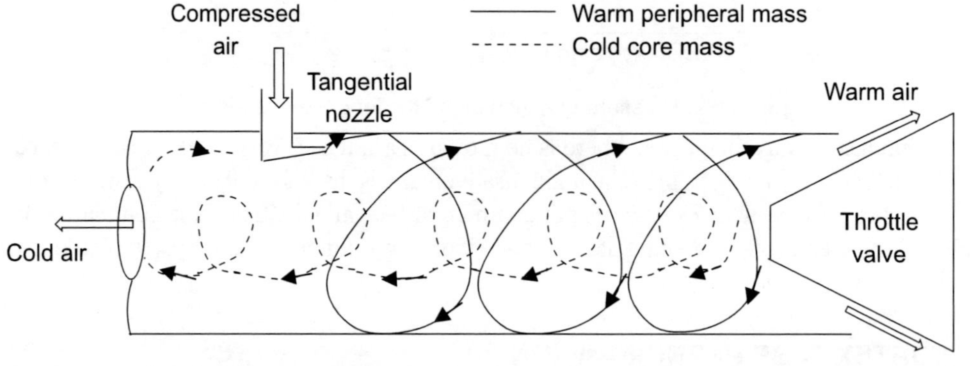

Figure 7.13: Schematic of a vortex tube refrigeration system

It is observed that if the throttle valve is fully closed, all air (warm and cold) will strike back and there will be no cooling effect. The opening of the throttle valve causes warm air to go out being separated and the cold air is received at the orifice. Cold air flow rate and temperature both decrease. The temperature is found to be minimum when the cold air flow rate is 25% to 30% of the total flow rate. Further decrease in flow rate increases the temperature of the cold air.

In summary, peripheral mass gets warmer due to following reasons:

1. When a rotary or swirling motion is imparted to a mass of air, some of the air molecules gain higher kinetic energy and move to periphery and rest with less kinetic energy remains in the core. Thus, there is a division of molecules on the basis of kinetic energy. The peripheral mass having higher kinetic energy will have higher temperature and the core mass having lower kinetic energy will have lower temperature.
2. Increase in temperature of the peripheral mass is also due to the viscous dissipation of energy (frictional energy) between the mass and the tube wall.
3. Compression of the peripheral mass arising out of centrifugal force due to the swirling/ rotational motion of the compressed air.

Core mass gets cooler due to following reasons:

1. Due to transfer of momentum from core to periphery, core mass becomes cooler.
2. Due to rotary motion, there develops a pressure gradient in radial direction. The periphery is at higher pressure due to maximum centrifugal force and the whole mass in a compressive state. Thus, the core mass does work on peripheral mass, i.e., peripheral mass acts as a sink to the core mass. So the core mass becomes cool. But in turbine, work is done due to expansion/reduction of pressure and the cooling effect is produced.

This refrigeration system is suitable for spot cooling. During machining of materials, where coolant is not used, vortex tube refrigeration system is used. The costumes of miners, spray painters can also use this refrigeration system for cooling purpose.

7.6 MAGNETIC REFRIGERATION SYSTEM

Emil Gabriel Warburg, a German physicist was first to observe, in 1881, the magnetocaloric effect in an iron sample, which got heated a few milikelvins when moved into a magnetic field and cooled down again when removed out of it. The magnetocaloric effect is a phenomenon in which a temperature change of a suitable material is caused by exposing the material to a changing magnetic field. This is also known as *adiabatic demagnetization.* This thermal response of a solid to the application or removal of magnetic field is maximized when the solid is near its magnetic ordering temperature. Temperature very near to absolute zero may be obtained by adiabatic demagnetization of certain paramagnetic salts. Currently, alloys of gadolinium producing 3-4 per tesla of change in magnetic field is used for magnetic refrigeration.

The basic principle is represented in Fig. 7.14. Each atom of the paramagnetic salt may be considered to be a tiny magnet. Normally, the atoms or the magnets are randomly oriented such that the net magnetic force is zero. When exposed to a strong magnetic field, the atoms align themselves to the direction of the magnetic field. Now if the magnetic field is suddenly removed, the atoms come back to their original random orientation. This requires work to be done at the expense of the internal energy. As a result the salt cools.

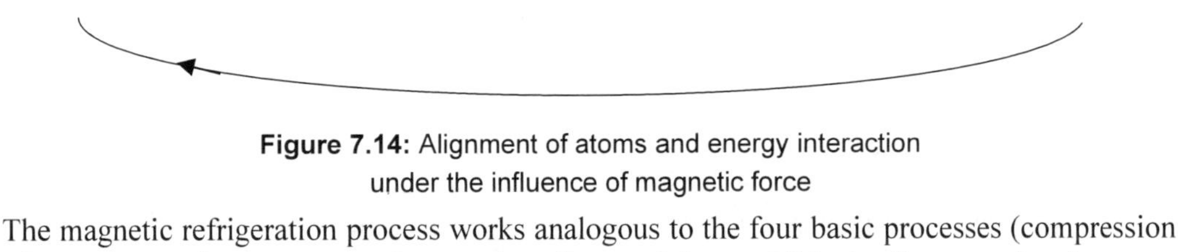

Figure 7.14: Alignment of atoms and energy interaction under the influence of magnetic force

The magnetic refrigeration process works analogous to the four basic processes (compression of gas, heat rejection, expansion or throttling and heat addition) of conventional vapour compression cycle as shown in Fig. (7.15).

Adiabatic magnetisation: A magnetocaloric substance is placed in an insulated environment. The increasing external magnetic field ($+H$) causes the magnetic dipoles of the atoms to align in the magnetic field direction and the atoms lose their energy. Since there is no heat lost due to the adiabatic condition, the energy given up by the atoms is stored as heat in the magnetocaloric material and increases the temperature ($T + \Delta T$).

Heat rejection: Heat stored is removed by passing some coolant, may be liquid helium or gas. The magnetic field is held constant so that reabsorption of heat by the dipoles can be prevented. When gets cooled (T), the cooling medium is removed.

Adiabatic demagnetisation: The substance is returned to another adiabatic condition so that total energy remains constant. As the magnetic field is removed, the atoms again get back to their original random orientation by the expense of the internal energy. Thermal energy transfers to the magnetic energy. Thus, the material gets cooled ($T - \Delta T$).

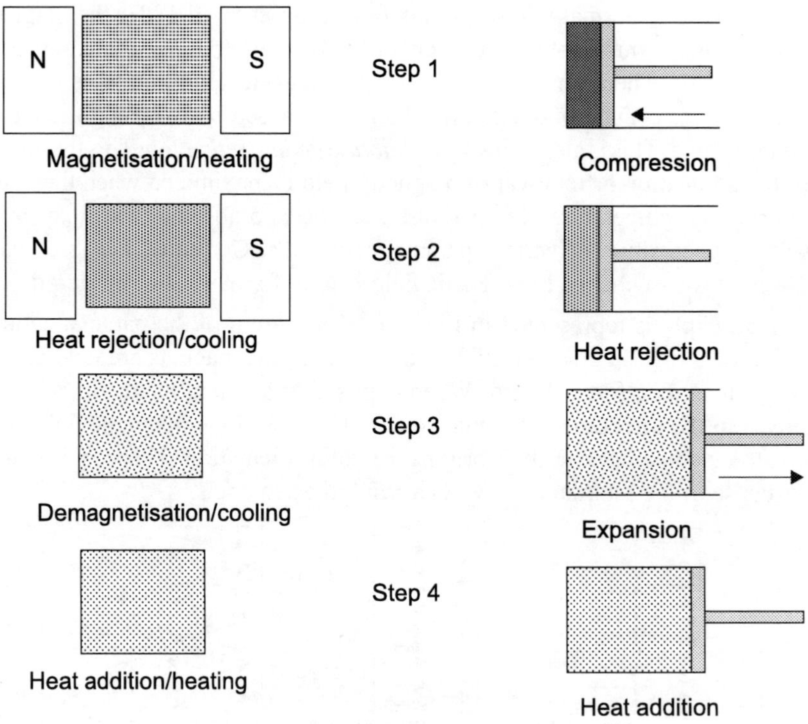

Figure 7.15: Comparison among the processes of magnetic refrigeration with that of the vapour compression refrigeration

Heat addition: When this material comes in contact with the environment to be cooled, it absorbs the heat from the surroundings and the temperature of the magnetocaloric material becomes T and the cycle restarts. However, this follows the processes of a reverse Brayton cycle. If the rotor in the magnetic refrigeration system rotates fast, the temperatures of the processes during heat rejection and heat addition cannot vary significantly. Thus, the reverse Brayton cycle approaches reverse Carnot cycle.

Magnetocaloric material: The most important factor for magnetic refrigeration system is the availability of suitable magnetocaloric material. Paramagnetic salt, such as cerium magnesium nitrate, is the most useful refrigerant. Eventually, paramagnetic salts become either diamagnetic or ferromagnetic, limiting the lowest temperature.

Presently, widely used magnetocaloric material is gadolinium (Gd) and its alloys with silicon (Si_2) and germanium (Ge_2). Besides this, lanthanide (La) and its alloys with aluminium, cobalt, nickel are of use as magnetocaloric material. These are very expensive because they are the rare earth metals. Heusler alloys or nickel and manganese based alloys (Ni_2Mn-X, where, X may be Ga, Co, In, Al, Sb) can be used as magnetocaloric material for magnetic refrigeration system.

Disadvantages of this system

1. Non-availability of good magnetocaloric material with Curie point at room temperature and above. (Curie temperature or Curie point is the temperature above which certain magnetic materials undergo a sharp change in their magnetic properties or lose their permanent magnetic properties. It is just like the critical points for refrigerants. If the Curie point is at room temperature or above, maintaining a temperature lower than this will not change the material's magnetic properties.)
2. In this system the problem is that, unlike heat transfer process in conventional refrigeration system, the transport mechanism for heat is the slow molecular diffusion. Therefore filigree porous structures are considered where a heat transport fluid takes the heat and transports it away from the material's surface.
3. The irreversibilities like magnetic hysteresis loss and transport fluid frictions are to be made minimum.

Advantages of this system

1. The temperature nearing 0 K (even up to 0.5 K) can be obtained.
2. The COP, in the range of 80% of RC cycle, can be obtained if losses can be made minimum. This value of COP is, obviously, higher in comparison to the other conventional refrigeration systems.

7.7 PCM BASED COOLING TECHNOLOGY

In this technology, phase change materials (PCMs) are used to obtain desired cooling amount. They absorb large amount of energy as latent heat as they undergo a reversible phase transition from solid to liquid. Very common example is storing of food and beverages in ice. If food can be preserved at 10°C, it can be done so till the ice reaches 10°C. For this ice has to absorb (latent heat of 80 kJ/kg + C_P of water 4.19 kJ/kg-K × ΔT 10°C) 121.9 kJ/kg amount of heat. So, how long 1 kg of ice can preserve the food at 10°C can be calculated. But in designing PCM thermal control systems for specific applications a number of factors other than heat of fusion must be considered. These include thermodynamics, heat transfer and various physical and thermodynamic properties of PCMs should be considered. In selection of a PCM for a specific application, the most important criterion is the operational temperature range of the components to be cooled. The PCM must have a melting point temperature well within this range to ensure that unintentional undercooling or overheating will not damage the components. Property characteristics are as suggested here. Heat of fusion – High; Thermal conductivity – High; Specific Heat – High; Density – High; Volume change during melting – Low; Vapour Pressure – Low; Melting and Freezing Behaviour – Dependable and Reversible. It should be non-toxic, compatible with container material, easily available and of low cost.

When to use PCM: Suppose a component peak temperature is 100°C. It is to be cooled down to 40°C. It can be achieved by using only radiator. But the radiator size and mass will be higher. Instead a PCM material having melting temperature 60°C can be used for initial heat absorption. It absorbs the latent heat of fusion and may take some amount of sensible heat. Thus, reaching a temperature of 65°C. For this required mass of PCM is to be calculated. By this the component will reach a temperature of 70°C. Then cooling from 70°C to 40°C is done in the radiator. Thus, the radiator size and mass is reduced very much. So before using a PCM the first question to be answered is whether the added mass of the thermal storage (PCM) is less than the mass saved by reducing the radiator size. For this the knowledge of the nature of heating and cooling (duty cycle) is required.

Possible uses of PCM: PCM based cooling technology or sometimes called solid heat sink is used for cooling Computer electronics, cellular phones, spacecraft and avionics (thermal control), microwave power transistor, solar energy storage, shoe cooler, etc. In space applications, the thermal load is considered to be cyclic. In this arrangement, when the component generates heat during the "On" portion of its cycle, the energy is stored via phase change in the PCM. During the "Off" portion the latent heat of fusion is removed by radiator, heat pipe, or other means to refreeze the PCM in preparation for the next "On" portion. The alternate melting and freezing of the PCM enables the component to operate very nearly isothermally at all times. (Fig. 7.16).

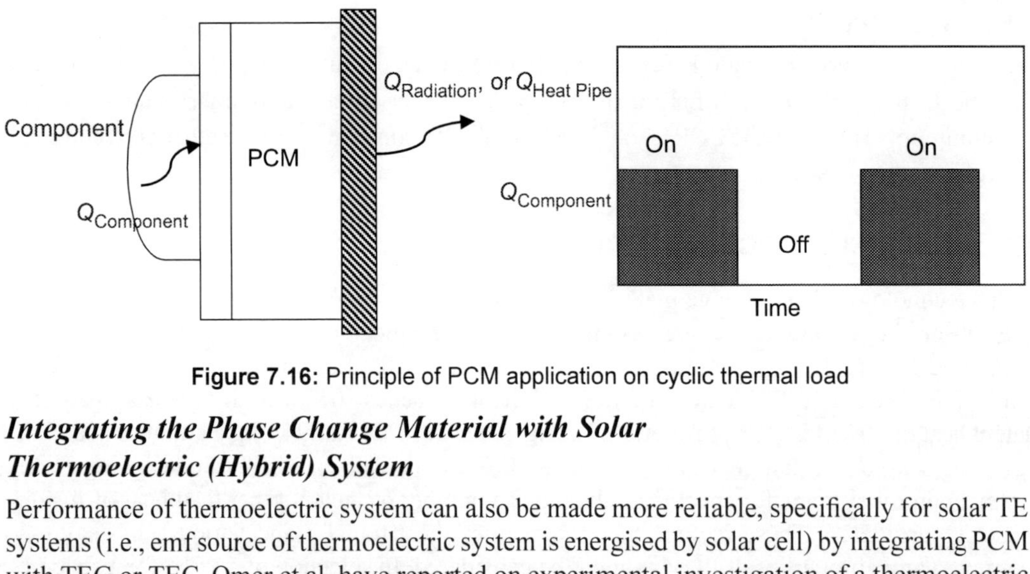

Figure 7.16: Principle of PCM application on cyclic thermal load

Integrating the Phase Change Material with Solar Thermoelectric (Hybrid) System

Performance of thermoelectric system can also be made more reliable, specifically for solar TE systems (i.e., emf source of thermoelectric system is energised by solar cell) by integrating PCM with TEG or TEC. Omer et al. have reported on experimental investigation of a thermoelectric refrigeration system integrated with PCM along with thermal diode. Heat is rejected by PCM during cooling process and gets solidified. Total heat rejection by PCM will be summation of its sensible heat and the latent heat of fusion. During 'Off period' PCM absorbs latent heat and sensible heat from the refrigerated/cooling cabinet and keeps it at the desired temperature. Thermal diode is used so that heat will not be leaked to the PCM when there will be power off. Studies reveal that a salt, mixture of sodium nitrate and potassium nitrate, may be used with thermoelectric generator to act as the heat source as its melting temperature is nearly 270°C.

7.8 ADSORPTION REFRIGERATION SYSTEM (ARS)

Like vapour absorption refrigeration system, this is also a heat engine driven refrigeration system. The basic ARS consists of four main components: (1) an adsorbent bed, (2) a condenser, (3) an expansion valve, and (4) an evaporator. An adsorbent bed contains highly porous bodies called adsorbent with very large internal surface (such as silica gel, activated carbon, zeolite, etc.). The adsorbent has high adsorptive property to certain gases (such as water, ammonia, methanol, etc.) called adsorbate or refrigerant. The adsorbent is packed on a metal surface which is required to conduct the heat transferred from/to a heat transfer fluid that flows in the bed. The functions of other three components are similar as they do in the vapour compression refrigeration system. The adsorbent bed is, sometimes, called the *thermally driven compressor* which circulates the refrigerant in an adsorption cycle by periodical switching between heating and cooling heat transfer fluids.

Four ideal thermodynamic processes that constitute the adsorption refrigeration cycle is described below. (Refer. Fig. 7.17 for schematic diagram and Fig. 7.18 for thermodynamic process on P-T plane)

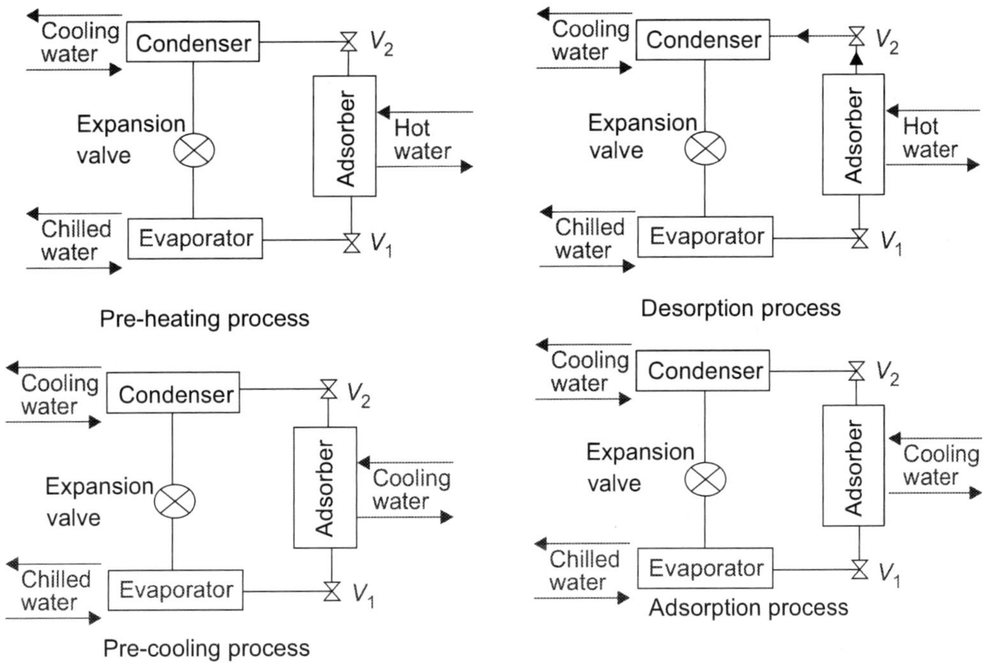

Figure 7.17: Schematic of an adsorption refrigeration system

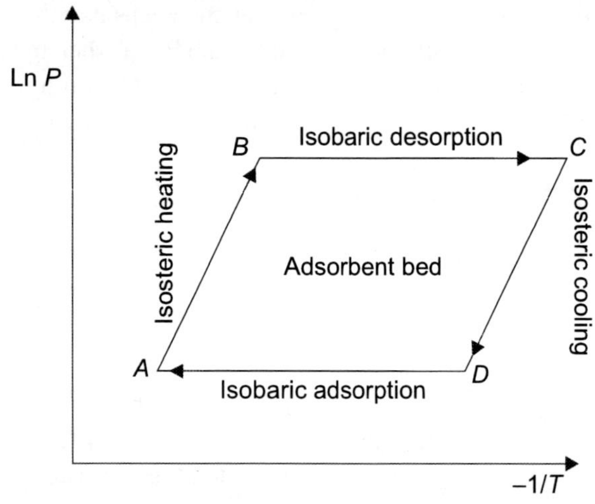

Figure 7.18: Representation of adsorption refrigeration system on *P-T* plane

Pre-heating process: Before beginning of this process, the valves between the adsorbent bed and both of the condenser and evaporator (V_1 and V_2) are kept closed. Thus, the adsorbent bed becomes a closed system. Then the bed is sensibly heated to certain temperature T_1 by any non-conventional heat source (solar, geothermal, etc.) or waste heat (industry exhaust, etc.) directly or indirectly through the HTFs. Since the adsorbent bed is closed and the adsorbate or refrigerant in adsorbed phase remains constant at the maximum concentration, the temperature and pressure in the adsorbent bed rise. This process lasts until the vapour pressure in the adsorbent bed reaches the condenser pressure.

Desorption process: After the adsorbent bed reaches the condenser pressure valve (V_2) is opened. The heating process of adsorbent bed is still going on. This induces adsorbed refrigerant vapour to get vaporised from the solid surface of the adsorbent bed. This vaporisation process is called desorption. The refrigerant vapour flows to the condenser and condenses there. This process continues till the adsorbent bed reaches the minimum concentration level. The vapour pressure of the bed is considered to remain equal with that of the condenser during this process.

Pre-cooling process: After the bed reaches the maximum temperature, the valve (V_2) is closed. Now the adsorbent bed becomes a closed system. Cooling water is made to flow through the bed. Cooling process reduces the vapour pressure in the bed to that of the evaporator. During this period high pressure liquid refrigerant from the condenser flows through the expansion valve and its pressure reduces to evaporator pressure. The liquid and vapour mixture of refrigerant flows to the evaporator.

Adsorption process: During this process the valve between the adsorber and evaporator (V_1) is opened. The refrigerant vapour generated due to the heat absorbed by the evaporator from the conditioned space (refrigerating effect) is directed to the adsorbent bed. The vapour is adsorbed in the adsorbent bed. The adsorption of adsorbate or refrigerant increases with the decrease in temperature. That is why the bed temperature is kept low during this process by continuing the circulation of coolant through the bed. The coolant takes up both (a) the residual heat of the pre-heating and desorption process, and (b) the heat generated due to adsorption process.

Thus, the adsorption refrigeration cycle is completed. COP of the system is calculated from the following equation.

$$COP = Q_{evap}/(Q_{Pre\text{-}cooling} + Q_{Desorption}) = [1 - (T_2/T_1)] / [(T_2/T_1) - 1]$$

Properties of material pairs: The design and performance of an adsorption refrigeration system primarily depends upon the working pair of materials. A good adsorbent should have wider range of adsorption capacity with temperature variation, higher heat and mass transfer properties along with thermal stability. The refrigerant or adsorbate should have the usual desirable properties as discussed earlier. In this system, normally used adsorbent and adsorbate pair is silica gel and water. The other pairs are zeolite-water, activated carbon and methanol, activated carbon – ammonia, etc. Generally, on the basis of the driving force involved in the adsorption process, the adsorbent-adsorbate pairs have been classified into three categories. (a) physical; (b) chemical; and (c) composite.

Physical adsorbent-adsorbate pair: Silica gel-water, zeolite-water, activated carbon-ammonia, activated carbon-methanol, etc., are the physical adsorbent-adsorbate pairs. In these pairs van der Waal's force is responsible to contain adsorbate. Silica gel is an amorphous silicon dioxide and has a granular, vitreous and highly porous form. It has a pore diameter in the range of 2.0 to 3. Nm and the pore volume 0.3 to 0.4 cm^3/gm. Specific surface area is 400 to 700 m^2/gm. The maximum adsorption capacity at equilibrium is between 0.35 and 0.4 kg/kg of silica gel. The adsorption heat is 2500 to 2800 kJ/kg and the desorption temperature could be maximum 120°C. The main advantage of this adsorbent is that its regeneration temperature is around 85°C which makes it suitable for solar energy use and low temperature waste heat sources.

Zeolites are microporous alumina silicate crystals composed of alkali or alkali soil. This pair desorption temperature ranges from 70 to 250°C. Its adsorption heat is 3300 to 4200 kJ/kg. This higher adsorption heat leads to lower COP in comparison to silica gel-water system.

Activated carbon has a large specific area for adsorption between 800 to 1500 m^2/gm. Its adsorption heat is nearly 1800 to 2000 kJ/kg which is lower than the other two. This leads to higher COP of the system in comparison to other two. Activated carbon-methanol system has a higher specific adsorption rate of 0.55 kg/kg of adsorbent. It has a regeneration temperature of around 100°C. The demerit of activated carbon system is that it has low thermal conductivity.

Chemical adsorbent-adsorbate pair: The strong chemical bond between the adsorbent and the adsorbate takes place in chemical adsorption process. The uptake in the chemical adsorbents is not limited by the available surface area of the material which leads to higher mass transfer in comparison to physical adsorbents. Metal chlorides are commonly used as chemical adsorbents. Some of the examples are calcium chloride, strontium chloride, magnesium chloride, barium chloride, etc. The specific adsorption capacity approaches 1 kg/kg of adsorbent and desorption temperature is 40 to 80°C. However the stability of this system is lower due to agglomeration and swelling phenomena of the adsorbent bed. This instability reduces the heat and mass transfer which limits the cooling capacity of the chemical adsorption system.

Composite adsorbent-adsorbate pair: Composite adsorbents are made by combining the physical and chemical adsorbents in a proportion so that the drawbacks of both the systems can be minimized. In such composites, physical adsorbents (porous media) improve the heat and mass transfer characteristics of the chemical adsorbents while limiting their swelling characteristic and chemical adsorbents increase the uptake of the adsorbent pair. Common example is silica gel with metal chloride. The silica gel is immersed in a chloride salt solution and then dried up remove water. Performance of the composite adsorbents can be enhanced by (i) coating the adsorbent over the heat transfer metal surfaces of the adsorber so that thermal contact resistance can be eliminated, and (ii) by using novel adsorbent materials known as metal organic frameworks (MOFs) which are highly crystalline porous materials and have large inter-particle voids that provide an attractive adsorption characteristic compared to common adsorbents.

EXERCISE

1. List the effects observed when two dissimilar metals are connected and junctions are held across an emf.
2. List the effects observed when two dissimilar metals are connected and junctions are held at two different temperatures.
3. What is the significance of Thompson effect in thermoelectric refrigeration?
4. State the principle of thermoelectric refrigeration system.
5. Define 'figure of merit' and its significance.
6. Discuss the factors affecting 'figure of merit'.
7. Discuss the advantages and disadvantages of TE refrigeration over VC refrigeration system.
8. What replaces the current of TE system in VC system?
9. Why the semiconductors are considered to be best materials for TE refrigeration?
10. What are the desirable properties for a thermoelectric material?
11. State some material combinations used for thermoelectric refrigeration system.
12. Derive an expression for amount of current to obtain maximum COP.
13. Derive an expression for amount of current to obtain maximum cooling effect.
14. Why sudden expansion of liquid at high pressure causes reduction in temperature of liquid?

    [**Ans:** Some liquid flashes into vapour as the liquid pressure suddenly decreases in adiabatic condition. Because the saturated temperature of liquid, corresponding to the decreased pressure, decreases. The liquid becomes vapour by absorbing latent heat from the surrounding liquid, thereby decreasing the temperature of the liquid.]
15. Is it always true that whenever a fluid is throttled, its temperature decreases? State the criteria for a fluid to be cooled when throttled.
16. Why is it required to liquify a gas? Write the basic principle of gas liquification. Write some of the applications of liquified gases.

MULTIPLE–CHOICE QUESTIONS

1. Refrigeration is produced in a thermoelectric system due to

(a) Seebeck effect (b) Peltier effect

(c) Joulean effect (d) Thompson effect

2. EMF or electricity is produced in a thermoelectric system due to

(a) Seebeck effect (b) Peltier effect

(c) Joulean effect (d) Thompson effect

3. Thermoelectric effect is more observed in

(a) conductors (b) insulators

(c) semiconductors (d) superconductors

4. Which element is considered the working fluid for thermoelectric refrigeration system?

(a) Electron (b) Proton

(c) Neutron (d) Nucleus

5. Performance of thermoelectric refrigeration system mostly depends upon

(a) current passed through it

(b) figure of merit

(c) temperature difference between hot and cold end

(d) all of the above

6. It is said to be cryogenic study when the temperature is below

(a) –100°C (b) –125°C

(c) –150°C (d) –200°C

7. Expansion through turbine is said to be —— expansion process.

(a) restrained (b) unrestrained

(c) open (d) centrifuged

8. Liquefaction of gases follows the inference drawn from porous plug experiment conducted by

(a) Kelvin (b) Kelvin-Plank

(c) Dalton (d) Joule-Thompson

9. When a gas is throttled, the pressure and temperature of the gas

(a) always decreases

(b) always increases

(c) pressure decreases but temperature may decrease or increase

(d) pressure increases but temperature may decrease or increase

10. In the expression $\mu = (\partial T/\partial P)h$, μ is the Joule-Thompson coefficient. Cooling is produced when

(a) $\mu = 0$, (b) $\mu > 0$,

(c) $\mu < 0$, (d) $\mu = 1$

ANSWERS

1. (b) **2.** (a) **3.** (c) **4.** (a) **5.** (b) **6.** (c) **7.** (a) **8.** (d) **9.** (c) **10.** (b)

8

Psychrometry

8.1 INTRODUCTION

Air conditioning is the science and practice of creating a desirable controlled climate in indoor space. This desirable controlled climate has become an essential part of life due to globalisation of industrial, agricultural activities and drastic changes in the climatic conditions. For example, a desirable controlled climate is essential to grow apple tree in a hot climatic condition. Besides we need desirable controlled climate to preserve food, medicines and other essential products for long-term use. Air conditioning has a number of applications in industries, agriculture, health sector, space research, etc. 'Comfortable stay' is the most popular application of air conditioning. These applications, in detail, are discussed in Chapter 11. However, desirable conditioning of air is done by controlling the following factors.

1. Temperature
2. Humidity
3. Air movement
4. Ventilation
5. Cleaning (pure and odour free)

Besides the above 5 factors, there is a need for varying the constituents of atmospheric air. Normal atmospheric air is a mixture of dry air and water vapours. Dry air consists of 21% of oxygen, 78% of nitrogen, 0.04% of carbon dioxide, gases like argon, 0.9% and other gases like helium, neon, etc., are 0.06%. In some cases, constituents of air need to be changed for preserving of some articles. Take the case of apple preservation. Apple is to be stored at a temperature of 0°C to 1°C and 90% relative humidity. At the same time to keep it fresh for a longer period the natural enzyme activity should be retarded so that rate of ripening of the apple gets retarded. This can be done in an air-tight refrigerated room which is sealed after the fruit is placed in it. The oxygen content in the store room is reduced from 21% to 2.5% and carbon dioxide level is increased from 0.25% to 2.5%. In this context it can be said that it is difficult to preserve non-living products like meat, fish, etc., because they are more susceptible to spoilage. Therefore, they are normally stored at a temperature below –15°C.

It is essential to know the thermodynamic properties of available atmospheric air which is a mixture of water vapours and dry air so that the above said parameters can be controlled by different thermodynamic processes.

8.2 PSYCHROMETRY

The available air in the atmosphere is considered to be a mixture of dry air and water vapours, otherwise called 'moist air' in the study of air conditioning. The study of the thermodynamic properties of the moist air is called 'psychrometry'. The water vapour has its partial pressure in the moist air. So the total pressure of moist air (i.e., atmospheric pressure), from Dalton's law of partial pressure, is the summation of the partial pressure of dry air and that of the water vapour.

$$P_{atm} = P_a + P_v \quad [8.1]$$

Atmospheric pressure at sea level is considered to be at 1.0 bar or 101.325 kPa. As the altitude of a place goes on increasing, the pressure and temperature goes on decreasing. This can be expressed by the following empirical relations.

Temperature, $t = t_o - 0.0065z$; t is in °C, t_o is 15°C and z is the altitude in m.

Pressure, $P = P_o\,(1 - 0.0065z/T_o)^{5.256}$; T_o is in K, P is in kPa and P_o is the atmospheric pressure at sea level, i.e., 101.325 kPa.

If the partial pressure of water vapour corresponds to the saturation pressure of water vapour at the mixture temperature, the vapour is said to be saturated. Saturated air is a mixture of dry air and saturated water vapour. The partial pressure of water vapour in the moist air is very low (i.e., from steam tables the saturated vapour pressure at atmospheric temperature 20°C is nearly 2.3 kPa). So the water vapour remains in superheated vapour state in the air even at a lower temperature. And for this reason, the water vapour is assumed to be an ideal gas in moist air and accordingly property calculations are carried out.

8.2.1 Parameters in Psychrometry

- **Dry bulb temperature (DBT):** It is the actual temperature of moist air as recorded by a thermometer and denoted by t.
- **Specific humidity/humidity ratio:** It is the mass of water vapour (in kg) per kg of dry air and expressed as

$$(\omega) = m_v/m_a \quad [8.2]$$

Its unit, for this textbook, is expressed in kg/kga

From universal gas law, $PV = mRT$; $m_v = P_vV/R_vT$ and $m_a = P_aV/R_aT$

So $\omega = (P_v/R_v) / (P_a/R_a)$

$R_v = 0.461$ kJ/kg-K, $R_a = 0.287$ kJ/kg–K

So $\omega = 0.622P_v/P_a = 0.622P_v/(P_{\text{total}} - P_v)$ [8.3]

P_{total} = Constant and its value is the atmospheric pressure of that place.

So it can stated that $\omega \;\alpha\; P_v$.

Sometimes, humidity ratio at saturation (ω_S) is explained as follows. As water vapour is continuously added to the unsaturated moist air at 2 (vapour pressure P_{v2}), then the partial pressure of water vapour goes on increasing to P_{v1}, state point 1, and so on (Fig. 8.1). The maximum partial vapour pressure will be at point '*S*' where the moist air becomes saturated (vapour pressure is $P_{v,\text{sat}}$). Thus, at state '*S*' the partial pressure of water vapour is $P_{v,\text{sat}}$, the saturated vapour pressure at temperature *t*, and it is the maximum partial vapour pressure that water vapour at this temperature can achieve. The moist air at this state is called 'saturated air'. $\omega_S = 0.622\, f_S P_{v,\text{sat.}}/(P_{\text{total}} - P_{v,\text{sat.}})$; where f_S is a function of temperature. For approximate calculation, with negligible error, the value of f_S can be taken 1.

So $$\omega_S = 0.622\, P_{v,\text{sat.}}/(P_{\text{total}} - P_{v,\text{sat.}}) \quad [8.4]$$

- **Saturated air:** The air is said to be saturated when the partial vapour pressure in the moist air becomes equal to the saturation pressure of water vapour at the moist air temperature. When the air gets saturated, any decrease in temperature results in condensation of the water vapours into liquid. The same is also explained in the above chapter 'humidity ratio at saturation'.
- **Degree of saturation (μ):** This represents the capacity of moist air to absorb water vapour. It is defined as the ratio of humidity ratio at temperature *t* to the humidity ratio of saturated air at same temperature *t* and expressed as (ω/ω_S).

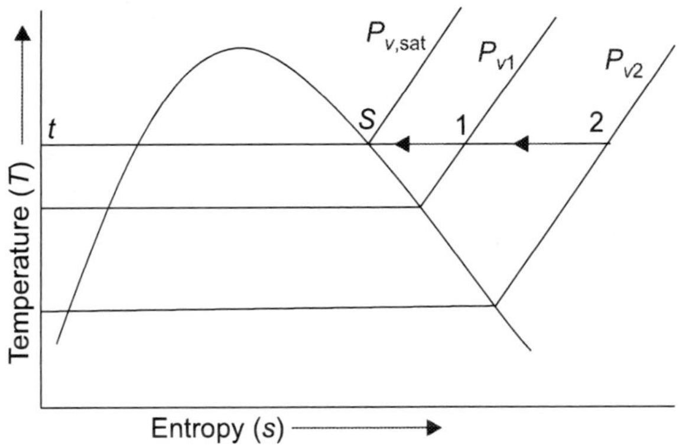

Figure 8.1: For saturated air at state '*s*', the partial pressure of water vapour is equal to the saturation pressure of water

- **Relative humidity (ϕ):** It is the ratio of the mass of water vapour present in a fixed volume of moist air to the mass of water vapour present in the same volume if the moist air is saturated at the same temperature and expressed as $\phi = m_v/m_{v,\text{sat}} = P_v/P_{v,\text{sat}}$ (from universal gas law). Now $\omega = 0.622 \times \phi P_{v,\text{sat}}/P_a$
- **Dew point temperature (DPT):** This is the temperature to which the temperature of the unsaturated moist air is to be brought down at the constant pressure so that the water vapour present in the moist air starts condensing, i.e., first dew drop will be formed. This is shown in the figure and denoted as t_{dpt}. Before condensation specific humidity and partial vapour pressure remain unchanged.
- **Thermodynamic wet bulb temperature:** (Fig. 8.2:) Let us take a water filled long trough. Put it in an insulated chamber. Moist air is made to flow over it. As the unsaturated air flows over the long trough, water from the trough vaporises and mixes with it. Thus at the outlet of the chamber the air becomes saturated. This is an adiabatic

saturation process of air. The latent heat of vaporisation is partly drawn from air and partly drawn from water itself. Thus, air and water both get cooled. After certain time, this saturator reaches a steady state condition. The temperature of water in trough after reaching the steady state is called *thermodynamic wet bulb temperature* and this temperature is same as that of the saturated air at the exit. This experiment shows that for any state of unsaturated moist air, there exists a temperature (t^*) at which the air becomes adiabatically saturated by the evaporation of water into air, at exactly the same temperature (t^*) into the air. This is a thermodynamic property of moist air. Energy balance for the moist air can be expressed as

Initial enthalpy (h) + enthalpy added due to evaporated water $(\omega^* - \omega)\ h_f^*$ = Exit enthalpy (h^*) [8.5]

where, superscript * is the values at the adiabatic saturation temperature t^*.

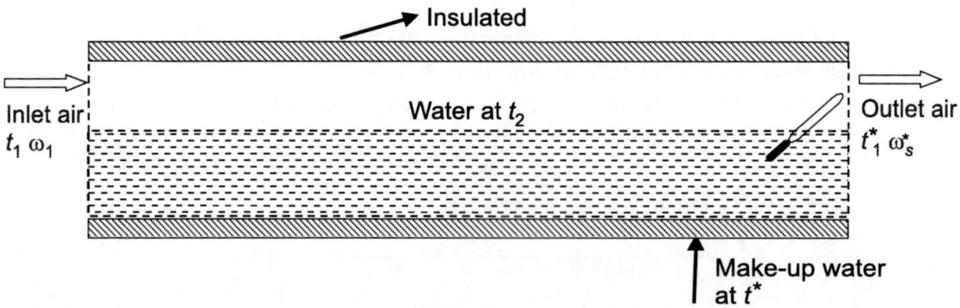

Figure 8.2: Process of adiabatically saturating the air

This method of measuring the WBT is difficult in practice. Hence, alternately wet bulb temperature of moist air is measured as described.

Wet bulb temperature (WBT) is the temperature t' recorded by a thermometer whose bulb is wrapped with a water soaked wick, in rapidly moving air (air velocity should be around 3.5 m/s). This wet bulb thermometer is normally available in combination with dry bulb thermometer and called psychrometer. Thermodynamic analysis shows that $t' \approx t^*$.

WBT is of importance in evaporative cooling because it is the minimum temperature to which air can be cooled in evaporative cooling. This also can tell how much air is saturated.

As per the 'straight line law' the condition of the moist air moves towards the saturation line at the temperature of the wet surface. This is true for air-water vapour mixture. Because this process is governed by a non-dimensional number called Lewis number (Le) and expressed as $h_c/h_D C_{Pa} = \alpha/D$, where, α = thermal diffusivity and D = mass transfer diffusivity and the value of this number for air-water vapour mixture is nearly equal to 1. Since this an evaporation process, the moist air moves along the constant enthalpy line (or same as constant WBT line). In the adiabatic saturator moist air gets saturated and reaches state point 2 (Fig. 8.3a) on saturated line whereas in case of wet bulb thermometer exit air may reach a state point 2 (Fig. 8.3b) due to lack of time and contact area to get saturated. But the wet surface temperature in both the cases remains same. So the measured temperature of sump water in case of adiabatic saturator (t^*) and that of wick (t') are same.

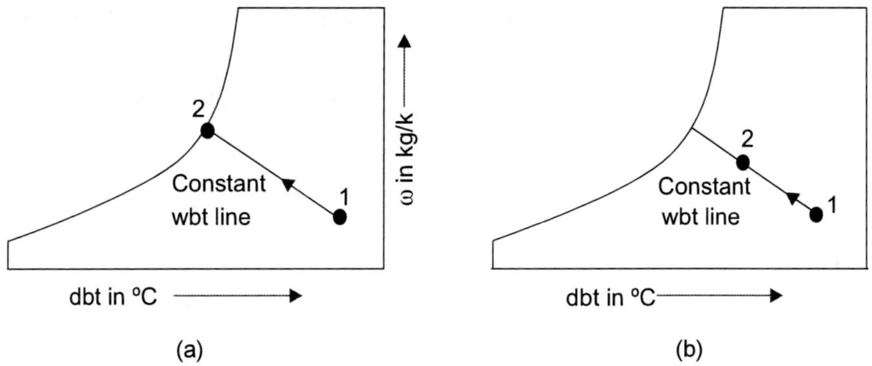

Figure 8.3: (a) 1-2 Path of moist air in adiabatic saturator; (b)1-2 in wet bulb thermometer

This statement is not valid for a mixture of air and vapour other than water.

- **Enthalpy of moist air:** This is calculated as the summation of the enthalpy of dry air and that of the water vapour present in the moist air.

Enthalpy of the dry air,

$$h_a = C_{Pa.} \times t = 1.005 \times t \qquad [8.6]$$

Enthalpy of water vapour can be calculated by three methods (Fig. 8.4)

(i) $C_{Pv.} \times t + h_{fg,\text{ water}}$ (at 0°C) (Process line 1-2*g*-0) [8.7a]

(ii) $C_{PW} t_{\text{dpt}} + h_{fg}(\text{dpt}) + C_{PV}(t - t_{\text{dpt}})$ (process line 1-1*g*-1*f*-1*R*) [8.7b]

(iii) $h_g(t)$ (Enthalpy of superheated vapour at temperature *t*, from steam tables)

Out of these three methods first one is most reliable.

$$\begin{aligned}\text{So enthalpy of moist air } &= C_{Pa}t_{.} + \omega\{(C_{Pv}t) + h_{fg,\text{ water}}\text{ (at 0°C)}\}\\ &= t\,(C_{Pa} + \omega \times C_{Pv}) + \omega \times h_{fg,\text{ water}}\text{ (at 0°C)}\\ &= 1.005 \times t + \omega\,(1.88t + 2500) = (1.005 + 1.88\omega)t + 2500\omega \qquad [8.8]\end{aligned}$$

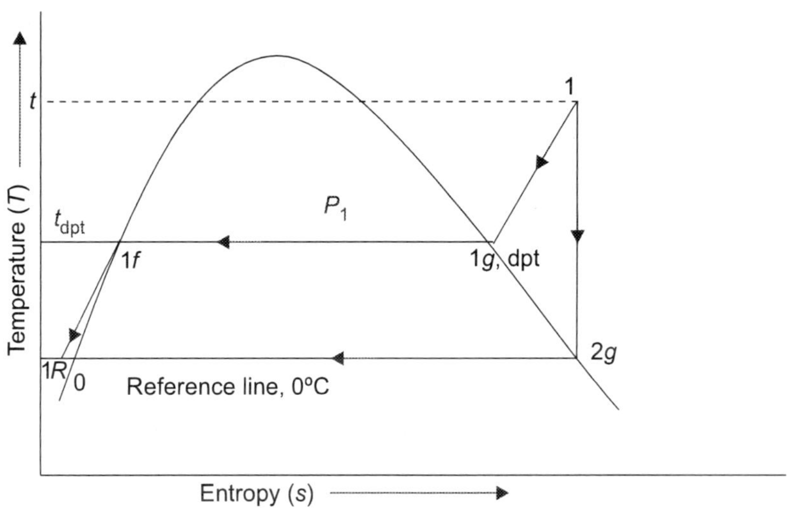

Figure 8.4: Shows the calculation of water vapour enthalpy.

To have easy calculation of enthalpy of moist air an average value of $\omega = 0.009$ kg of w.v. per kg of d.a. (for a usual range of air conditioning system) is considered.

Hence, the value of $(C_{Pa.} + \omega \times C_{Pv}) = 1.005 + 0.009 \times 1.88 = 1.0216$ kJ/kg-K

This value is defined as the '*humid specific heat*'.

Thus, the enthalpy of moist air is expressed as $h = 1.0216 \times t + \omega \times h_{fg,\,\text{water}}$ (at 0°C)

$$= 1.0216 \times t + \omega \times 2500 \text{ kJ/kga.} \quad [8.9]$$

This expression is mostly applicable for standard atmospheric condition, i.e. around 50% RH and 20°C.

Or it can be calculated from $h = 1.005 \times t + \omega \times h_g(t)$ from steam tables and this is more accurate than other expressions.

- **Latent heat of moist air:** Atmospheric air is a mixture of many non-condensable gases (i.e., non-condensable in the range of air conditioning application) like oxygen, carbon dioxide, nitrogen, etc., and water vapour. Hence, it is only the water vapour which liberate latent heat during its change of phase. So latent heat of moist air can be calculated as humidity ratio × latent heat of water vapour at that pressure and temperature.
- **Specific volume of moist air (v):** It is defined as the volume of the moist air per kg of dry air and this is calculated from Amagat-Leduc's law (for ideal mixture of gases). It states that the total volume of the mixture is the summation of the individual volume of a gas as if it exists alone at the mixture pressure and temperature. So the total volume of the moist air

$V = V_a + V_v = m_a R_a T/P + m_v R_v T/P = (T/P) m_a \{R_a + \omega R_v\}$ (since $m_v = \omega m_a$) [8.10]

So $V/m_a = \{R_a T/P\}\{(1 + (0.462/0.287)\omega\}$ (Putting the values of R_a and R_v)

V/m_a is the specific volume of the moist air (as per definition) $v = \{R_a T/P\}\{1 + 1.609\omega\}$ [8.11]

From perfect gas law approximation specific volume of moist air can also be taken equal to the volume occupied by 1 kg of dry air at P_a and T or volume occupied by the mass of water vapour (i.e., numerical value of specific humidity) at P_v and T.

Mathematically this can be expressed as

$$v = R_a T/P_a \quad [8.12]$$

or

$$v = m_v R_v T/P_v \quad [8.13]$$

Specific volume of moist air can calculated from following three relations.

(a) $v = \{R_a T/P\}\{1 + 1.609\,\omega\}$

(b) $v = R_a T/P_a$

(c) $v = m_v R_v T/P_v$

- Density of moist air (ρ) can be calculated from $m = m_a + m_v$. Dividing both sides by V_a
  $m/V_a = m_a/V_a + m_v/V_a\, \rho = \rho_a + (m_v/m_a)(m_a/V_a)$
  or $\rho = \rho_a + \omega\rho_a$, or, $\rho = \rho_a(1 + \omega)$ kg/m^3 of *da* [8.14]

8.3 CONSTRUCTION OF PSYCHROMETRIC CHART

Moist air is a mixture of dry air and water vapour. Hence, as per the phase rule at least three intensive properties are required to fix the state point of any of the components. In the psychrometric chart, there are two properties (dry bulb temperature on the abscissa and humidity ratio/partial pressure on ordinate) shown and the third property is the total pressure basing on which the chart is being constructed. Thus, a psychrometric chart constructed is applicable for a particular total pressure (i.e., atmospheric pressure) based on which it is constructed as shown in Eq. 8.3. For any other total pressure (let us say at higher altitude where atmospheric pressure is different from that at sea level) another psychrometric chart is to be constructed.

Psychrometric chart is constructed on the basis of the saturated water vapour line on the *P-T* plane as shown in the Fig. 8.5. In air conditioning calculations water vapour in the atmospheric air is in a superheated state. Hence, the right side of the line is taken for consideration only and the ordinate is taken on the right side as shown in Fig. 8.6. Secondly, as it is shown in Eq. (8.3) that pressure is directly proportional to humidity ratio. The knowledge of humidity ratio is more important than the partial pressure of water vapour in moist air in air conditioning calculations. So humidity ratio is taken on the ordinate of the chart.

Constant specific humidity lines (Fig. 8.6): This value is calculated by using Eq. 8.3 and is a horizontal line drawn parallel to abscissa to intersect the saturation line.

The saturation curve (Fig. 8.6): The saturation curve line of 100% RH can be plotted from the data in the steam tables at atmospheric pressure. The saturated water vapour pressure at a particular temperature is read from the steam table. By using Eq. 8.4, specific humidity at saturation (ω_S) is calculated. The intersection point of temperature(dbt) and ω_S is a point on the saturation line. Similarly, for other temperatures (ω_S) values are calculated and plotted. By Joining all those points, saturation curve is generated.

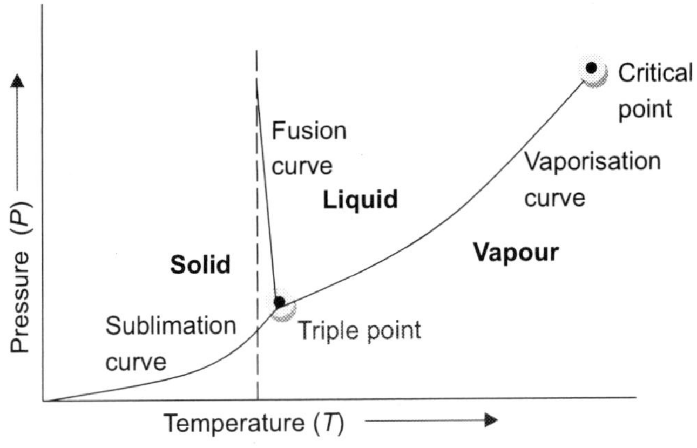

Figure 8.5: Equilibrium diagram on *P-T* plane of water

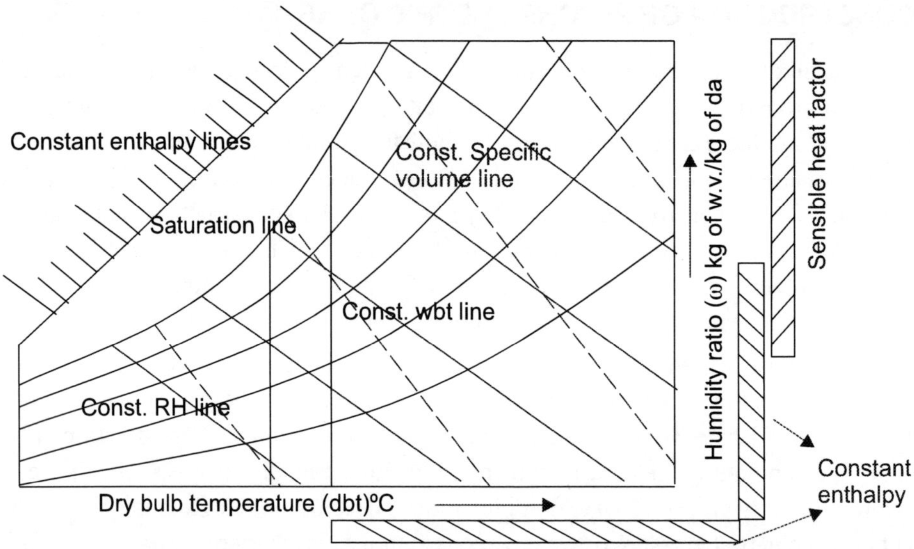

Figure 8.6: Psychrometric chart at an atmospheric pressure P_1

Constant relative humidity lines (Fig. 8.6): As shown in the figure, the linear distance between the abscissa and the saturation curve is linearly scaled by their respective value of relative humidity. For example, 50% relative humidity curve is generated by joining all the mid points of the linear distance between the dbt in abscissa and the corresponding point in the saturation curve. The reason is that relative humidity (ϕ) linearly varies with $P_v/P_{v,sat}$ as the formula indicates. On the 100% relative humidity line, i.e., saturation curve dbt, wbt and dpt are identical as per their definitions.

Constant wet bulb temperature lines: These lines are drawn on the basis of adiabatic saturation of moist air as discussed earlier. The energy balance equation for this adiabatic saturation process is initial enthalpy (h) + enthalpy added due to evaporated water $(\omega^* - \omega)$ h_f^* = Exit enthalpy (h^*)

or $$h - \omega\, h_f^* = h^* - \omega^*\, h_f^* = \text{Constant} \qquad [8.15]$$

Because all the values correspond to adiabatic saturation temperature and all the points on this constant wet bulb temperature line will have same value.

Let us consider the constant wet bulb temperature line of 25°C.

The initial point is the intersection point of saturation curve and the vertical line of 25°C dbt.

For 25°C, $\omega^* = 0.622 \times 3170/(101325–3170) = 0.0201$ kg/kga

$$h^* = (1.005 + 1.88\,\omega^*)t + 2500\omega^* = 76.32 \text{ kJ/kga}$$

$$h_f^* = 104.89 \text{ kJ/kg (from steam tables at 25°C)}$$

So the constant, C from equation $h^* - \omega^*\, h_f^*$ = Constant is 74.211

Then let us take the state point for 36°C on the constant wbt line.

As discussed above $74.211 = h - \omega\, h_f^* = (1.005 + 1.88\,\omega)t + 2500\omega - \omega\, h_f^*$

Substituting the respective values $\omega = 0.01544$ kg of wv/kg of da and thus the corresponding state point is fixed on the constant wbt line. Similarly, more points for other temperatures can be fixed. And join those points to get the constant WBT line.

The corresponding enthalpy (h) calculated to be 75.825 kJ/kga

Constant enthalpy lines: The previous calculation shows that there is a little difference between the enthalpy at saturation (at 25°C) and enthalpy at 36°C. This is called *enthalpy deviation* and is expressed as $D = h - h^* = (\omega - \omega^*)\, h_f^*$ [8.16]

In the previous calculation it was 76.32 – 75.825 = 0.495

Hence, the inference drawn is that constant enthalpy line can be drawn for different dbt values from enthalpy mathematical expression, but it is difficult to have separate lines from constant wbt lines as the difference is very small. But it can easily be demarcated if a big sheet is used. However, the enthalpy values at saturation can be read from the chart which coincides with the constant wbt line's initial point. Thus, the enthalpy of the moist air at certain state can be read by extending the constant wbt line passing through that state point and the intersection point of wbt line and constant enthalpy line reads the value of the specific enthalpy of that moist air.

Constant specific volume lines (Fig. 8.6): These lines are drawn by solving the perfect gas equation for the pressure (P_a) as a function of temperature and specific volume.

Let us discuss drawing a constant specific volume line of 0.8. This is a hit-and-trial method.

First assume a dry bulb temperature of 18°C. $P_{v,\text{sat}}$ of water vapour is 2.062 kPa.

So specific volume at saturated state will be

$$R_aT/(P_\text{a}) = 0.287 \times 291/(101.325 - 2.062) = 0.8413 \text{ m}^3\text{/kg}$$

Second assumption is 12°C ($P_{v,\text{sat}}$ = 1.401 kPa) and corresponding specific volume at saturated state is 0.82 m^3/kg

Third assumption 8°C. Similarly, corresponding specific volume at saturated state is 0.805 m^3/kg.

Then for 7°C the corresponding specific volume at saturated state is 0.8 m^3/kg.

The corresponding specific humidity (ω) from Eq. 8.3 is $0.622 \times P_\text{sat}/P_\text{da} = 0.00622$ kg/kga

Thus, the initial point of 0.8 constant specific volume line is fixed on the saturation curve from intersection point of dbt 7°C and $\omega = 0.00622$ kg /kga.

Second step: Assume another dbt in the unsaturated zone, may be 12°C.

From perfect gas equation, $P_a = 102350.6$ Pa, which is greater than 101325 Pa. So this assumption is wrong. Then assume another temperature 9°C. Then P_{da} is calculated to be 101273.25 Pa. And P_{wv} is 51.75 Pa. Then corresponding specific humidity is calculated to be 0.00032 kg of wv/kg of da. Thus, another point is fixed in the unsaturated zone.

With similar dbt assumption a number of points can be fixed. By joining these points constant specific volume line can be drawn. It can also be shown that at any point on this line $v = v_a = \omega v_v$

Sensible heat factor lines: (Discussed in the later part of elementary processes.) The SHF scale is shown on the right hand side of the psychrometric chart (Fig. 8.6). By knowing the initial state of air and the sensible heat factor, the process line can be drawn. The slope of the process line is calculated from $\tan\theta = (1/2451)\,(1/\text{SHR}-1)$ (discussed in the later part of elementary processes).

8.4 MEASURING THE PSYCHROMETRIC PARAMETERS

Measuring of the thermodynamic wet bulb temperature or the adiabatic saturation temperature of moist air is practically difficult. Hence, a good approximation of it is the wet bulb temperature, as defined above. The two properties, dry bulb temperature and wet bulb temperature, are measured by an instrument called 'psychrometer'. It consists of two thermometers. Refer to Fig. 8.7.

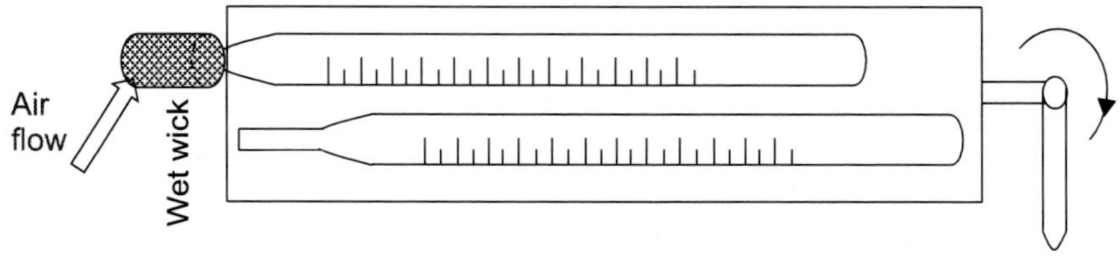

Figure 8.7: Sling psychrometer

One measures the dry bulb temperature and the other one whose bulb is wrapped with wetted wick measures the wet bulb temperature. The instrument is fitted with a swivel handle and is rotated by hand for at least one minute. The air velocity (should be around 3.5 m/s) past the wet wick causes the maximum rate of evaporation so that the air in contact with the wet bulb gets saturated. The same can be achieved by drawing the air over the wet wick by a hand operated aspirator or blower. Now the reading on this thermometer is the 'wet bulb temperature'. The difference between dbt and wbt is called the *wet bulb depression*. As the unsaturated air is blown through the wet wick, water evaporates into the air stream. The amount of evaporation is proportional to the difference in vapour pressure of water vapour in unsaturated air and the water in the wick. Due to evaporation, the remaining water in wick becomes cooled and wet bulb thermometer measures that temperature. So WBT is always less than DBT and become equal when air gets saturated. From this it is clear that there remains a direct relationship among WBT, RH and DPT. If there is more water vapour in air the vapour pressure is less and low rate of evaporation will be there. So, cooling of wick water is less and WBT is high. Thus, when WBT is high it indicates RH is high and DPT is also high. This can be verified with psychrometric chart.

Another instrument is called *Hygrometer*. It measures the relative humidity of the moist air. The active element is a horse hair, whose length varies with the change in the relative humidity. This change in length is calibrated and displayed using suitable instrument so that relative humidity of moist air can directly be read. The electrical hygrometer uses the 'change in resistance or capacity of electrical device with the change in relative humidity of moist air' property.

Example 8.1

The environmental air is at 35°C dbt and 22°C wbt. An air duct having its surface temperature 18°C is passes through this environment. Will the duct sweat?

Ans: From psychrometric chart, it is found that the dew point temperature of this environmental air is 16°C. Since the surface temperature is higher than the dpt of the environmental air, the duct will not sweat.

Example 8.2

The environmental air is at 35°C dbt and 60% RH. An air duct having its surface temperature 22°C passes through this environment. Will the duct sweat?

Ans: From psychrometric chart, it is found that the dew point temperature of this environmental air is 26.1°C. Since the surface temperature is lower than the dpt of the environmental air, the duct will sweat.

Example 8.3

A sample of moist air is at 32°C dbt. Partial pressure of water vapour in the air is 2.5 kPa. What is the dpt of the air?

Ans: From the steam tables, saturation temperature of steam at 2.5 kPa is 21°C and this is the dpt of moist air.

Example 8.4

Calculate the specific volume of moist air which is at total pressure of 90 kPa and dbt 30°C. Its specific humidity is 0.015 kg/kg of da.

Ans: From relation $\omega = 0.622\ P_v/(P - P_V)$, we get partial pressure of w.v. $P_V = 2.118$ kPa

and $\quad P_a = 90 - 2.118 = 87.882$ kPa

(a) From relation, $v = \{R_a T/P\}\{1 + 1.609\omega\} = (0.287 \times 303/90)(1 + 1.609 \times 0.015)$

$= 0.99\ \text{m}^3/\ \text{kg of da}$

(b) From relation, $v = R_a T/P_a$, $v = 0.287 \times 303/87.882 = 0.99\ \text{m}^3/\text{kg of da}$

(c) From relation, $v = m_v R_v T/P_v = 0.015 \times 0.4615 \times 303/2.118 = 0.99\ \text{m}^3/\text{kg of da}$

Example 8.5

A sample of moist air at a pressure of 736 mm of Hg. Its dbt is 30°C and dpt is 20°C. Determine its humidity ratio, relative humidity, degree of saturation, specific enthalpy, and specific volume.

Ans: Since the pressure is not 1.0 bar or 101.325 kPa, the psychrometric chart available which is constructed for 1.0 bar cannot be used. So, analytical method, which uses the perfect gas relations, should be followed.

As per the definition of dpt, it is the saturation temperature of water vapour at that corresponding pressure. So from steam tables P_{sat} at 20°C = 2.367 kPa. But P_{sat} at 30°C = 4.302 kPa.

Total pressure = (736 /760) × 101.325 = 98.125 kPa

From the relationship, $\omega = 0.622 P_{\text{w.v.}}/P_{\text{d.a.}} = 0.622 P_{\text{w.v.}}/(P_{\text{total}} - P_{\text{w.v.}})$ and $\phi = P_{\text{w.v}}/P_{\text{w.v.sat}}$

Specific humidity, $\omega = 0.622 \times 2.367/(98.125 - 2.367) = 0.015375$ kg of w.v./kg of da

Humidity at saturation, $\omega_S = 0.622 \text{P}_{\text{wv, sat.}}/(P_{\text{total}} - P_{\text{wv,sat.}}) = 0.622 \times 4.302/\ (98.125 - 4.302)$

$= 0.0285$ kg of w.v./kg of da

Degree of saturation, $(\mu) = \omega/\omega_S = 0.015375/0.0285 = 0.5395$

Relative humidity, $\phi = 2.367/4.302 = 0.5502 = 55.02\%$

Specific enthalpy $= 1.005 \times t_{\text{dbt}} + \omega\ (1.88 t_{\text{dbt}} + 2500) = 1.005 \times 30 + 0.015375\ (1.88 \times 30 + 2500)$

$= 69.45$ kJ / kg of da

Specific volume of dry air, $v_{da} = R_{da}\ T/P_{da} = 0.2871 \times 303/(98.125 - 2.367) = 0.90845\ m^3/kg$ of da

Example 8.6

A sample of moist air at standard atmospheric pressure (1 bar) has dbt 40°C and relative humidity 80%. Find the specific humidity, degree of saturation, dew point temperature, specific enthalpy, and specific volume.

Ans: The sample air is at 1 bar. Hence, the properties can be read from the available psychrometric chart. But the chart is small enough to accommodate a wide range of temperature and humidity. So properties are calculated by analytical method.

From steam tables P_{sat} at 40°C = 7.482 kPa.

RH = 0.8 = $P_{wv}/P_{sat} = P_{wv}/7.482$; So P_{wv} = 5.9856 kPa

Specific humidity, ω = 0.622 × 5.9856/(101.325 – 5.9856) = 0.039 kg of w.v./kg of da

Humidity at saturation, $\omega_S = 0.622P_{wv,sat.}/(P_{total} - P_{wv,sat.})$ = 0.622 × 7.482/ (101.325 – 7.482)
= 0.0496 kg of w.v./kg of da

Degree of saturation, (μ) = ω/ω_S = 0.039/0.0496 = 0.786

Dew point temperature is obtained from the steam tables, i.e., saturation temperature corresponding to 5.9856 kPa is 36.0°C.

Specific enthalpy = $1.005 \times t_{dbt} + \omega\,(1.88t_{dbt} + 2500)$ = 1.005 × 40 + 0.039(1.88 × 40 + 2500)
= 140.633 kJ / kg of da

Specific volume, $v = R_{da}\ T/P_{da} = 0.2871 \times 313/(101.325 - 5.9856) = 0.9454\ m^3/kg$ of da

Example 8.7

dbt and wbt of a sample of moist air are 35°C and 25°C. The air is at standard atmospheric pressure (1bar). Calculate the relative humidity, specific humidity, degree of saturation, dew point temperature, specific enthalpy and specific volume of the moist air.

Ans: WBT has no direct relation with any of the property of the moist air. Hence, one of the mostly used correlation, Carrier's equation is used to calculate the partial vapour pressure of the moist air.

$$P_V = P_V' - [(P - P_V')(t - t')1.8 / \{2800 - 1.3(1.8t + 32)\}]$$

where, P is the total pressure, P_V' is the saturated vapour pressure of water vapour at wbt, t is the dbt and t' is wbt in °C, pressures may be of any unit but consistent.

From steam tables at 25°C, P_{sat} = 3.211 kPa. Now substituting the values in the above equation P_V = 2.5512 kPa which is also P_{wv} as has been used previously for prtial pressure of water vapour.

P_{sat} at 35°C from steam tables is 5.702 kPa.

So relative humidity, $\phi = P_{wv}/P_{sat}$ = 2.5512/5.702 = 0.4474 = 44.74%

Specific humidity, ω = 0.622 × 2.5512/(101.325 – 2.5512) = 0.01606 kg of w.v./kg of da

Humidity at saturation, $\omega_S = 0.622\ P_{wv,\ sat.}/(P_{total} - P_{wv,sat.}) = 0.622 \times 5.702/(101.325 - 5.702)$ $= 0.0371$ kg of w.v./kg of da

Degree of saturation, $(\mu) = \omega\ /\omega_S = 0.01606/0.0371 = 0.433$

Dew point temperature is obtained from the steam tables, i.e., saturation temperature corresponding to 2.5572 kPa is 21.5°C.

Specific enthalpy $= 1.005 \times t_{dbt} + \omega\ (1.88t_{dbt} + 2500) = 1.005 \times 35 + 0.01606\ (1.88 \times 35 + 2500)$ $= 76.382$ kJ / kg of da

Specific volume, $v = R_{da}\ T/P_{da} = 0.2871 \times 308/(101.325 - 2.5512) = 0.8952\ m^3$/kg of da

The above properties can also be obtained directly from psychrometric chart of standard atmospheric pressure (1 bar). The state point is the intersection point of 35°C dbt and 25°C wbt (as shown in Fig. 8.6). The values obtained are as follows.

Relative humidity is 44.9%, specific humidity is 0.016 kg of wv/kg of da, dpt is 21.3°C, specific enthalpy is 76.8 kJ/kg of da, specific volume is 0.895 m^3/kg of da.

So $Q_{12} = 12 \times 139.89 \times (12 \times 57.555 + 0.2774 \times 83.9) = 964.75$ kW = 274.33 TR.

So condensate rate is 0.2774 kg/s and cooling capacity is 274.33 TR.

Example 8.8

This is a case of maintaining the relative humidity of indoor air so that there will be no condensation on the glass surfaces of the room during winter.

Indoor air of a room with glazed windows is maintained at 22°C. Outdoor temperature is 10°C. What should be the maximum relative humidity to be maintained inside the room so that there is no condensation on the glazed surfaces.

Ans: Neglecting the thermal resistance of the glazed surfaces, it is assumed that the outdoor temperature is equal to inside temperature of the glazed surface, i.e., 10°C. (Now the new question is what should be the RH at 22°C dbt so that its DPT is 10°C.)

Analytical solution: Refer Example 8.3.

Partial pressure of water vapour (P_{WV}) in the moist air is the saturation pressure of water vapour at dpt, i.e., 1.227 kPa (from steam tables).

Saturation pressure (P_{WS}) at 22°C is (from steam tables) 2.642 kPa.

So the relative humidity is 1.227/2.642 = 0.4644, i.e., 46.44%.

Hence, the maximum relative humidity of the indoor air to be maintained is 46.5%.

Alternatively: From psychrometric chart: The total pressure is 1.0 bar (atmospheric pressure). Hence, available psychrometric chart can be used. (Refer Fig. 8.8.)

Dew point temperature is same as that dbt at saturation. So intersecting point of constant humidity ratio line corresponding to 10°C dbt and 22°C dbt is the desired state point. The RH line passing through that point is 46%.

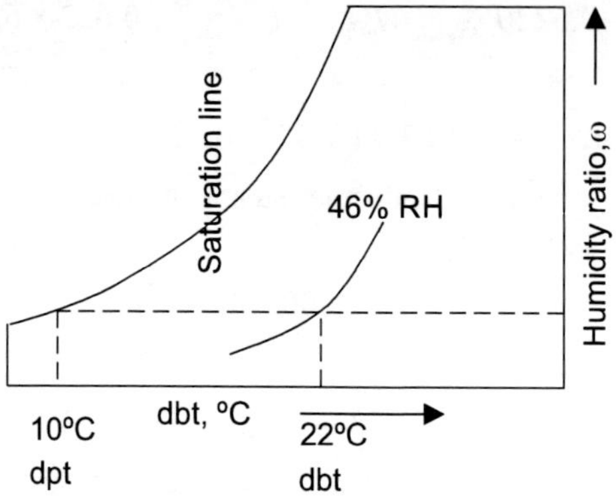

Figure 8.8: Example 8.8

Example 8.9

A 3.0 m^3 tank contains saturated air at 25°C and 97 kPa. Determine (a) the mass of dry air, (b) the specific humidity, and (c) specific enthalpy of air.

Ans: At 25°C, saturated vapour pressure is 3.1698 kPa.

Specific humidity at saturation, $\omega_S = 0.622 \times 3.1698/(97.0 - 3.1698) = 0.021$ kg/kga

Specific volume of air, $v_a = R_aT/P_a = 0.287 \times 298/\ (97.0 - 3.1698) = 0.9115\ m^3/kga$

So $m_a = 3.0/0.9115 = 3.29$ kg

Specific enthalpy of moist air, $h = 1.005 \times 25 + 0.021 \times 2546.5$ (i.e., h_g at 25) = 78.6 kJ/kga

Example 8.10: A sample of moist air has dbt 30°C and wbt 25°C. Barometric pressure is 101 kPa.

(a) Calculate the humidity ratio and the relative humidity of the sample moist air (without using the psychrometric chart or **dbt and wbt correlations**).

(b) Calculate the humidity ratio and the enthalpy of air if it is adiabatically saturated.

Ans: (a) wbt of moist air is the temperature attained by the air after getting adiabatically saturated.

When moist air is adiabatically saturated, the relationship between saturated air and unsaturated air is expressed by $h_S - h = (\omega_S - \omega) \times h_{f(\text{at wbt})}$

$h = 1.005 \times t + \omega \text{ X } h_{g(\text{at dbt})}$

Now substituting the values $76.453 - (1.005 \times 30 + \omega \times 2556.3) = (0.02015 - \omega) \times 104.89$

This gives $\omega = 0.01802$ kg/kga

So $0.01802 = 0.622 \times P_v/(101 - P_v)$; This gives $P_v = 2.8436$ kPa

Relative humidity $= P_v/P_{S(\text{at dbt})} = 2.8436/4.246 = 66.97\%$

(b) Adiabatically saturated air implies the humidity ratio and enthalpy at wbt.

At 25°C, $P_{vS} = 3.169$ kPa;

So $\omega_{sat} = 0.622 \times P_{vS}/(P - P_{VS})$

$= 0.622 \times 3.169/(101 - 3.169) = 0.02015$ kg/kga

Enthalpy, $h_S = 1.005\ (25) + 0.02015 \times (h_{g(25)}) = 25.125 + 0.02015 \times 2547.3 = 76.453$ kJ/kga

Example 8.11

A sample of moist air has dbt 30°C and humidity ratio 0.015 kg/kga. Calculate the (a) enthalpy, (b) dpt of the moist air for barometric pressures 85 kPa and 101 kPa.

Ans: Enthalpy = 1.005 × 30°C + ω × $h_{g(\text{at } 30°\text{C})}$ = 1.005 × 30 + 0.015 × 2556.3 = 68.5 kJ/kga

Again 0.015 = 0.622 × $P_v/(85 - P_v)$; This gives P_v = 2.002 kPa

From steam table Saturation temperature corresponding to 2.002 kPa is 17.5°C.

So dpt of the sample air is 17.5°C

Similarly, for barometric pressure (total pressure) of 101 kPa

Enthalpy remains same, i.e., 68.5 kJ/kga

And P_v = 2.378 kPa

From steam table, corresponding saturation temperature is 20.5°C.

So dpt of the sample air at 101 kPa barometric pressure is 20.5°C.

Example 8.12

15 m^3/s of moist air at 35°C dbt and 24°C wbt at 101 kPa enters a cooling tower and pass through a spray of water at 38°C with a flow rate of 20 kg/s. Air leaves the cooling tower at 31°C being saturated. (A) To what temperature the water is cooled? (B) How much kg of make-up water is needed per second?

Ans: Specific enthalpy of the air entering the cooling tower (h_1) = 1.005 × 35 + ω_1 × $h_{g(35)}$.

Specific humidity(ω_1) can be calculated using Carrier's equation or by the method shown in Example 8.10.

At 24°C, P_{vS} = 3.0 kPa; So $\omega_S = 0.622 \times P_{vS}/(P\text{-}P_{VS})$

= 0.622 × 3.0/(101 – 3.0) = 0.019 kg/kga

Enthalpy, h_S = 1.005 (24) + 0.019 × ($h_{g(24)}$) = 24.12 + 0.019 × 2545.5 = 72.4845 kJ/kga

$h = 1.005 \times t + \omega \times h_{g(\text{at dbt})}$

For adiabatic saturation of moist air, $h_S - h = (\omega_S - \omega) \times h_{f(\text{at wbt})}$

Now substituting the values 72.4845 – (1.005 × 35 + ω_1 × 2565.3) = (0.019 – ω_1) × 100.5

This gives ω_1 = 0.01436 kg/kga.

[Using Carrier's equation, $P_V = P_V' - [(P - P_V')(t - t')1.8/\{2800 - 1.3(1.8t + 32)\}]$;

where P is the total pressure, P_V' is the saturated vapour pressure of water vapour at wbt, t is the dbt and t' is wbt in °C, pressures may be of any unit but consistent.

3.0 – [(101.0 – 3.0) (35 –24) 1.8/{2800 – 1.3(1.8 × 35 + 32)}] = 2.275 kPa

Specific humidity, ω_1 = 0.622 × 2.275/(101.0 – 2.275) = 0.01433 kg/kga]

Specific enthalpy of air, h_1 = 1.005 × 35 + 0.01436 × 2565.3 = 72.013 kJ/kga

Specific volume of moist air, $v = R_aT/P_a$ = 0.287 × 308/(101 –2.28) = 0.8954 m^3/kga

So mass of air = 15.0 / 0.8954 = 16.752 kg/s

Humidity ratio of exit air, which is at 31°C and saturated, ω_S = 0.622 × 4.45/(101 – 4.45) = 0.02867

Specific enthalpy exit air, h_2 = 1.005 × 31 + 0.02867 × 2558.1 = 104.5 kJ/kga

Total heat taken by air = 16.752 × (104.5 – 72.013) = 544.22 kW

Mass of water carried away by air = $m_a \times (\omega_S - \omega_1)$ = 16.752 × (0.02867 – 0.01436) = 0.24 kg/s

So the make-up water required is 0.24 kg/s

Temperature of water coming out of the cooling tower

= 38 – [544.22/(4.19 × 20)] = 31.5°C.

(More accurate calculation can be done by considering the mass transfer.)

Example 8.13

Atmospheric air (101 kPa) at 27°C and 50% RH enters an air conditioning unit and leaves it at 13°C and 90% RH. Calculate the refrigerating capacity and water removal rate from the air if the air flow rate is 3.5 m^3/s.

Ans: Relative humidity 50% = P_v/P_S

From steam tables, P_S at 27°C is 3.6 kPa and this gives P_v = 1.8 kPa

Specific humidity, ω_1 = 0.622 × 1.8 / (101 – 1.8) = 0.0113 kg/kga

Specific enthalpy of incoming air, h_1 = 1.005 × 27 + 0.0113 × $h_{g(27)}$(2550.5) ≈ 56.0 kJ/kga

Similarly for dbt 13°C and RH 90%

P_S at 13°C is 0.01514 and this gives, P_v = 0.9 × 0.01514 = 1.363 kPa.

So ω_2 = 0.622 × 1.363/(101 – 1.363) = 0.00851

Specific enthalpy of outgoing air, h_2 = 1.005 × 13 + 0.00851 × $h_{g(13)}$(2521.6) = 34.524 kJ/kga

Specific volume of air at inlet, $v = R_aT/P_a$ = 0.287 × 300 /(101 – 1.8) = 0.868 m^3/kga

Mass of air = 3.5 / 0.868 = 4.032 kg/s

Cooling capacity = 4.032 × (56 – 34.524) = 86.6 kW

Water removal rate from air, $m_a(\omega_1 - \omega_2)$ = 4.032 × (0.0113 – 0.00851) = 0.01125 kg/s

Example 8.14

The compressor of an air conditioning unit compresses air at 28°C, 50% RH and 101 kPa to 400 kPa and is then sent to an inter-cooler. Calculate the minimum temperature to which air should be cooled so that there will be no condensation.

Ans: In this question, dpt of air is to be calculated below which condensation will start.

Specific humidity of the sample air remains constant even after compression.

0.5 = $P_{v1}/P_{S(28)} = P_{v1}$/3.815 kPa. This gives P_{v1} = 1.91 kPa

Specific humidity, ω_1 = 0.622 × 1.91/ (101 – 1.91) = 0.012 kg/kga

After compression 0.012 = $P_{v2}/(400 - P_{v2})$;

This gives P_{v2} = 7.571 kPa and corresponding saturated temperature (from steam table) is 40.5°C.

This is the dpt of the air after compression and this is the minimum temperature to which air should be cooled to avoid condensation.

EXERCISE

1. What do you mean by latent heat of moist air?
2. What is the condition when dbt, wbt and dpt of moist air is same?
3. What information wet bulb depression gives?

4. What is the minimum temperature that can be achieved by evaporative cooling?

5. How many independent properties are required to fix the state point of a two-component mixture. What are those properties in a psychrometric chart?

6. Why a psychrometric chart prepared for 1.0 bar total pressure cannot be used for any other total pressure without modification?

7. Specific volume shown in a psychrometric chart is of dry air or of water vapour?

8. What is amount of heat is required to raise the temperature of 0.5 m^3/s of air at 10°C and 80% RH to 30°C.

9. The atmospheric conditions are 25°C dbt and specific humidity of 10 gm/kg of da. Determine its dew point temperature (take $p_{atm} = 1$ bar).

10. Prove that the partial pressure of water vapour in the atmospheric air remains constant as long as the specific humidity remains constant.

11. What is the difference between wet bulb temperature and thermodynamic wet bulb temperature?

MULTIPLE–CHOICE QUESTIONS

1. Atmospheric air contains oxygen and carbon dioxide — and — respectively.

(a) 21% and 0.4% (b) 21% and 0.04%
(c) 20% and 1.4% (d) 20.1% and 1.14%

2. For air conditioning calculation, air is considered to be a mixture of

(a) O_2 and N_2 (b) dry air and water vapour
(c) water vapour and N_2 (d) dry air and N_2

3. Water vapour in atmospheric air is assumed to behave as perfect gas because water vapour is at

(a) high pressure (b) low pressure
(c) high temperature (d) low temperature

4. In evaporative cooling minimum temperature that can be attained by the moist air is its

(a) dpt (b) dbt
(c) wbt (d) water temperature

5. Moist air has dbt 25°C, wbt 25°C. What will be its dpt?

(a) 15°C (b) 20°C
(c) 25°C (d) 22°C

6. Hygrometer measures — of moist air.

(a) wbt (b) degree of saturation,
(c) humidity ratio (d) relative humidity

7. Psychrometer measures — of moist air.

(a) dbt
(b) wbt
(c) both (a) and (b)
(d) specific humidity

8. One psychrometric chart can be used for psychrometric calculations for

(a) one pressure
(b) different pressures
(c) one humidity ratio
(d) one temperature

9. Independent parameters considered for constructing psychrometric chart are

(a) temperature and partial pressure of WV
(b) total pressure and temperature
(c) temperature and humidity ratio
(d) total pressure, temperature and partial pressure of WV

10. Saturation line on the psychrometric chart is that of

(a) water vapour
(b) dry air
(c) moist air
(d) none of the above

11. On a pshychrometric chart constant enthalpy line and extended constant wbt line

(a) almost coincide
(b) are wide apart
(c) vertical to each other
(d) 45° to each other

12. Unit of specific enthalpy of moist air is

(a) kJ/kg
(b) kJ/kg of dry air
(c) kJ/kg of moist air
(d) kJ/kg of saturated air

13. Partial pressure of water vapour is — to specific humidity.

(a) directly proportional
(b) inversely proportional
(c) reciprocal
(d) equal

14. The value of humid specific heat is — kJ/kg-K.

(a) 1.216
(b) 1.0216
(c) 1.00216
(d) 1.005

15. Volume occupied by the water vapour (V_v)in moist air is calculated from

(a) $m_v\ R_v\ T/P$
(b) $m_v\ R_v\ T/P_v$,
(c) $m_v\ R_v\ T_v/P_v$
(d) $m\ R\ T/P$

16. During winter dew is formed due to

(a) presence of more WV
(b) Lower temperature at same partial pressure of WV
(c) Lower temperature at reduced partial pressure of WV
(d) Lower temperature at high partial pressure of WV

17. Consider the following statements: A psychrometer measures

1. wet bulb temperature
2. dew point temperature
3. dry bulb temperature

Of these statements, which statement is correct?

(a) 1 and 2

(b) 2 and 3

(c) 1 and 3

18. If the specific heats of dry air and water vapour are 1.00 kJ/kg-K and 1.88 kJ/kg-K, respectively and the humidity ratio is 0.011, then the specific heat of moist air at 25°C and 50% relative humidity will be

(a) 1.0207 kJ/kg-K (b) 1.869 kJ/kg-K

(c) 1.891 kJ/kg-K (d) 0.9793 kJ/kg-K

19. When the wet bulb and dry bulb temperatures are equal, which of the following statements is/are correct?

1. Air is fully saturated.
2. Dew point temperature is reached.
3. Partial pressure of vapour is equal to the total pressure.
4. Humidity ratio is 100%.

(a) 1 and 2 (b) 1 only

(c) 1, 2 and 3 (d) 2 and 3

20. Moist air exists at a pressure of 1.01 bar. The partial pressure and saturation pressure of water vapour are 0.01 bar and 0.02 bar, respectively. What are the relative humidity and humidity ratio of the moist air, respectively?

(a) 50% and 0.00622 (b) 100% and 0.0126

(c) 50% and 0.0126 (d) 100% and 0.00622

ANSWERS

1. (b) **2.** (b) **3.** (b) **4.** (c) **5.** (c) **6.** (d) **7.** (c) **8.** (a) **9.** (d) **10.** (a)

11. (a) **12** (b) **13** (a) **14.** (b) **15.** (a) **16.** (b) **17.** (c) **18** (a) **19** (b) **20.** (a)

9

Elementary Psychrometric Processes

9.1 INTRODUCTION

Available atmospheric air is not always suitable, when directly used, for fulfilling our purpose (may be for comfort or for any other industrial applications). Hence, this moist air has to pass through certain thermodynamic processes to attain its desired state. In these thermodynamic processes the moist air passes through from available state to desired state are called 'psychrometric processes'. These processes can be represented on a psychrometric chart. These psychrometric processes involve both mass (water vapour) and heat transfer. Some of these processes are described below.

9.2 DIFFERENT PSYCHROMETRIC PROCESSES

This involves addition or removal of heat to heat or cool the air, which is called *sensible heating* or *sensible cooling* and addition or removal of moisture, which is called *humidification* or *dehumidification*. Combination of any of the two processes is also carried out whenever it is required. For example, cooling and dehumidification during summer in humid climate or heating and humidification during winter.

9.2.1 Sensible Heating/Cooling

When the moist air is passed over a heating/cooling coil there is only change of its dbt without any change of humidity ratio. This process is called sensible heating (O-2)/Cooling (O-1) as shown in Fig. 9.1. The amount of heat transfer during the process is calculated as follows.

Mass conservation (dry air), $m_{a1} = m_{a2} = m_a$. [9.1]

Mass conservation (water vapour), $m_{v1} = m_{v2} = m_v$. So $\omega_1 = \omega_2 = \omega$ [9.2]

Energy conservation, $m_a \times h_1 + q_{12} = m_a \times h_2$, $q_{12} = m_a (h_2 - h_1)$ [9.3]

q_{12} is the sensible heat transfer during the process.

Putting the empirical relation of enthalpy from Eq. 8.9,

$$q_{12} = m_a \times (1.0216 \times t_2 + \omega \times 2500 - 1.0216 \times t_1 + \omega \times 2500) = m_a \times 1.0216 \times (t_2 - t_1)$$

Air flow rate is normally measured in volume flow rate (cubic metre per minute, cmm).

So $m_a = \rho Q_V/60 = 1.2Q_V/60$ (standard air density, at 1 bar and 20°C, is assumed to be 1.2 kg/m^3)

$$q_{12} = 1.2Q_V/60 \times 1.0216 \times (t_2 - t_1) = 0.0204\, Q_V \times \Delta t \qquad [9.4]$$

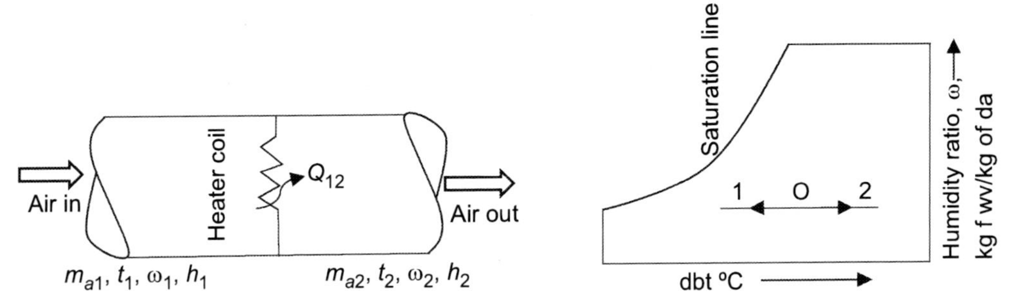

Figure 9.1: Sensible heating/cooling

9.2.2 Pure Humidification/Dehumidification

When water vapour is added (by spraying water) or removed (by using desiccants or absorbents) from the moist air, the process is called pure humidification/dehumidification as shown in Fig. 9.2. The amount of heat transfer during the process is calculated as follows.

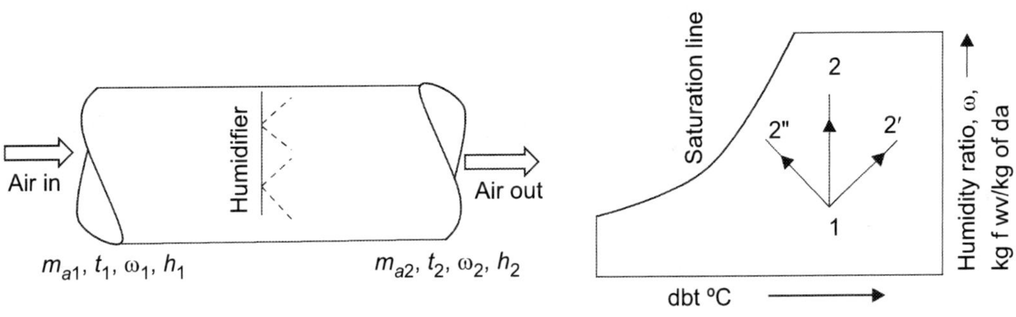

Figure 9.2: Humidification process

Mass conservation (dry air), $m_{a1} = m_{a2} = m_a$ [9.5]

Mass conservation (water vapour), $m_a\omega_1 + m_{v\text{added}} = m_a\,\omega_2$, So $\omega_2 - \omega_1 = \omega_{\text{added}}$ [9.6]

Energy conservation, $m_a \times h_1 + \omega_{\text{added}} h_w = m_a \times h_2$ [9.7]

Substituting the Eq. (8.8) for enthalpy, $h = 1.005 \times t_{\text{dbt}} + \omega \times h_g$ (at t_{dbt})

$1.005 \times (t_2 - t_1) + \omega_2\, h_g(t_2) - \omega_1\, h_g(t_1) = (\omega_2 - \omega_1) \times h_w$

$1.005 \times (t_2 - t_1) + \omega_2 \times (h_g(t_2) - h_g(t_1)) = (\omega_2 - \omega_1) \times (h_w - h_g(t_1))$

$h_g(t_2) - h_g(t_1) = 2500 + 1.88\, t_2 - 2500 - 1.88\, t_1$

The equation will be $(1.005 + 1.88\,\omega_2)(t_2 - t_1) = (\omega_2 - \omega_1) \times (h_w - h_g(t_1))$

Alternatively, $(1.005 + 1.88\,\omega_1)(t_2 - t_1) = (\omega_2 - \omega_1) \times (h_w - h_g(t_2))$ [9.8]

This equation indicates:

- If $h_w > h_g(t_1)$, $t_2 > t_1$, i.e., air will be sensibly heated and humidified. This usually happens when steam is added to the air.
- If $h_w < h_g(t_1)$, $t_2 < t_1$, i.e., air will be sensibly cooled and humidified. This usually happens when water is sprayed into the air. The water droplets get vaporised by taking the latent heat from the air, thus decreasing the temperature of the air.
- If $h_w = h_g(t_1)$, $t_2 = t_1$, i.e., air is purely humidified and there is no change in the air temperature. This usually occurs (approximately) when the water vapour is added to the room air due to perspiration, infiltration, etc., as the water vapour is at room temperature.

During this process latent heat transfer, $Q_L = m_a \times (h_2 - h_1)$ [9.9]

t_2 being equal to t_1 for pure humidification

$Q_L = m_a \times (1.005 \times (t_1 - t_1) + \omega_2\, h_g(t_1) - \omega_1\, h_g(t_1)) = m_a \times (\omega_2 - \omega_1) \times (2500 + 1.88\, t_1)$ [9.10]

For standard air, $m_a = \rho Q_V/60 = 1.2Q_V/60$

So $Q_L = 1.2Q_V/60 \times \omega_2 - \omega_1) \times (2500 + 1.88\, t_{dbt,1})$

1.88 t_1, being very small for air conditioning range of calculation, is neglected.

So $Q_L = 1.2Q_V/60 \times \omega_2 - \omega_1) \times (2500) = 50\, Q_V\, \Delta\omega$ [9.11]

9.2.3 Combined Heating/Cooling with Humidification/Dehumidification Process

The different combined processes are as follows.

(a) Heating and humidification: This process is required during winter months when the air is cool and dry. Here heat pumps are used instead of refrigeration systems and humidifiers (water sprayers) are used to keep the relative humidity at comfort level. This process can be accomplished in two different ways.

(i) by injecting saturated steam into a stream of air flowing through a duct and represented by line 1-2 in Fig. 9.3(a), and

(ii) by first sensibly heating the air (line 1-3) followed by spraying water at wet bulb temperature resulting in evaporative cooling (line 3-2) and humidification (Fig. 9.3 b and c).

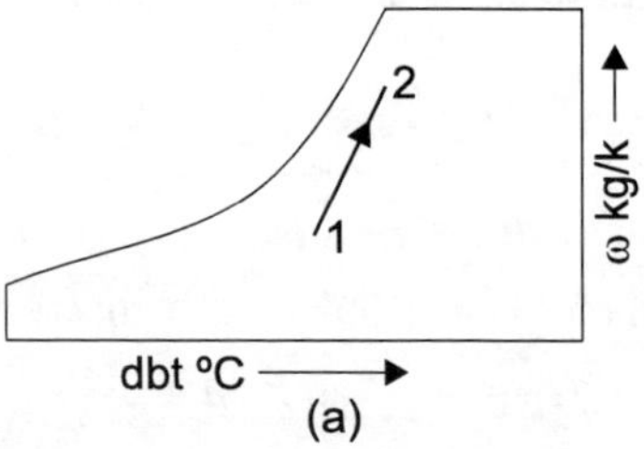

Figure 9.3: (a) Steam injection method

(b) Cooling and dehumidification is necessary during summer months in coastal zones where the air is hot and humid. Here it needs a refrigeration system, i.e., a cooling coil which cools and dehumidifies the air when it flows over it as shown in Fig. 9.4 (detail is given in Section 9.2.6).

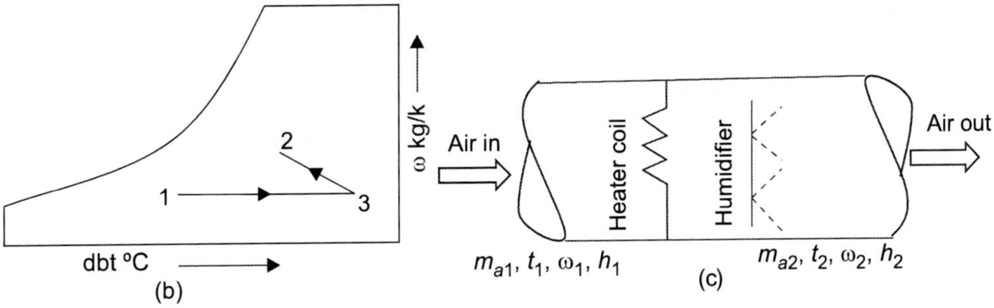

Figure 9.3: (b) Heating and humidification process, (c) Schematic of heating and humidification

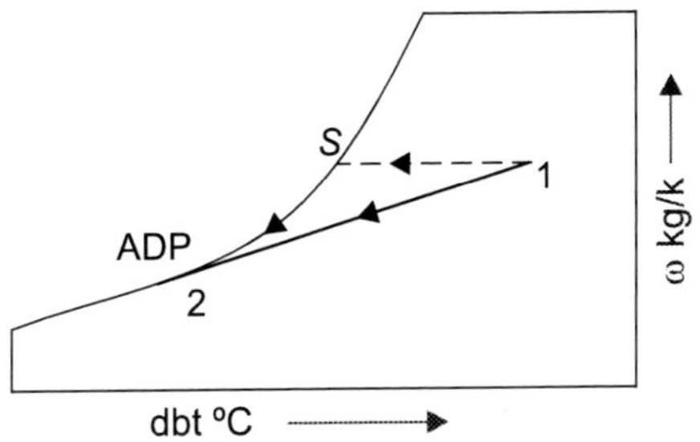

Figure 9.4: Cooling and dehumidification (1-2)

(c) Cooling and humidification is necessary during summer months in desert areas where the air is hot and dry. Here it requires evaporative cooling. Hence, the air coolers, otherwise called desert coolers, are used which adds water (thus increasing the relative humidity) to the air as the air flows through it and cools the air to its wet bulb temperature (the minimum temperature it can attain) as shown in Fig. 9.5.

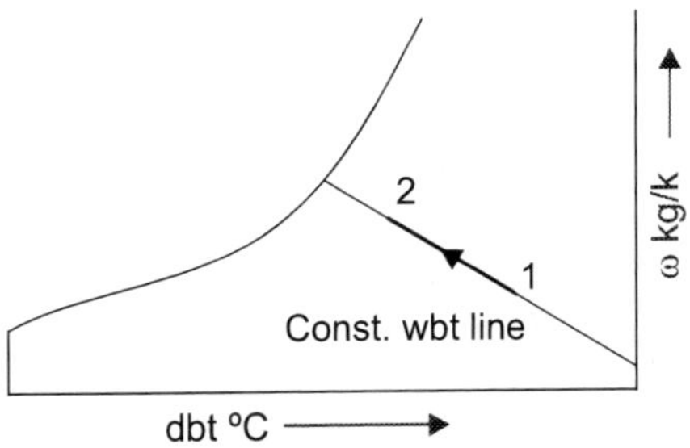

Figure 9.5: Evaporative cooling process, 1-2

(d) Heating and dehumidification process is not usually carried out. However, during these processes, the energy transfer is by adding the two processes discussed in Sections 9.2.1 and 9.2.2. Here an example of heating and humidification process is considered. It is assumed that this combined process occurs separately as shown in Fig. 9.6. First 1-3 is sensible heating and 3-2 is pure humidification. The resultant process is 1-2. Now the summation of the energy transfer during individual process is the net energy transfer during the combined process. This method is used only to simplify the energy transfer calculation.

Mass conservation (water vapour): $m_a\omega_1 + m_{v,\text{added}} = m_a\,\omega_2$

Energy conservation: $m_a \times h_1 + m_{w,\text{added}} \times h_w + q_{12} = m_a \times h_2,\ q_{12}$

$$= m_a (h_2 - h_1) - m_a \times (\omega_2 - \omega_1) h_w$$

q_{12} is the total heat transfer during the process.

q_{12} can be expressed as the summation of two processes, 1 to 3 (sensible heating) and 3 to 2 (pure humidification) as shown in Fig. 9.6.

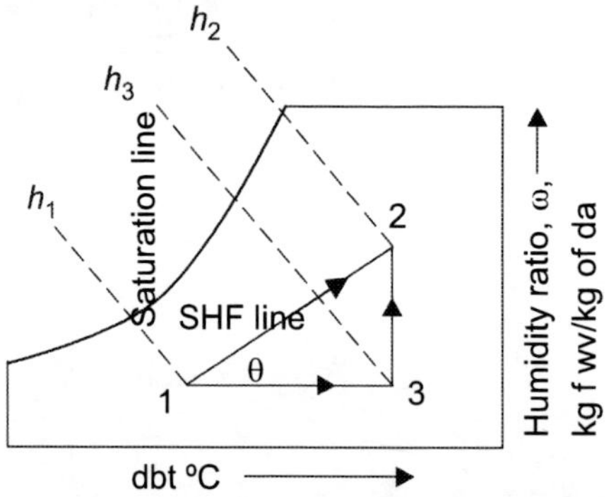

Figure 9.6: Defines the SHF line during combined process

$$Q_{\text{total}} = Q_S + Q_L = 0.0204 Q_V \Delta t_{\text{dbt}} + 50 Q_V \Delta\omega = Q_V (0.0204 \Delta t_{\text{dbt}} + 50 \Delta\omega) \quad [9.12]$$

Here we define a term 'sensible heat ratio' (SHR) which is expressed as the ratio of sensible heat and the total heat.

$$\text{SHR} = Q_S / (Q_S + Q_L) \quad [9.13]$$

$$= 0.0204 Q_V \Delta t / Q_V (0.0204 \Delta t + 50 \Delta\omega)$$

$$= 1/[\,1 + \{50/0.0204 \times (\Delta\omega/\Delta t)\}] = 1/\{1 + 2451(\Delta\omega/\Delta t)\}$$

On psychrometric chart $\Delta\omega/\Delta t$ is the slope of the process line, i.e., $\tan\theta$

$$\text{So SHR} = 1/(1 + 2451 \tan\theta) \quad [9.14]$$

$$\tan\theta = (1/2451)(1/\text{SHR} - 1) \quad [9.15]$$

In the psychrometric chart this process line can be drawn in terms of 'sensible heat ratio' or 'sensible heat factor' line.

9.2.4 Chemical Dehumidification

During rainy season dbt of air is not so high, but the relative humidity is very high. So simple dehumidification of air will decrease the latent heat load on the cooling coil. Air can be dehumidified by passing it over the sorbents or desiccants that have very high affinity for moisture. The moisture gets absorbed or adsorbed by the sorbents like silica gel, activated alumina, etc. And during the process, it gives up its latent heat, raising the dry bulb temperature of air. It moves along the constant enthalpy (≈ wbt) line. The process is shown on the line 1-2 in Fig. 9.7.

9.2.5 Adiabatic Mixing of Two Streams of Air

Let one stream of moist air '1' at temperature t_1, specific humidity ω_1 and specific enthalpy $h_1$adiabatically mixes with another moist air stream '2' at temperature t_2, specific humidity ω_2

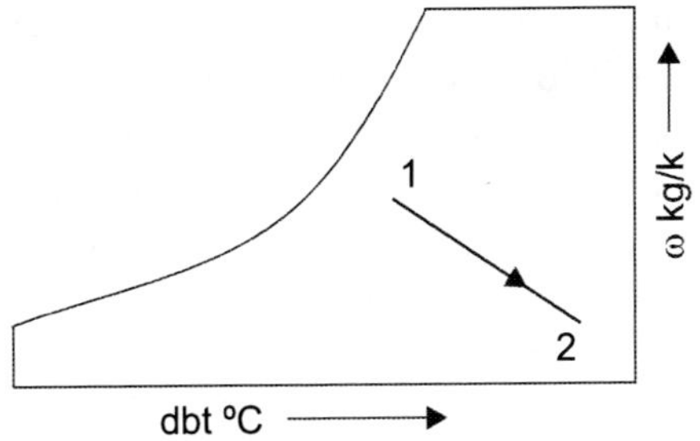

Figure 9.7: Chemical dehumidification process (1-2)

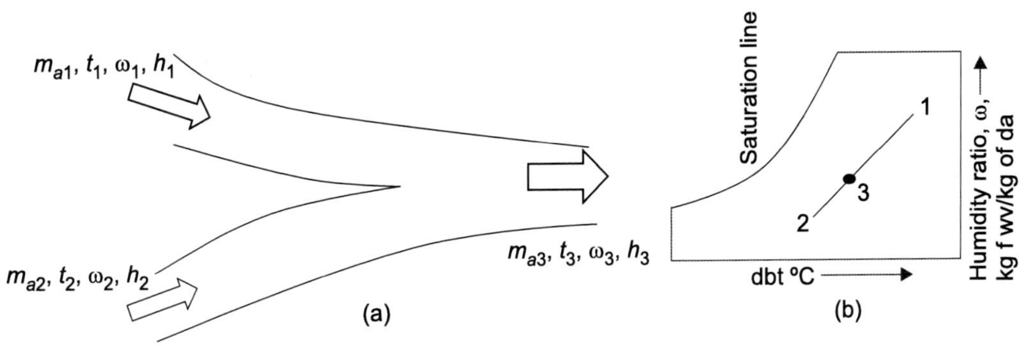

Figure 9.8: Adiabatic mixing of two streams (a) schematic view, (b) state points on psychrometric chart

and specific enthalpy h_2 [Fig. 9.8 (a) and (b)]. The resultant stream properties are calculated as follows.

Conservation of mass (dry air), $m_{a1} + m_{a2} = m_{a3}$ [9.16]

Conservation of mass (water vapour), $m_{a1}\,\omega_1 + m_{a2}\,\omega_2 = m_{a3}\,\omega_3$ [9.17]

Conservation of energy, $m_{a1}\,h_1 + m_{a2}\,h_2 = m_{a3}\,h_3$

So $h_3 = (m_{a1}\,h_1 + m_{a2}\,h_2) / m_{a3}$ [9.18]

Substituting h as $C_{P,\text{humid}} \times t + \omega h_{fg}$ (0°C) kJ /kg of a and rearranging

$t_3 = (m_{a1}t_1 + m_{a2}t_2)/m_{a3} + h_{fg}\,(0°\text{C})/C_{P,\text{humid}}\,\{(m_{a1}\omega_1/m_{a3}) + (m_{a2}\,\omega_2\,/m_{a3}) - \omega_3\}$ [9.19]

The second term in the above expression is negligible.

So $t_3 \approx (m_{a1}t_1 + m_{a2}t_2)/m_{a3}$ [9.20]

The symbol of approximation is due to the assumption that $C_{P,\text{humid}}$ is same for all the streams and the second term is negligible.

If this mixing process is plotted on the psychrometric chart, the resulting state '3' will lie on the line joining the two states of '1'and '2' and in the inverse ratio of the mass of air m_{a1} and m_{a2}.

9.2.6 Cooling and Dehumidification while Air Passes Over a Cooling Coil

When moist air is made to pass over a cooling coil whose surface temperature is maintained below the dew point temperature of the incoming air, the water vapour in the immediate vicinity of the coil attains the dew point temperature and condenses. If sufficient time and large coil surface area is provided all of the incoming air gets in contact with the coil surface for a longer period and it will attain the coil temperature. Thus, incoming air is will leave the coil at coil surface temperature and at saturated state. It can be imagined that first the air sensibly cooled to saturated state at constant humidity ratio to point '*S*' (Fig. 9.9). Then the water vapour in the air condenses along the saturated line. Thus, cooling and dehumidification process occur. These two processes occur simultaneously. Hence, the resultant of these two processes is represented by a single process line or referred to as *coil conditioning line* which starts from '1' and moves towards the cooling coil temperature on saturation line in a straight path and leaves the coil at the average temperature of the cooling coil. This is called *straight line law*. This law states that when air transfers heat and mass to or from wetted surface the condition of the air drives towards the saturation line at the temperature of the wetted surface. The straight-line law asserts that point 2 lies on a straight line drawn between point 1 and the saturation curve at the wetted surface temperature. The point where this process line intersects the saturation curve on the psychrometric chart is called the *apparatus dew point* or coil ADP and the corresponding temperature is called t_{ADP}. This t_{ADP} is also the average temperature of the cooling coil surface because the temperature at each point of the cooling coil is not same.

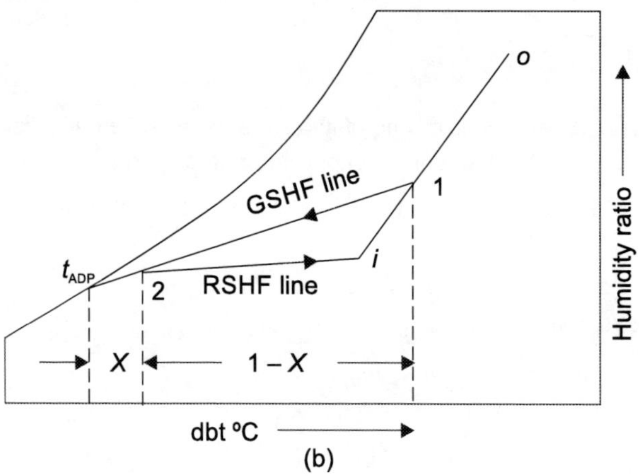

Figure 9.9: Cooling and dehumidification process with bypass factor

9.2.7 Bypass Factor of a Coil

Moist air is made to pass over a process coil (heating/cooling) for conditioning purpose. As it passes over the coil, some of the air comes in direct contact with the coil surface, some passes over at the close vicinity of the coil and some may pass over unaffected. Accordingly, the temperature of the air will be different for different layers. However, it is assumed that a fraction of moist air attains the temperature of the coil surface and rest amount of air passes the coil unaffected or bypassed. Hence an average value or bulk mean temperature of the air stream is considered and that is less than the temperature of the coil surface. The ratio of mass of moist air bypassed to the total mass of moist air is called the *bypass factor* (X) of the process coil.

From the principle of energy conservation of moist air that passes over the process coil:

Energy of moist air that bypasses the coil : $Xm_a C_{P,\text{humid}} t_1$

Energy of moist air that attains the coil surface temperature: $(1–X)\, m_a C_{P,\text{humid}} t_{\text{ADP}}$

Energy of moist air that leaves the coil: $m_a C_{P,\text{humid}} t_2$

Now $Xm_a C_{P,\text{humid}} t_1 + (1–X)\, m_a C_{P,\text{humid}} t_{\text{ADP}} = m_a C_{P,\text{humid}} t_2$

This gives $X = (t_2 - t_{\text{ADP}}) / (t_1 - t_{\text{ADP}})$ [9.21a]

Also can be expressed as $X = (\omega_2 - \omega_{\text{ADP}}) / (\omega_1 - \omega_{\text{ADP}})$ [9.21b]

Figure 9.9 shows t_2 is the temperature of air leaving the coil, t_1 is the temperature of air entering the coil, t_{ADP} is the apparatus dew point and it is the average temperature of the coil. $(1–X)$ is known as *contact factor*.

9.3 AIR WASHER

Air washer, sometimes referred to as spray washer, is a system in which psychrometric processes like heating, cooling, humidification, or dehumidification can be carried out. A year-round air conditioning can be obtained with the help of this device.

It has a chamber with a sump at its bottom (Fig. 9.10). The water from the sump (may be hot, cold, or at room temperature) is pumped to the nozzles from where it is sprayed into the chamber such that whole chamber becomes full of water droplets. The air is drawn in horizontally through a filter. There are drift eliminators at the exit to arrest the water particle being carried away with the air. The whole chamber may be assumed to be insulated.

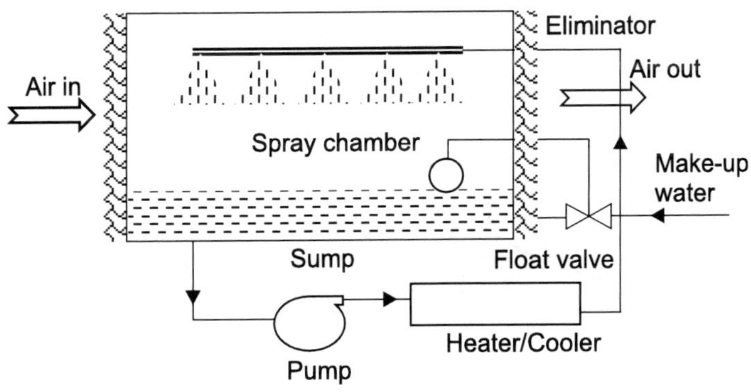

Figure 9.10: Air washer

The outgoing moist air will be heated, cooled, humidified or dehumidified depending upon the temperature of the water sprayed (Refer to Fig. 9.11).

Assume water temperature is t_W and air temperature is t_1.

Case I: If $t_W > t_1$, i.e., if hot water, from the sump, is sprayed over the air in the chamber, the outgoing air will be **heated and humidified**, shown by line *O-A*.

Case II: If $t_W = t_1$, i.e., if the sump water temperature is same as the dry bulb temperature of incoming air, there will be no sensible heating of air. There will be pure humidification as shown by line *O-B*. This process is difficult to achieve as discussed earlier. The sump water is to be heated to keep the enthalpy of water and moist air same because with the addition of water vapour to moist air, the enthalpy of the later increases and the sump water temperature decreases.

Case III: If $t_{wbt} < t_W < t_1$, i.e., if the sprayed water temperature is less than the dry bulb temperature of air but more than its wet bulb temperature, there will be cooling and humidification and the process is shown by line *O-C*. In this case, enthalpy of moist air increases. So to keep the temperature of sprayed water in the range, sump water is to be heated.

Case IV: If $t_W = t_{wbt}$, i.e., If the sump water temperature is equal to the wet bulb temperature of the incoming air, the air will be adiabatically saturated and the process will be along the constant wet bulb temperature line. The moist air and sump water temperature will be same and equal to t_{wbt} of the incoming air. If the air washer is infinitely long, moist air will be saturated. For a finite system, air will attain a state point behind the saturation point, but on the constant wet bulb temperature line. This is shown as a line *O-D*.

Case V: If $t_{wbt} > t_W > t_{dpt}$, if sump water temperature lies between dew point and wet bulb temperature of the incoming air, cooling and humidification will occur as shown by the line *O-E*. But the enthalpy of air decreases and water gets cooled.

Case VI: If $t_{dpt} = t_W$, i.e., if dew point temperature of incoming air is same as that of the water in the sump, the humidity ratio of the moist air will remain same, but the enthalpy of air decreases causing the decrease in temperature of air, i.e., sensibly cooled as shown by the line *O-F.*

Case VII: If $t_W < t_{dpt}$, i.e., if sump water temperature is less than the dew point temperature of the incoming air, the air is cooled and dehumidified. This is what happens in case of a cooling coil. This is shown by the line *O-G*.

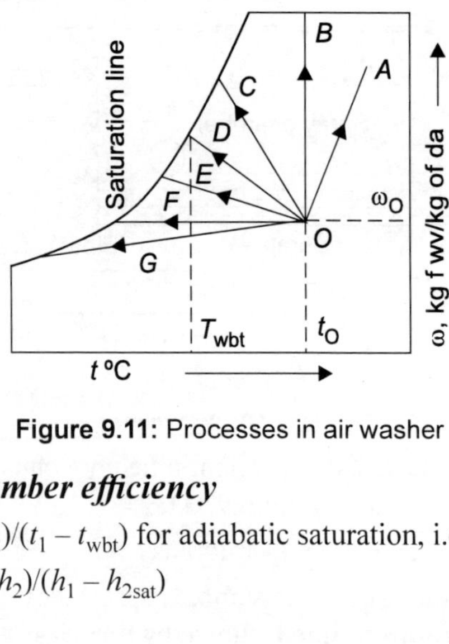

Figure 9.11: Processes in air washer

Air washer or spray chamber efficiency

It is expressed as $\eta = (t_1 - t_2)/(t_1 - t_{wbt})$ for adiabatic saturation, i.e., for $t_w = t_{wbt}$ [9.22]

Or else in general $\eta = (h_1 - h_2)/(h_1 - h_{2sat})$ [9.23]

Example 9.1

What will be the outlet condition of 10 cmm of moist air at 10°C and 40% RH at standard atmospheric pressure when it is heated by a heating coil of 2.5 kW?

Ans: This is a 'sensible heating' process.

So the empirical equation (Eq. 9.4) $q_{12} = 0.0204\ Q_V \times \Delta t_{dbt}$ gives t_2, i.e., outlet dbt of air is 2.5/(10 × 0.0204) + 10 = 22.25°C

The empirical relation is derived for standard air, i.e., $\rho = 1.2$ kg/m^3

To be more precise, this can be calculated from the first principle.

Enthalpy of moist air at inlet, $h_{in} = m\{(1.005 + 1.88\omega)t_{in} + 2500\omega$

Referring Example 8.6, specific humidity (ω) is calculated to be 0.003068 kg of wv/kg of da

Density of moist air is calculated from the equation, $\rho = \rho_a (1 + \omega)$ kg/m^3 of da

$\rho_a = P_d/R_aT = (101.325 - 0.49)/(0.287 \times 283) = 1.2415$ kg/m^3

So $\rho = 1.2415 (1 + 0.003068) = 1.2453$ kg/m^3

Mass of moist air, $m = (10/60) \times 1.2453 = 0.20755$ kg/s

Now $h_{out} - h_{in} = 0.20755(1.005 + 1.88 \times 0.003068)(t_{out} - t_{in}) = 2.5$ kW

$0.20978\ (t_{out} - 10) = 2.5$, So $t_{out} = 21.917$°C

The outlet dbt is 22.25°C as calculated above by empirical relation as above. The error in using empirical is within an acceptable range.

Example 9.2

Moist air at 42°C and 28% RH at standard atmospheric pressure flows at the rate 12 kg/s through an insulated duct. Water at 22°C is sprayed over the incoming air at a rate of 0.05 kg/s. What will be outlet state of moist air when all the water sprayed is evaporated.

Ans: Specific humidity (ω_1) of the incoming air is 0.0157 kg of wv/kg of da as calculated in earlier examples. So the specific enthalpy of moist air,

$h_{in} = (1.005 + 1.88\omega)t_{in} + 2500\omega = (1.005 + 1.88 \times 0.0157) \times 42 + 2500 \times 0.0157$

$= 82.7$ kJ/kg of da

h_f of water at 22°C is 92.2 kJ/ kg which is less than h_g of water at 42°C, i.e., 2577.9 kJ/kg.

So the moist air will be sensibly cooled.

Water vapour conservation, $m_a (\omega_2 - \omega_1) = m_{wv} = 0.05$, so $\omega_2 = (0.05/12) + 0.0157 = 0.0199$

Enthalpy of moist air at outlet, $h_2 = (1.005 + 1.88\omega_2)t_2 + 2500\omega_2$

But $(m_a + m_v)h_2 = m_ah_1 + m_vh_f(22°C) = 12 \times 82.7 + 0.05 \times 92.2 = 997.01$

So $h_2 = 997.01/12 = 83.084$ kJ/kg of da, Substituting this value in above equation

$t_2 = 31.98$°C.

Example 9.3

A room is to be maintained at 21°C dbt and 45% RH. Outdoor air is at 27°C dbt and 20°C wbt. What is the (a) sensible cooling load, (b) latent cooling load, (c) total cooling load for the fresh air flow rate of 1.5 kg/s.

Ans: Analytical method: Refer Examples 8.6 and 8.7 to calculate other properties of moist air analytically.

From steam tables, P_{Vsat} at 21°C is 2.48 kPa; at 27°C is 3.6 kPa; at 20°C is 2.34 kPa.

For indoor air, (RH) $0.45 = P_V/P_{Vsat} = P_V/2.48$; So $P_V = 1.116$ kPa

So specific humidity, ω = 0.622 × 1.116 / (101.325 – 1.116) = 0.00693 kg/kg

For outdoor air: From Carrier's equation, substituting the respective values, P_V is 1.883 kPa.

So specific humidity = 0.622 × 1.883 / (101.325 – 1.883) = 0.01178 kg/kg

Sensible heat cooling load, $Q_S = 0.0204 Q_V \Delta t = 0.0204\ (1.5/1.2) \times 60 \times (27 - 21) = 9.18$ kW

Latent heat cooling load, $Q_L = 50\ Q_V\ \Delta\omega = 50\ (1.5/1.2) \times 60 \times 0.01178 - 0.00693)$

$= 18.1875$ kW

Total cooling load = 9.18 + 18.1875 = 27.3675 kW.

(Moist air density is assumed 1.2 kg/m^3)

Alternatively: Initial enthalpy of moist air, 1.0216 × *t* (27°C) + ω (0.01178) × 2500 = 57.0332 kJ/kg

Final enthalpy, is 1.0216 × *t* (21OC) + ω (0.00693) × 2500 = 38.78 kJ/kg

So total cooling load is 1.5 × (57.0332 – 38.78) = 27.38 kW

Alternatively: From psychrometric chart (as shown in Fig. 9.12), fix the state points of moist air (*i* for indoor air and *o* for outdoor air) and read the respective values from constant enthalpy line. Procedure for reading of enthalpy is discussed in Chapter 8.

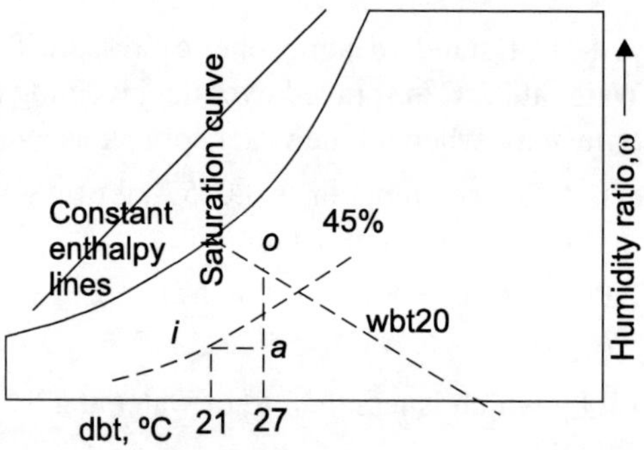

Figure 9.12: Example 9.3

Sensible heat cooling load = mass of air × ($h_a \times h_i$)

Latent heat cooling load = mass of air × ($h_o - h_a$)

Total cooling load = mass of air × ($h_o - h_i$)

For check: Sensible heat + Latent heat = Total heat

Note: There may be little variation (maximum 5%) amongst all these three methods of determining the enthalpy values because of the assumptions in analytical calculation and optical errors in using the chart.

Example 9.4

Moist air with a flow rate of 6 kga/s is at 18°C and 40% RH and 1.0 bar pressure. It is heated by 30 kW of heater and steam at 100°C at a rate of 0.1 kg/s is added to it. Calculate the outlet state and sensible and latent heat added.

Ans: Initial specific humidity of moist air, as calculated, is ω_1 = 0.622(0.4 × 0.02062)/(1 – 0.4 × 0.02060) = 0.005586 kg of wv/kg of da.

Specific enthalpy, $h_1 = (1.005 + 1.88\ \omega_1)t_1 + 2500\omega_1 = 32.244$ kJ/kg of da

Mass balance for water vapour, $m_a(\omega_2 - \omega_1) = 0.1$ kg/s, so $\omega_2 = 0.02225$

Total enthalpy of outgoing moist air, $H_2 = m_a h_1$ + heat added + $m_v(h_g(t_{\text{steam}}))$

$$= 6 \times 32.224 + 30 + 0.1(2675.44) = 490.888 \text{ kJ}$$

So $h_2 = 81.815$ kJ/kg of da = $(1.005 + 1.88\ \omega_2)t_2 + 2500\omega_2$

Substituting the values, $t_2 = 25.02$°C

Sensible heat transfer, $Q_S = m_a\ (1.005 + 1.88\omega_1)(t_2 - t_1) = 6.0(1.005 + 1.88 \times 0.005586)(25.02 - 18) = 42.77$ kW

Latent heat transfer = $m_a(\omega_2 - \omega_1)(2500 + 1.88\ t_1) = 6(0.02225 - 0.005586)(2500 + 1.88 \times 18) = 253.34$ kW

Total heat transfer = 42.77 + 253.34 = 296.11 kW

Check: total heat transfer = $m_a(h_2 - h_1) = 6 \times (81.815 - 32.224) = 297.546$ kW

Example 9.5

Moist air at 42°C, 70% RH and 1 bar pressure enters a cooling coil and leaves the coil at 20°C and saturated. The condensate also leaves at 20°C. Mass flow rate of air is 12 kga/s. Determine the condensate rate and the cooling capacity.

Ans: Specific humidity of the incoming air, $\omega_1 = 0.622(0.7 \times 0.082)/(1\text{–}0.7 \times 0.082) = 0.037877$ kg of wv/kg of da.

Specific enthalpy, $h_1 = (1.005 + 1.88\ \omega_1)t_1 + 2500\omega_1 = 139.89$ kJ/kg of da

Specific humidity of outgoing air, $\omega_2 = \omega_{\text{Sat}}(20°\text{C}) = 0.01476$

Specific enthalpy, $h_2 = (1.005 + 1.88\ \omega_2)t_2 + 2500\omega_2 = 57.555$ kJ/kg of da

Mass balance of water vapour, $m_w = m_{\text{da}}(\omega_1 - \omega_2) = 12(0.037877\text{–}0.01476) = 0.2774$ kgw/s

Energy balance, $m_a h_1 = m_a h_2 + Q_{12} + m_w h_f\,(20°\text{C})$

So $Q_{12} = 12 \times 139.89 - (12 \times 57.555 + 0.2774 \times 83.9) = 964.75$ kW = 274.33 TR

So condensate rate is 0.2774 kg/s and cooling capacity is 274.33 TR

Example 9.6

10 kga/s of moist air at 40°C and 60% RH adiabatically mixes with 4 kga/s of moist air at 28°C and 18°C dpt. Calculate the final state of the mixed air. Pressure is 1 bar.

Ans: For the first stream $\omega_1 = 0.622 \times 0.6 \times 0.07375/(1\text{–}0.6 \times 0.07375) = 0.0288$

Specific enthalpy, $h_1 = (1.005 + 1.88\omega_1)t_1 + 2500 \times \omega_1 = 114.366$

Similarly, for second stream, $\omega_2 = 0.622 \times 0.02062/(1\text{–}0.02062) = 0.0131$ (0.02062 bar is the saturated pressure at 18°C) and $h_2 = 61.58$

So $h_3 = (m_{a1}\ h_1 + m_{a2}\ h_2)\ /\ m_{a3} = (10 \times 114.366 + 4 \times 61.58)/(10 + 4) = 99.284$ kJ/kg

Specific humidity, $\omega_3 = (m_{a1}\omega_1 + m_{a2}\omega_2)/m_{a3} = (10 \times 0.0288 + 4 \times 0.0131)/14 = 0.02431$

So $t_3 \approx (m_{a1}t_1 + m_{a2}t_2)/m_{a3} = (10 \times 40 + 4 \times 28)/14 = 36.57$°C

This state point can also be determined from the psychrometric chart. Fix state points 1 and 2. Join the two points. State point 3 will be on the line in reverse proportion of the masses, i.e., m_{da1}/m_{da2} = length of point 2 to 3/length of points 1 to 3.

Adiabatic mixing with bondensation

Example 9.7

4 kga/s of moist air at 0°C and 90% RH is adiabatically mixed with 1 kga/s of moist air at 35°C and 70% RH. Find the state of the mixed air.

Ans: Like the previous example, ω_3, h_3 and t_3 are calculated to be 0.007783 kg of wv/kg of da, 26. 8252 kJ/kg of da and 7.226°C, respectively.

But at 7.226°C of dry bulb temperature, $\omega_S = 0.0063378$ which is less than ω_3. So the mixed air will be in 'fog region'. For this, the equation used is

$$h_4 = h_3 - (\omega_3 - \omega_4)\, h_{f4} = 26.8252 - (0.007783 - \omega_4)\, 4.1868 \times t_4$$

By trial and error, $t_4 = 8.88$°C and $\omega_4 = 0.0071$ kg of wv/kg of da

So h_4 is calculated to be 26.7998 kJ/kg of da

And condensate rate, $m_{a3}(\omega_3 - \omega_4) = 5(0.007783 - 0.0071) = 0.003414$ kg/s

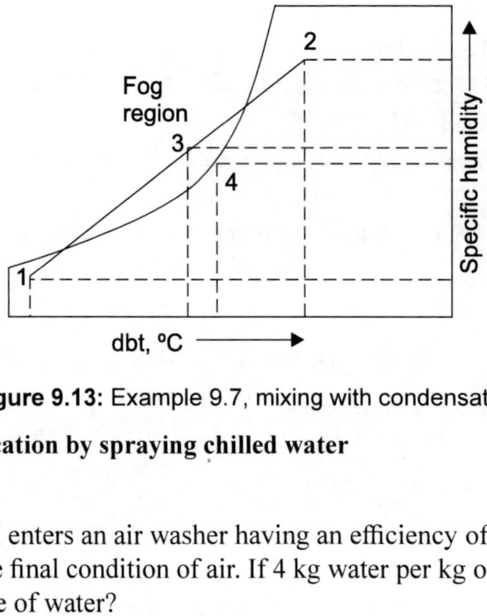

Figure 9.13: Example 9.7, mixing with condensation

Cooling and dehumidification by spraying chilled water

Example 9.8

Air at 23°C and 50% RH enters an air washer having an efficiency of 70% and is sprayed with water at 5°C. What is the final condition of air. If 4 kg water per kg of air is sprayed, what will be the rise in temperature of water?

Ans: P_{sat} at 23°C, from steam tables, is 2.837 kPa.

From given data $0.5 = P_{wv}/2.837$

So $P_v = 1.4185$ kPa and corresponding DPT is 12.5°C.

Hence, it is a cooling and dehumidification process as $t_w < t_{dpt}$

The process is represented by the process line *O-G* as shown in Fig. 9.13.

Specific humidity $\omega_1 = 0.622 \times 1.4185/(101.325 - 1.4185) = 0.00883$ kg of wv/ kg of da

So $h_1 = (1.005 + 1.88\omega_1)t_1 + 2500\omega_1 = 45.572$ kJ/kg of da

Now ω_{sat} at 5°C $= 0.622 \times 0.872/(101.325 - 0.872) = 0.0054$ kg of wv / kg of da

Again h_{sat} at 5°C $= (1.005 + 1.88\omega_{sat})t_{sat} + 2500\omega_{sat} = 18.576$ kJ/kg of da

Air washer efficiency = 0.7 = $(23 - t_2)/(23 - 5)$; Gives t_2 = 10.4°C;

Similarly, ω_2 = 0.00643 kg of wv/kg of da

Again 0.7 = $(45.572 - h_2)/(45.572 - 18\ 576)$; Gives h_2 = 26.675 kJ/kg of da

So heat lost by the air = 45.572– 26.675 = 18.897 kJ/kg of da = Heat gained by water

Temperature rise in water = 18.897/(4 × 4.18) = 1.13°C.

Cooling and dehumidification by desiccant and spray chamber

Example 9.9

Outdoor air is at 40°C and 70% RH, 1.0 atmospheric pressure. It is first dehumidified by a desiccant wheel and its temperature rises to 80°C. Then the air exchanged heat with the outdoor air in a heat exchanger having effectiveness 0.9. Then it is passed through a spray chamber having humidification efficiency 0.75. Find the exit state of the air from the spray chamber.

Ans: Refer Fig. 9.14.

P_{Sat} at 40°C is 0.07384 bar.

The specific humidity of outdoor air, ω_1 = 0.622 × 0.7 × 0.07384/(1.032 – 0.7 × 0.07384) = 0.034

Specific enthalpy, $h_1 = (1.005 + 1.88\omega_1)t_1 + 2500 \times \omega_1$ = 127.5 kJ/kg of da

The adsorption in the desiccant wheel is a constant enthalpy process. So h_2 = 127.5 kJ/kg of da

So from specific enthalpy equation ω_2 calculated to 0.01777 taking t_2 = 80°C

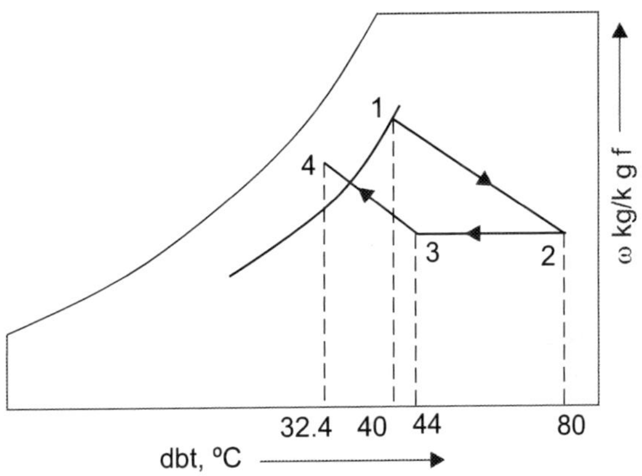

Figure 9.14: Cooling and dehumidification process with desiccant and spray chamber

Now the dehumidified air enters the heat exchanger.

$0.9 = (t_2 - t_3)/(t_2 - t_1)$, So t_3 = 44°C and corresponding RH (from psychrometric chart) is 30%.

Then the air passes through the spray chamber (or a desert cooler can be used for further cooling and humidification). So the conditioning process is along the constant wbt line. wbt of air at 3 is 28.5°C.

Spray chamber efficiency = 0.75 = $(t_3 - t_4)/(t_3 - t_{wbt})$, so t_4 = 32.4°C and the corresponding RH is 74%.

9.4 HOW A ROOM IS AIR CONDITIONED

Indoor air of a room gains heat, in the form of sensible heat, and moisture, in form of latent heat, from the outdoor air and from the occupants (may be human or articles in the conditioned space). These sensible heat gain and latent heat gain are discussed in chapter 10 'Cooling Load Calculation'. When the indoor air is to be conditioned to attain the desired state, it has to pass through one or combined psychrometric processes as mentioned above. However, the terms used in conditioning the indoor air are given below.

Room sensible heat (RSH): The heat gained by the indoor air due to heat conduction from roof, wall, electric equipment, etc., increases the room air temperature, i.e., dbt only. This is called RSH gain by the indoor air.

Room latent heat (RLH): The heat gained by the indoor air due to release of latent heat by the moisture from perspiration, tea/coffee making, etc.

Room total heat (RTH): RSH + RLH

Room sensible heat factor (RSHF): This is expressed as RSHF = RSH/RTH and RSHF line is drawn with a slope oftan θ as discussed in Section 9.2.3. This is shown in Fig. 9.6.

Outdoor air sensible heat (OASH): Sensible heat carried by the outdoor air by virtue of its dbt.

Outdoor air latent heat (OALH): Latent heat carried by the outdoor air due to the moisture present.

When outdoor air mixes with return air from indoor, the mixed air will have:

RSH + OASH = TSH (total sensible heat)

RLH + OALH = TLH (total latent heat); GTH (grand total heat) = TSH + TLH

Grand sensible heat factor (GSHF) = TSH/GTH and GSHF line is drawn in same way as RSHF.

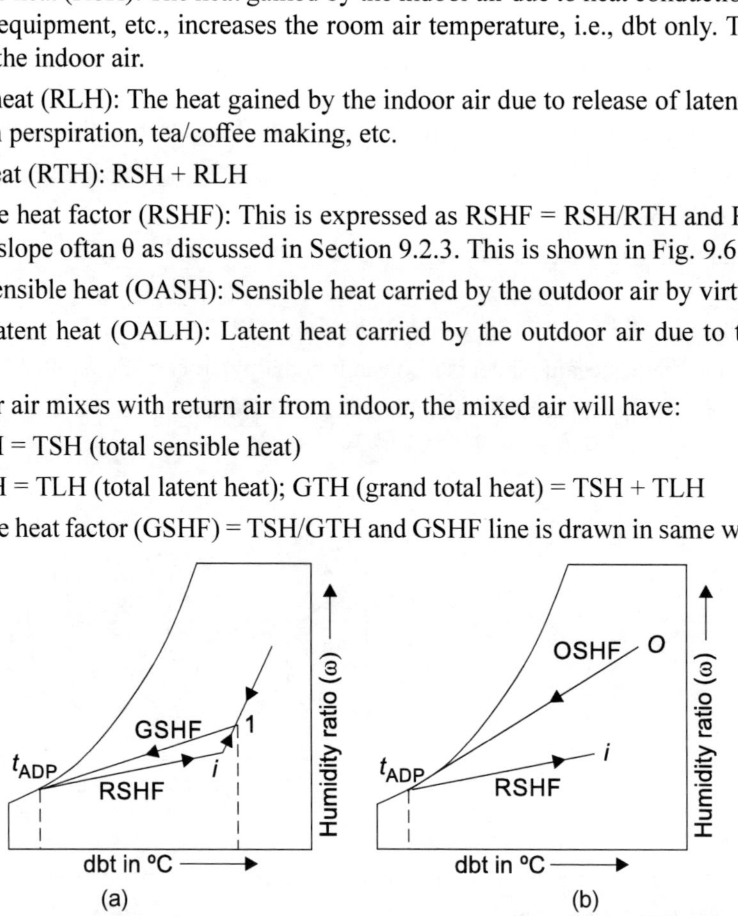

Figure 9.15: Psychrometric processes (a) ventilation without bypass factor, (b) 100% outdoor air without bypass factor

Three possible psychrometric processes for air conditioning a room are shown in Fig. 9.15 (a), (b) and (c). Fig. 9.15 (a) shows the process of conditioning the room with ventilation with zero bypass factor. '*O*' is the outdoor air state, '*I*' is the indoor and '1' is the mixed air state passing over the conditioning coil. It leaves the conditioning coil at t_{ADP}, so the supply air state is same as t_{ADP}. Supply air attains state '*I*' along RSHF line receiving the room total heat. Similarly, Fig. 9.15 (b) shows 100% ventilation (desirable for operation theatres) with zero bypass factor.

Outdoor air at '*O*' passes over the cooling coil and leaves it at t_{ADP}. So the supply air state to the room will also be at t_{ADP}. Then it moves along the RSHF line taking up the RTH and leaves the room at '*I*'. Fig. 9.15 (c) shows the process with ventilation and non-zero bypass factor of coil. Here the mixed air moves towards t_{ADP} but comes out of conditioning coil at '2' due to bypass factor. So the supply air state will be '2'. Then it moves along the RSHF line receiving RTH of the room and goes out as return air at '*I*'.

In this figure bypass factor of the coil is also shown as $X = (t_2 - t_{ADP})/(t_1 - t_{ADP})$.

Summer air conditioning with ventilation (air needs cooling and dehumidification)

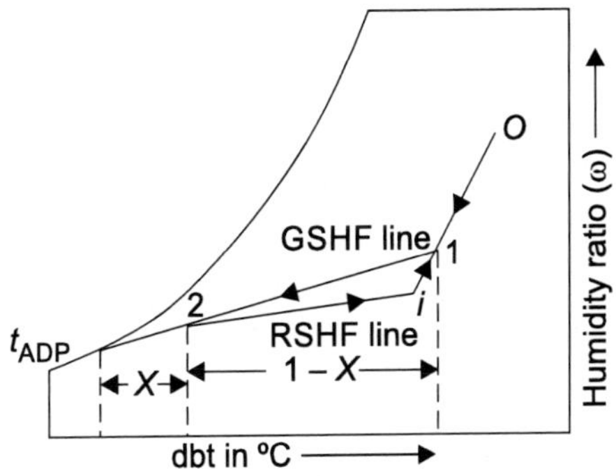

Figure 9.15: (c) Psychrometric processes: ventilation with bypass factor

Example 9.10

The outdoor air of a city in summer season is dbt 40°C and RH 70%. Indoor air condition is 24°C and 50% RH. The RSH and RLH gain are 40 kW and 10 kW, respectively. Bypass factor of the cooling coil is 0.15. Ventilation air is 20% of the total air. Calculate the required ADP of the cooling coil and its refrigerating capacity.

Ans: At the given condition of outdoor air it is difficult to get the state points on psychrometric chart as its size is small. So the different property values of moist air have been calculated analytically for 1.0 bar atmospheric pressure.

For outdoor air (Refer Fig. 9.16)

P_{Sat} at 40°C = 0.07384 bar, So $P_V = 0.7 \times 0.07384 = 0.05169$ bar

$\omega_O = 0.0339$ kg of wv/kg of da; $h_O = 1.005 \times t_{dbt} + \omega\,(1.88t_{dbt} + 2500) = 127.5$ kJ/kg of da

P_{Sat} at 24°C = 0.03003 bar, So $P_V = 0.5 \times 0.03003 = 0.0150$ bar

$\omega_i = 0.00948$ kg of wv /kg of da; $h_i = 48.25$ kJ/kg of da

$\omega_1 = 0.2\,\omega_O + 0.8\omega_i = 0.014364$ kg of *v*/kg of da; $h_1 = 0.2h_O + 0.8h_i = 64.1$ kJ/kg of da

$t_1 = 0.2t_O + 0.8t_i = 27.2$°C

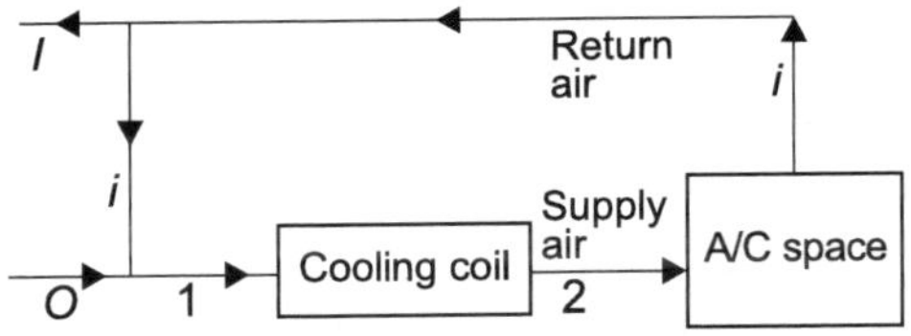

(a)

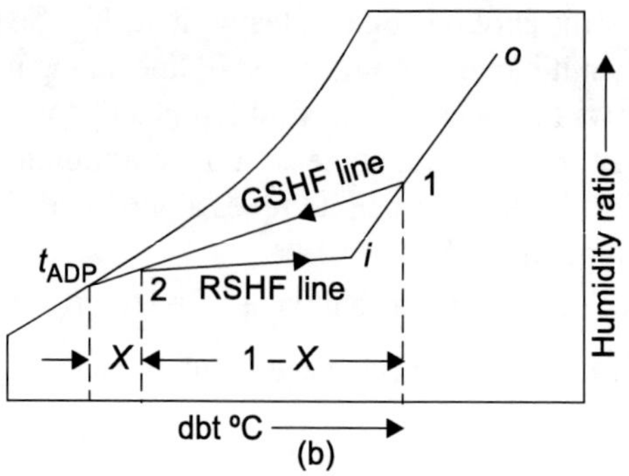

Figure 9.16: Example Problem 9.10. Cooling and dehumidification with bypass factor, (a) schematic of the process (b) psychrometric processes on the chart

RSH = 40 kW = 0.0204 (24 – t_2)$Q_{V,S}$; RLH = 10 kW = 50 (0.00948 – ω_2)$Q_{V,S}$

RSH/RLH = 4 = (0.0204/50){(24 – t_2)/(0.00948 – ω_2)}

So ω_2 = 0.007032 + 0.000102t_2 ...(1)

BPF = (t_2 – t_{ADP})/(t_1 – t_{ADP}) = 0.15; So t_2 = 4.08 + 0.85 t_{ADP} ...(2)

(ω_2 – ω_{ADP})/ (ω_1 – ω_{ADP}) = 0.15; So ω_{ADP} = (ω_2 – 0.002155)/(0.85) ...(3)

There are four variables and three equations.

Hence solution is done by trial and error.

Assume t_{ADP} is 10°C. Substituting this value in above equations, we get

t_2 = 12.58°C, ω_2 = 0.008315, and ω_{ADP} = 0.00725

Now the corresponding t_{ADP} from psychrometric chart is 9°C (does not match with assumed value).

Next assumption, t_{ADP} = 9°C

The corresponding values, t_2 = 11.73, ω_2 = 0.00823, ω_{ADP} = 0.00715

This value of ω_{ADP} gives t_{ADP} = 9°C. (It matches the assumed value.) So t_{ADP} = 9°C

So h_2 = 1.005 × t_2 + ω_2 (1.88t_2 + 2500) = 32.545 kJ/kg of da

Mass of supply (Total) air = RTH / (h_i – h_2) = (40 + 10) kW/ (48.25 – 32.545) = 3.183 kg/s

Mass of ventilating air = 3.183 kg/s / 5 = 0.6366 kg/s

Refrigerating capacity of the cooling coil = 3.183 (h_1 – h_2) = 3.183 (64.1 – 32.545) = 100.44 kW

For a check: Load on the cooling coil = Ventilation load + RTH = 0.6366 (h_O – h_i) + 50 = 100.45 kW

Alternative solution: This can also be solved with the psychrometric chart.

1. Outdoor air and indoor air state points are located on the psychrometric chart as '*O*' and '*i*'.
2. State point '1' is located by taking the inverse of air mass ratio, i.e., 1:4.
3. Draw the RSHF line through '*i*' and slope tanθ = 0.408 (1/SHR -1).
4. Draw a line through '1' by trial and error such that it cuts RSHF line at a point where the BPF (0.15) condition is satisfied.
5. This line through '1', where it cuts the saturation line is ADP. Then other properties can be evaluated.

Example 9.11

Outdoor air is 35°C and 70% RH. Indoor air condition is 24°C and 50% RH. RSH gain and RLH gain of the room are 300 kW and 100 kW, respectively. 20% fresh air is mixed with 80% of return air. If the minimum temperature of supply air is need to be 12°C, find the refrigerating capacity and bypass factor of the cooling coil.

Ans: From steam tables P_{Sat} at 35°C is 0.05628 bar, So P_V from RH relation 0.0394 bar.

P_{Sat} at 25°C is 0.03003 bar, so P_V is 0.0150 bar

AS shown in the above examples

$\omega_O = 0.02551$ kg of v/kg of da; $h_O = 1.005 \times t + \omega\ (1.88t + 2500) = 100.6286$ kJ/kga

$\omega_i = 0.00948$ kg of wv /kg of da; $h_i = 48.25$ kJ/kga

$\omega_1 = 0.2\ \omega_O + 0.8\omega_i = 0.012686$ kg of v/kg of da; $h_1 = 0.2\ h_O + 0.8\ h_i = 58.72572$ kJ/kga

$t_1 = 0.2t_O + 0.8t_i = 27.0$°C; t_2 (given) 12°C

$h_2 = 1.005 \times t_{dbt} + \omega\ (1.88t_{dbt} + 2500) = 31.857$ kJ/kga

RSH = 300 kW = $0.0204\ (24 - t_2)Q_{VS}$; RLH = 100 kW = $50\ (0.00948 - \omega_2)Q_{VS}$

RSH/RLH = 3 = $(0.0204/50)\{(24 - 12)/(0.00948 - \omega_2)\}$;

So $\omega_2 = 0.007848$ kg of wv/kga

Now from triangle similarity $(\omega_1 - \omega_2)/(t_1 - t_2) = (\omega_1 - \omega_S)\ /(t_1 - t_S)$

Substituting the respective values $0.0003225 = (0.012686 - \omega_S)/(27 - t_S)$

$\omega_S = 0.00398 + 0.0003225t_S$

By trial and error let us find the value of t_S.

Assume $t_S = 5$; $\omega_S = 0.0056$

From equation, $\omega_S = 0.622\ P_S/(P - P_S)$; $P_S = 0.00892$ bar

The corresponding saturation temperature from steam table is 5.5 (does not match)

Next assumption is 7; $\omega_S = 0.00624$ and $P_S = 0.01$ bar

The corresponding saturation temperature is nearly 7 and it does match.

So t_S or t_{ADP} is taken as 7°C.

Supply air quantity is calculated from m = RTH/$(h_i - h_2)$ = 400/(48.25 – 31.857) = 24.4 kg/s

Refrigerating capacity, $m(h_1 - h_2) = 655.6$ kW

BPF of coil, $(t_2 - t_{ADP})/(t_1 - t_{ADP}) = 0.25$ (of course, a higher value).

Alternatively, this can be solved with the help of psychrometric chart as already been discussed. Find the state points *O*, *i* and 1 on the chart. Draw the RSHF line through *i*. This will intersect the vertical line 12°C and the state point will be 2. Then join the points 1 and 2 and extend it so that it intersects the saturation line at ADP. Thus, state point t_{ADP} is determined. Then other data can be calculated.

Example 9.12

Atmospheric air at 101.3 kPa, 35°C and 70% RH is to be conditioned to 20°C and 60% RH. This is achieved first by cooling, followed by dehumidification and then by reheating. The cooling coil ADP is 10°C. The bypass factor of the heating coil is 0.25. Determine (a) cooling coil bypass factor and cooling coil capacity, (b) heating coil surface temperature and heating coil capacity, and (c) water removal rate.

Ans: Figure 9.17 shows the conditioning process on psychrometric chart.

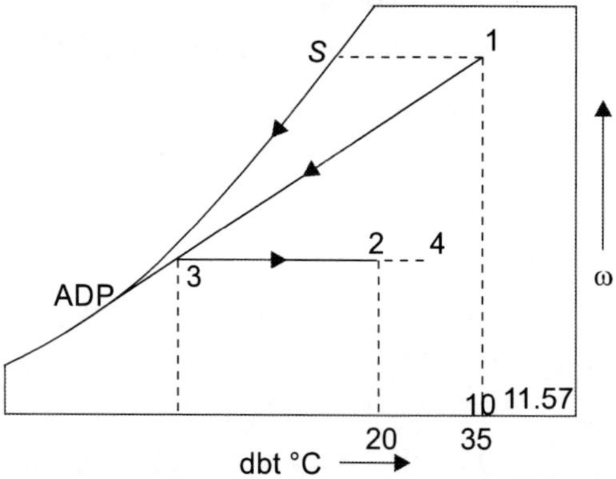

Figure 9.17: Example 9.12

Property values at state point 1: P_S at 35°C is 5.628 kPa (from steam tables)

RH 0.7 = P_v/P_S; P_v = 3.94 kPa;

So specific humidity, $\omega_1 = 0.622 \times 3.94/\ (101.3 - 3.94) = 0.0252$ kg/kga

Specific enthalpy, $h_1 = 1.005 \times 35 + 0.0252 \times 2565.3$ (i.e., h_g at 35) = 99.82 kJ/kga

Specific volume, $v_1 = R_aT/P_a = 0.287 \times 308/(101.3 - 3.94) = 0.908$

So mass flow rate of air, $m_a = 300/0.908 = 330.4$ kga/min

Similarly property values at state point 2: P_S at 20°C is 2.339 kPa

Specific humidity, $\omega_2 = 0.00874$ kg/kga; specific enthalpy, $h_2 = 42.283$ kJ/kga

Similarly, property values at state point ADP: P_S at 10°C is 1.228 kPa

Specific humidity $\omega_{ADP} = 0.007633$ kg/kga; specific enthalpy $h_{ADP} = 29.284$ kJ/kga

Cooling coil bypass factor, $(\omega_3 - \omega_S)/(\omega_1 - \omega_S)$

$$= (0.00874 - 0.007633)/(0.0252 - 0.007633) = 0.063$$

Again, $0.063 = (h_3 - h_{ADP})/(h_1 - h_{ADP}) = (h_3 - 29.284)/(99.82 - 29.284)$; $h_3 = 33.73$ kJ/kga

Again, $0.063 = (t_3 - 10)/(t_1 - 10)$; $t_3 = 11.575$°C

Cooling capacity = (330.4/60) × (99.82 – 33.73) = 364.0 kW = 103.4 TR

Heating coil bypass factor is 0.25 and t_4 is the heating coil surface temperature.

So $0.25 = (t_4 - t_2)/(t_4 - t_3)$; $t_4 = 22.8$°C

Heating coil capacity, $(m_a/60) \times (h_2 - h_3) = (330.4/60) \times (42.283 - 33.73) = 47.1$ kW

Water removal rate, $m_a \times (\omega_2 - \omega_1) = 330.4 \times (0.0252 - 0.00874) = 5.438$ kg/min

Drying Process: (Heating and Dehumidification)

Problem 9.13

Hot air at 138 kPa, 65°C and 50% RH enters a seed drier and comes out at 103.5 kPa, 60°C and being Saturated. Calculate the loss of water and amount of heat transfer per kg of dry air from the seed.

Ans: Property values at state point 1: P_S at 65°C is 25.03 kPa (from steam tables).

So, $0.5 = P_v/25.03$; $P_v = 12.515$

So $\omega_1 = 0.622 \times 12.515/(138 - 12.515) = 0.062$ kg/kga

and $h_1 = 1.005 \times 65 + 0.062 \times 2618.3$ (i.e., h_g at 65) = 227.66 kJ/kga

Property values at state point 2: P_S at 60°C is 19.94 kPa (from steam tables).

So, $\omega_2 = 0.622 \times 19.94/(103.5 - 19.94 = 0.148$ kg/kga

and $h_2 = 1.005 \times 60 + 0.148 \times 2609.6$ (i.e., h_g at 60) = 446.521 kJ/kga

So, loss of water, $m_a \times (\omega_2 - \omega_1) = 1 \times (0.148 - 0.062) = 0.086$ kg/kga

And heat transfer rate, $m_a \times (h_2 - h_1) = 1 \times (446.521 - 227.66) = 218.86$ kJ/kga

Example 9.14

Outdoor condition: dbt 32°C and 50% RH and indoor condition: 24°C and 50% RH. Outdoor air, after passing over a cooling coil, mixes with 65% of return air. Cooling coil ADP is 10°C and its bypass factor is 0.1. Room total heat load is 18 kW. Calculate (a) the state of the supply air, (b) cooling load on the coil, and (c) RSH and RLH.

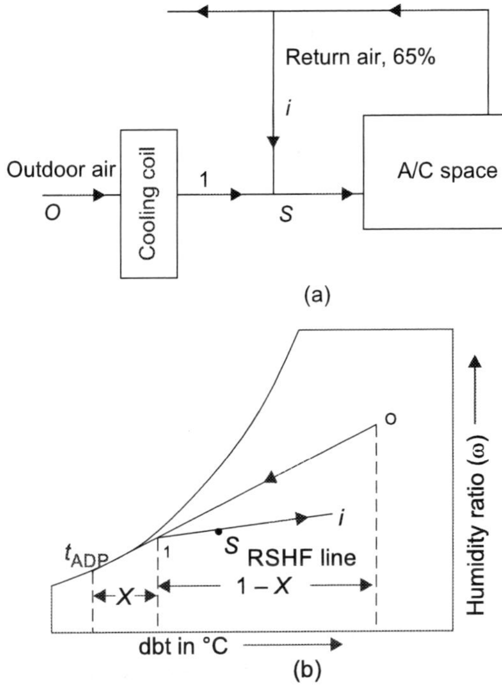

Figure 9.18: Mixing of return air with outdoor air after cooling coil
(a) schematic, (b) processes on psychrometric chart

Ans: Figures 9.18 (a) and (b) show the schematic of the air conditioning system and the state points on psychrometric chart.

At 32°C and 50% RH; $\omega_O = 0.015$ kg/kga, $h_O = 70.63$ kJ/kga

At 24°C and 50% RH; $\omega_i = 0.00934$ kg/kga, $h_i = 47.89$ kJ/kga

(Calculations are done as shown in Examples 8....)

At ADP 10ºC, $\omega_{ADP} = 0.622 \times 1.228/(101.325 - 1.228) = 0.00763$ kg/kga

For cooling coil:

Bypass factor, $0.1 = (t_1 - 10)/(32 - 10)$; So $t_1 = 12.2$°C

Similarly, $0.1 = (\omega_1 - 0.00763)/(0.015 - 0.00763)$; $\omega_1 = 0.00837$ kg/kga

Specific enthalpy, $h_1 = 1.005 \times 12.2 + 0.00837 \times 2523.5 = 33.38$ kJ/kg

For mixing of return air and conditioned air from cooling coil:

$65/35 = (12.2 - t_S)/(t_S - 24)$; $t_S = 19.87$ºC

$65/35 = (0.00837 - \omega_S)/(\omega_S - 0.00934)$; $\omega_S = 0.009$ kg/kga

$65/35 = (33.38 - h_S)/(h_S - 47.89)$; $h_S = 42.81$ kJ/kg

Total cooling load, 18 kW $= m_S \times (h_i - h_S) = m_S \times (47.89 - 42.81) = 3.5433$ kg/s

So outdoor air, $m_O = 0.35 \times 3.5433 = 1.24$ kg/s

Load on cooling coil $= 1.24 \times (h_O - h_1) = 1.24 \times (70.63 - 33.38) = 46.2$ kW.

RSH $= 0.0204 \times (Q_V)(t_i - t_S) = 0.0204 \times (3.5433 \times 0.85 \times 60) \times (24 - 19.87) = 15.225$ kW

(Specific volume of air $= 0.287 \times 293/(101.325 - 2.339) = 0.85$)

RLH $= 50 \times (Q_V) \times (\omega_i - \omega_S) = 50 \times (3.5433 \times 0.85 \times 60)(0.00934 - 0.009) = 3.072$ kW

Example 9.15

Outdoor condition: dbt 32ºC and 50% RH and indoor condition: 24ºC and 50% RH. Outdoor air mixes with 65% of return air. Cooling coil ADP is 10ºC and its bypass factor is 0.1. Room total heat load is 18 kW. Calculate (a) the state of the supply air, (b) cooling load on the coil, and (c) RSH and RLH.

Ans: Refer Figs. 9.16 a and b.

Air at state '*O*': At 32ºC and 50% RH; $\omega_O = 0.015$ kg/kga, $h_O = 70.63$ kJ/kga

Air at state '*I*': At 24ºC and 50% RH; $\omega_i = 0.00934$ kg/kga, $h_i = 47.89$ kJ/kga

Mixed air at state '1': $m_O/m_i = 7/13 = (t_i - t_1)/(t_1 - t_O) = (\omega_i - \omega_1)/(\omega_1 - \omega_O) = (h_i - h_1)/(h_1 - h_O)$

Substituting the values, $t_1 = 26.8$°C, $\omega_1 = 0.01132$ kg/kga, $h_1 = 55.85$ kJ/kga

Air at state '2': $0.1 = (t_2 - t_{ADP})/(t_1 - t_{ADP}) = (\omega_2 - \omega_{ADP})/(\omega_1 - \omega_{ADP})$

Substituting the values, $t_2 = 11.68$ºC, $\omega_2 = 0.008$

Specific enthalpy, $h_2 = 1.005 \times t_2 + \omega_2 \times (2500 + 1.88 \times t_2)$

$= 1.005 \times 11.68 + 0.008 \times (2500 + 1.88 \times 11.68) = 31.914$ kJ/kg

Room total load = 18 kW $= m_{a1}$ X $(h_i - h_2) = m_{a1} \times (47.89 - 31.914)$

So mass flow rate of air, $m_{a1} = 1.1267$ kg/s

Load on cooling coil, $m_{a1} \times (h_1 - h_2) = 1.1267 \times (55.85 - 31.914) = 26.97$ kW

In the previous example with the same data, load on the cooling coil is 46.2 kW, whereas in this case it is 26.97 kW. It shows conditioning of air after mixing of return air and outdoor air saves much amount of energy.

Winter Air Conditioning: (Heating and Humidification)

Example 9.16: (Spray of hot steam for heating and humidification)

Atmospheric air at 101 kPa, 15°C and 20% RH is humidified by saturated steam at 100°C. Air flow rate is 0.36 kg/s and steam added is 0.0025 kg/s. Calculate the dbt and the wbt of the air leaving the humidifier.

Ans: Saturation pressure of water vapour at 15°C is 1.705 kPa.

$0.2 = P_v/P_S = P_v/1.705$. This gives $P_v = 0.341$ kPa

Specific humidity of atmospheric air, $\omega_1 = 0.622 \times 0.341/(101 - 0.341) = 0.0034$ kg/kga

Specific enthalpy, $h_1 = 1.005 \times 15 + 0.0034 \times 2528.9$ (i.e., $h_{g(15)}$) = 23.67 kJ/kga

At 100°C, steam, $h_g = 2676.1$ kJ/kg and steam added is 0.0025 kg/s.

So, $(\omega_2 - \omega_1) = 0.0025/0.36 = 0.006944$ kg/kga; $\omega_2 = 0.010344$ kg/kga

$(1.005 + 1.88\ \omega_2)\ (t_2 - t_1) = (\omega_2 - \omega_1) \times (h_w - h_g(t_1))$

Let us substitute the numerical values in the above equation.

Humid specific heat = $(1.005 + 1.88 \times \omega_2) = 1.005 + 1.88 \times 0.010344 = 1.02445$ kJ/kga-°C

$1.02445\ (t_2 - t_1) = 0.006944\ (2676.1 - 2528.9)$; $t_2 - t_1 = 1$°C

So $t_1 = 15 + 1 = 16$°C.

To calculate the wbt, equation used is $h_S - h_2 = (\omega_S - \omega_2) \times h_f$ (at wbt)

Enthalpy, $h_2 = 1.005 \times 16 + 0.010344\ (2530.8) = 42.26$ kJ/kga

Now, by trial and error, the wbt is estimated to be 15.1°C

Also from psychrometric chart, wbt is found to be 15.1°C.

Note: ω_S at 16°C is 0.0114 kg/kga. It is more than ω_2 (0.010344 kg/kga). So the exit air is unsaturated. If more amount of steam is added and ω_2 value exceeds ω_S, the exit air is saturated and the excess water condenses out.

Example 9.17

35 m^3/min of moist air at 10°C and 70% RH enters an air conditioning system. Air passes over a heating coil and then humidified by saturated water vapour at 100°C. Total pressure of the system is 95 kPa. Air comes out at 20°C and 60% RH. Calculate (a) the temperature and RH of air leaving the heating section, (b) rate of heat transfer in heating section, (c) rate at which water is added during the humidification process.

Ans: For entry moist air

$0.7 = P_v/P_S$ (at 10) $= P_v/1.228$; $P_v = 0.86$ kPa

Specific humidity, $\omega_1 = 0.622 \times 0.86/(95 - 0.86) = 0.0057$

Enthalpy $h_1 = 1.005 \times 10 + 0.0057 \times 2519.2 = 24.41$ kJ/kga

For exit air

$0.6 = P_v/P_S$ (at 20) $= P_v/2.3392$; $P_v = 1.404$ kPa

Specific humidity, $\omega_2 = 0.622 \times 1.404/(95 - 1.404) = 0.00933$

Enthalpy, $h_2 = 1.005 \times 210 + 0.00933 \times 2537.4 = 43.774$ kJ/kga

Density of air, $\rho_1 = P_d/R_aT = (95 - 0.86)/(0.287 \times 283) = 1.16$ kg/m^3

So mass of air at inlet $= 1.16 \times 35/60 = 0.6767$ kg/s

From mass conservation of water vapour, $m \times \omega_1 + m_v = m \times \omega_2$

This gives mass of water vapour added, $m_v = 0.6767 \times (0.00933 - 0.0057)$

$= 0.002456$ kg/s of W.V.

Energy equation of air before (subscript i) and after humidification gives

$(m + m_v) \times h_2 - m_v \times h_v = m \times h_i$; substituting the values

$(0.6767 + 0.002456) \times 43.774 - 0.002456 \times 2675.6 = 0.6767 \times h_i$

So $h_i = 34.22$ kJ/kga

For sensible heating, humid specific heat can be calculated from

$(1.005 + 1.88 \times \omega_1) = 1.016$ kJ/kg-°C

$(h_i - h_1) = 1.016 \times (t_i - t_1)$; $(34.22 - 24.41) = 1.016 \times (t_i - 10)$; So $t_i = 19.65$°C

Again P_S at 19.65 = 2.27 kPa; So RH = (0.86/2.27) 100 = 37.9%

Rate of sensible heat transfer, $Q = 0.6767 \times (34.22 - 24.41) = 6.64$ kW = 398 kJ/min

Example 9.18

An air conditioning system consists of a heating section followed by an evaporative cooling section. Air, at a flow rate of 30 m^3/min, enters the system at 10°C, RH 70% and leaves the system at 20°C, 60% RH. Atmospheric pressure is at 100 kPa. Calculate (a) the temperature and RH of the air leaving the heating section, (b) the rate of heat transfer in heating section, and (c) the rate of water vapour added to air in evaporative cooler.

Ans: Refer Fig. 9.19.

From steam tables P_S at 10°C is 1.2281 kPa and P_S at 20°C is 2.3392 kPa

For state point 1, $P_{v1} = 0.7 \times 1.2281 = 0.85967$ and $\omega_1 = 0.622 \times 0.85967/(100 - 0.85967) = 0.0054$

And $h_1 = 1.005 \times 10 + 0.0054 \times 2519.2$ (i.e., h_g at 10) = 23.65 kJ/kga

Specific volume, $v_a = R_aT/P_a = 0.287 \times 283/(100 - 0.85967) = 0.82$ m^3/kga

So mass of air = 30/0.82 = 36.6 kga/min

For state point 2, $P_{v2} = 0.6 \times 2.3392 = 1.4035$ and $\omega_2 = 0.622 \times 1.4035/(100 - 1.4035) = 0.00885$

And $h_2 = 1.005 \times 20 + 0.00885 \times 2537.4$ (i.e., h_g at 20) = 42.556 kJ/kga

As the evaporative cooling occurs along the constant wbt line, which is also approximately same as constant enthalpy line, it can safely be assumed that $h_2 = h_3$

So h_3 can also be expressed as $(1.005 + 1.88\omega_3) + 2500\omega_3$;

But $\omega_1 = \omega_3 = 0.0054$

This gives $42.556 = (1.005 + 1.88 \times 0.0054) \times t_3 + 2500 \times 0.0054$; $t_3 = 28.6°C$

$RH = P_{v1}/P_{S(at\ 28.6)} = 0.85967/3.9453 = 21.8\%$

So heat transferred in heating section $= m_a \times (h_3 - h_1) = 36.6 \times (42.556 - 23.65) = 692$ kJ/min

During evaporative cooling, water vapour added $= m_a \times \Delta\omega = m_a(\omega_2 - \omega_3)$

$= 36.6 \times (0.00885 - 0.0054) = 0.1263$ kg of WV/min.

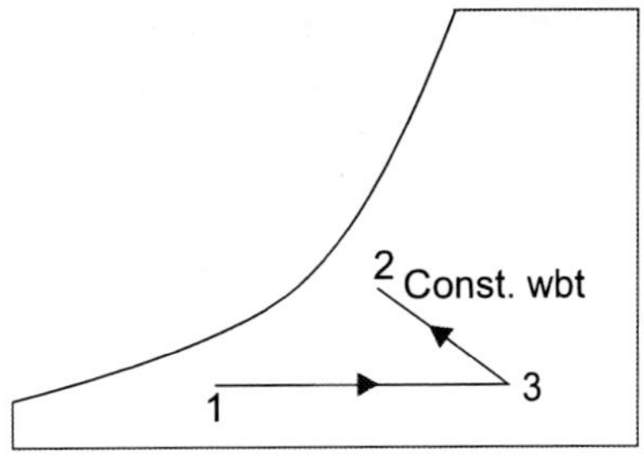

Figure 9.19: Example 9.18

Example 9.19

Moist air at 30°C and 80% RH is cooled to saturated air at 20°C. The total pressure is 101 kPa. Calculate the amount of water removed and cooling requirement per kg of dry air. The liquid water (condensed) leaves the system at 22°C.

Ans: $RH = 0.8 = P_v/P_S$;

From steam table, P_S is 4.2469 kPa at 30°C. So $P_v = 0.8 \times 4.2469 = 3.4$ kPa

Specific humidity $\omega_1 = 0.622 \times 3.4/(101 - 3.4) = 0.02167$ kg/kga

From steam table P_S of saturated air at 20°C is 2.3392 kPa;

Specific humidity at saturation, $\omega_S = 0.622 \times 2.3392/(101 - 2.3392) = 0.01475$ kg/kga

So water removed $= 0.02167 - 0.01475 = 0.00692$ kg/kga

Specific enthalpy of entry air $= 1.005 \times 30 + 0.02167 \times 2555.6(h_g$ at $30) = 85.53$ kJ/kga

Specific enthalpy of exit air $= 1.005 \times 20 + 0.01475 \times 2537.4(h_g$ at $20) = 57.527$ kJ/kga

Water going out at 22°C which carries heat of $Q_W = 0.00692 \times 92.281(h_f$ at $22) = 0.64$ kJ/kga

From energy balance around the cooling coil, $h_1 =$ Cooling load $Q + Q_W + h_2$

This gives the cooling requirement $Q = (85.53 - 57.527) - 0.64 = 27.36$ kJ/kga

EXERCISE

1. What happens to dbt of moist air during chemical dehumidification? Justify your answer.
2. What is the use of an air washer?
3. A sample of moist air has dbt 35°C and dpt 22°C. Calculate the RH of the sample air. If dbt is decreased by 5°C, what is the dpt and what is the value of RH? (dpt remains same)
4. Determine the humidity ratio of a sample of moist air whose dbt is 37°C and dpt is 22°C. What will be the humidity ratio if air gets saturated at that dbt.
5. Moist air at 10°C and 20% RH is heated to 22°C and 50% RH. Calculate the amount of heat transfer rate and the amount of water added if the mass flow rate of air is 2.0 kg/s. Atmospheric pressure is 100.0 kPa, (56 kJ/s, 48 kg /hr).
6. Moist air at 37°C and 10% RH is humidified while passing through a series of water laden wicks. Its temperature becomes 24°C. Calculate the relative humidity of the exit air and mass flow rate of water trough wicks per hour if the air flow rate is 900 kg/hr. Atmospheric pressure is 100 kPa, (50%, 4.9 kg/hr).

7. Moist air at 33°C and 80% RH is conditioned to 22°C and 50% RH. The air is first cooled to saturation, dehumidified and then it is sensibly heated. Calculate the heat removed and the water removed per hour and the heat added. Atmospheric pressure is 100.0 kPa. (134 kW, 127 kg/hr, 22 kW).

8. Define the terms: RSH, RLH, RTH, GTH, RSHF and GSHF. Show RSHF and GSHF on a psychrometric chart.

9. Atmospheric air at 1.013 bar enters an adiabatic saturation chamber with dry bulb temperature of 35°C. The adiabatic saturation temperature obtained at the end of experiment is 23°C. Find specific humidity, partial pressure of water vapour, relative humidity, dew point temperature and enthalpy of moist air.

10. Moist air enters a chamber at 5°C dbt and 2.5°C wbt at flow rate of 90 cmm. Barometer pressure is 101325 Pa while passing through the chamber, picks up 40.7 kw of sensible heat and also picks up 40 kg/hr of saturated steam 110°C. Determine the dbt and wbt of leaving state at 110°C.

11. What would be the final dbt, wbt and RH if 2 m^3 of air at 293 K dbt and 288 K wbt mixes with 3 m^3 of air at 273 K dbt and 80% RH?

12. Moist air at 35°C and 100% RH is entering a psychrometric device and leaving at 25°C and 100% RH. Name the appropriate device and justify it.

13. What is contact factor? Explain its usefulness.

14. 100 cmm of atmospheric air at 35°C dbt and 28°C wbt enters a cooling coil having ADP of 18°C and bypass factor 0.2. After the cooling, air enters an adiabatic dehumidification and comes out at 35°C. Find (a) amount of dehumidification in cooling coil in kg/hr, (b) amount of adiabatic dehumidification in kg/hr, and (c) total cooling coil capacity in term of TR.

15. Sketch and explain the principle of air washer.

16. Define humidity efficiency of an air washer.

17. A laboratory having an unusually large latent heat gain is required to be air conditioned. The design conditions and loads are as follows:

Outside conditions	: 40°C dbt, 27°C wbt
Inside conditions	: 25°C dbt, 50% RH
RSH	: 35 kW
RLH	: 20 kW
Cmmo	: 85 cmm

    Determine the followings:

    (i) Ventilation load

    (ii) Room and effective sensible heat factors

(iii) ADP and amount of re-heat for economical design

(iv) Supply of air quantity

(v) Conditioning of air entering and leaving coil and supply air temperature

(vi) Grand total heat assume a suitable BPF

MULTIPLE–CHOICE QUESTIONS

1. Sensible heating of moist air increases dry bulb temperature and decreases ——

(a) specific humidity (b) dew point temperature

(c) relative humidity (d) saturation temperature

2. Pure humidification process involves —— transfer.

(a) only mass (b) only energy

(c) mass and energy (d) none

3. Chemical dehumidification process involves

(a) increase in dbt (b) decrease in dbt

(c) increase in wbt (d) decrease in wbt

4. Chemical dehumidification process occurs along constant —— line.

(a) dbt (b) relative humidity

(c) specific humidity (d) enthalpy

5. Air Washer is a device that

(a) cleans the air (b) oxygenates the air

(c) humidifies the air (d) removes the air

6. Bypass factor of a coil is expressed by if m_b is the mass bypassed and m is total mass

(a) m_b/m_b (b) $(1–mb)/m$

(c) m/m_b (d) $m_b/(1–m)$

7. If two streams of air mix adiabatically, the state of mixed air will lie —— the line joining the two states

(a) perpendicular (b) at an angle of 60°

(c) on (d) at an angle of 45°

8. If two streams of air mix adiabatically, the state point of mixed air will lie on the line joining the two states and at a distance

(a) directly proportional to mass

(b) inversely proportional to mass

(c) directly proportional to specific humidity,

(d) inversely proportional to specific humidity

9. Bypass factor of a cooling coil —— cooling load.

(a) increases (b) decreases,

(c) keeps unchanged

10. The moist air going out of the air washer may be

(a) cooled (b) humidified

(c) both (d) thermodynamically unaffected

11. If the volume of moist air with 50% relative humidity is isothermally reduced to half of its original volume, then relative humidity of moist air becomes

(a) 25% (b) 60%

(c) 75% (d) 100%

12. Evaporative air-cooler is used effectively when

(a) dry bulb temperature is very close to the wet bulb temperature

(b) dry bulb temperature is high and relative humidity is high

(c) dry bulb temperature is low and relative humidity is high

(d) dry bulb temperature is high and the relative humidity is low

13. For which one of the following dbt, wbt and dpt has the same value?

(a) 0 per cent relative humidity line (b) 100 per cent relative humidity line

(c) 50 per cent relative humidity line (d) None of the above

14. If a sample of moist air of 50% relative humidity at atmospheric pressure is isothermally compressed to a pressure of two atmospheres, then

(a) its relative humidity will reduce to 25%

(b) its relative humidity will remain unchanged

(c) the sample of air will become saturated

(d) saturation pressure will increase to twice the value

15. A sample of moist air is at a temperature T and relative humidity 50%. Apart of the moisture is removed adiabatically by using an adsorbent. If the heat of adsorption is negligible, the resulting air will have the same

(a) dry bulb temperature but a lower wet bulb temperature

(b) wet bulb temperature but a higher dry bulb temperature

(c) dry bulb temperature but a higher wet bulb temperature

(d) wet bulb temperature but a lower dry bulb temperature

16. Moist air enters the cooling coil with mass flow rate of 10 kga/s at dry bulb temperature of 30°C and humidity ratio of 0.017 kg/kga. It leaves the cooling coil at dry bulb temperature of 16°C and humidity ratio of 0.008 kgw/kgda. If specific heat of humid air is 1.02 kJ/kgda-K and latent heat of water vapour is 2500 kJ/kgw. The sensible and latent heat transfer of cooling coil are, respectively

(a) 140 kW and 25000 kW
(b) 142.8 kW and 2.25 kW
(c) 142.8 kW and 225 kW
(d) 225 kW and 142.8 kW

17. In an air-conditioning plant the refrigeration load on the coil is 100 TR. Mass and enthalpy of air leaving the coil are 420 kg/minute and 40 kJ/kg, respectively. What will be the enthalpy of the air at the inlet to the coil under these conditions?

(a) 80 kJ/kg
(b) 90 kJ/kg
(c) 100 kJ/kg
(d) 102.5 kJ/kg

18. The state of air supplied by a cooling coil with a by-pass factor X lies on the psychometric chart at the

(a) intersection of RSHF line with saturation curve
(b) intersection of GSHF line with saturation curve
(c) point which divides RSHF line in proportion to X and $(1 - X)$
(d) point which divides GSHF line in proportion to X and $(I - X)$

19. In the case of a cooling coil with non-zero bypass factor, the apparatus, dew point temperature lies at the intersection point of

(a) room DB line with the saturation curve
(b) RSHF and GSHF lines
(c) RSHF and ESHF lines
(d) GSHF line with the saturation curve

20. Air conditioning has to be done for a hall whose RSH = 50 kW and RLH = 50 kW. There are no other sources of heat addition or leakages. What is the value of the RSHF?

(a) 0.25
(b) 0.5
(c) 0.75
(d) 1.00

21. Consider the following statements:

When a GSHF line is extended, it may strike the saturation curve at a point. This point is called

1. effective surface temperature
2. air saturation temperature
3. water boiling temperature
4. apparatus dew point

Which of the statements given above are correct?

(a) 1 and 2
(b) 1 and 4
(c) 2 and 3
(d) 3 and 4

22. Consider the following statements:

1. If air is heated in a closed chamber, its dew point will increase.
2. As relative humidity decreases, the difference between the wet bulb temperature and dew point will increase.
3. In spray humidification process, the enthalpy of air will decrease.
4. The dew point temperature is always an indication of moisture content of the air.

Which of these statements are correct?

(a) 1 and 2
(b) 2 and 4
(c) 1 and 3
(d) 3 and 4

23. For an air conditioned space, RTH = 100 kW, RSHF = 0.75, volume flow rate = 100 m^3/min, and indoor design specific humidity is 0.01 kg/kg of dry air. What is the specific humidity of the supply air?

(a) 0.010
(b) 0.0075
(c) 0.005
(d) 0.0025

24. In a spray washing system, if the temperature of water is higher than the dry bulb temperature of entering air, then the air is

(a) heated and dehumidified
(b) heated and humidified
(c) cooled and humidified
(d) cooled and dehumidified

25. Air (at atmospheric pressure) at a dry bulb temperature of 40°C and wet bulb temperature of 20°C is humidified in an air washer operating with continuous water recirculation. The wet bulb depression (i.e., the difference between the dry and wet bulb temperatures) at the exit is 25% of that at the inlet. The dry bulb temperature at the exit of the air washer is closest to

(a) 10°C
(b) 20°C
(c) 25°C
(d) 30°C

26. Consider the following statements:

Air washer can work as

1. Humidifier only
2. Dehumidifier only
3. Filter only

Which of the statements given above is/are correct?

(a) Only 1
(b) Only 2 and 3
(c) Only 1 and 3
(d) 1, 2 and 3

27. Bypass factor for a cooling coil

(a) increases with increase in velocity of air passing through it
(b) decreases with increase in velocity of air passing through it
(c) remains unchanged with increase in velocity of air passing through it
(d) may increase or decrease with increase in velocity of air passing through it depending upon the condition of air entering

28. The bypass factor of single cooling coil in an air conditioner is 0.7. The bypass factor, if three such cooling coils with the same apparatus dew point are kept one behind the other will be

(a) 0.210
(b) 0.292
(c) 0.343
(d) 0.412

29. Consider the following statements

1. Low value of the bypass factor for an air-conditioning equipment signifies higher performance of the equipment.
2. Bypass factor for an air-conditioning equipment signifies the fraction of ambient air mixed with the air to be conditioned.
3. Bypass factor for an air-conditioning equipment signifies the fraction of the air to be conditioned coming in contact with the conditioning surface.

Of these statements:

(a) 1 and 3 are correct
(b) 1 and 2 are correct
(c) 3 alone is correct
(d) 2 alone is correct

30. The atmospheric air at dry bulb temperature of 15°C enters a heating coil maintained at 40°C. The air leaves the heating coil at 25°C. The bypass factor of the heating coil is

(a) 0.375
(b) 0.4
(c) 0.6
(d) 0.67

31. In order to have a low bypass factor of a cooling coil, the fin spacing and the number of tube rows should be

(a) wide apart and high, respectively
(b) wide apart and low, respectively
(c) close and high, respectively
(d) close and low, respectively

32. In an air-conditioning process, 5 kJ/min heat is extracted from a room. If the sensible heat factor is 0.8, then the latent heat extracted will be

(a) 4 kJ/min
(b) 2 kJ/min
(c) 1 kJ/min
(d) 0.25 kJ/min

33. In order to cool and dehumidify a stream of moist air, it must be passed over a coil at a temperature

(a) which lies between the dry bulb and wet bulb temperatures of the incoming stream
(b) which lies between the wet bulb and dew point temperatures of the incoming stream
(c) which is lower than the dew point temperature of the incoming stream
(d) of adiabatic saturation of incoming stream

34. A cooling coil with a bypass factor of 0.1 and apparatus dew point (adp) of 12°C comes in contact with air having a dry bulb temperature of 38° C and dew point of 9° C.

Over the cooling coil, the air would undergo

(a) sensible cooling
(b) cooling and humidification
(c) cooling and dehumidification
(d) adiabatic saturation

35. When the air is passed through an insulated chamber having sprays of water maintained at a temperature higher than the dew point temperature of air entering but lower than its dry bulb temperature, then the air is said to be

(a) cooled and humidified
(b) cooled and dehumidified
(c) heated and humidified
(d) heated and dehumidified

36. For cooling and dehumidifying of unsaturated moist air, it must be passed over a coil at a temperature

(a) of adiabatic saturation of incoming stream

(b) which is lower than the dew point of incoming stream

(c) which lies between dry bulb and wet bulb temperatures

(d) which lies between wet bulb and dew point temperatures of the incoming stream

37. If air is passed through a solid chemical absorbent, the psychrometric process followed is

(a) heating and dehumidification with the bulb temperature remaining fairly constant

(b) cooling and dehumidification

(c) dehumidification with sharp rise in wet bulb temperature

(d) dehumidification at constant dry bulb temperature

38. Air at 35°C dbt and 25°C dew point temperature passes through the water shower whose temperature is maintained at 20°C. What is the process involved?

(a) Cooling and humidification

(b) Sensible cooling

(c) Cooling and dehumidification

(d) Heating and humidification

39. Consider the following statements: In chemical dehumidification process

1. dew point temperature decreases.
2. wet bulb temperature decreases.
3. dry bulb temperature increases.

Of these statements

(a) 1, 2 and 3 are correct

(b) 1 and 2 are correct

(c) 2 and 3 are correct

(d) 1 and 3 are correct

40. If moist air is sensibly cooled above its dew point, which of the following statements are correct?

1. Relative humidity decreases
2. Wet bulb temperature decreases
3. Wet bulb temperature increases
4. Humidity ratio remains constant

Select the correct answer using the codes given below:

Codes:

(a) 1 and 2

(b) 1 and 3

(c) 3 and 4

(d) 2 and 4

41. Consider the following statements:

In psychrometry, wet bulb temperature is a measure of enthalpy of moist air, so that in the psychrometric chart,

1. the constant enthalpy lines are also constant wet bulb temperature lines
2. the wet bulb and dry bulb temperatures are same at any condition

3. the wet bulb and dry bulb temperatures are equal at saturation condition.

Of these statements.

(a) 1 alone is correct
(b) 1 and 2 are correct
(c) 1 and 3 are correct
(d) 2 and 3 are correct

ANSWERS

1. (c) **2.** (c) **3.** (a) **4.** (d) **5.** (c) **6.** (a) **7.** (c) **8.** (b) **9.** (a) **10.** (c)
11. (d) **12.** (d) **13.** (b) **14.** (c) **15.** (b) **16.** (c) **17.** (b) **18.** (d) **19.** (d) **20.** (b)
21. (b) **22.** (a) **23.** (c) **24.** (b) **25.** (c) **26.** (d) **27.** (a) **28.** (c) **29.** (a) **30.** (c)
31. (c) **32.** (c) **33.** (c) **34.** (b) **35.** (a) **36.** (b) **37.** (a) **38.** (c) **39.** (d) **40.** (d)
41. (c)

10

Cooling Load Calculation

10.1 INTRODUCTION

'Cooling load' is the thermal load on the air conditioning system or on the cooling coil that should be removed to maintain the conditioned space temperature and humidity at desired value. This thermal load is the total heat, i.e., the summation of sensible heat and latent heat, gained by the space from different sources. The sources are external sources and internal sources. Below is given the list of sources (also refer Fig. 10.1) from which thermal load is gained by the space to be air conditioned. For heat pump application, where the room temperature is required to be maintained higher than the environment temperature, the thermal load on air conditioning system is called '*heating load*'.

10.2 SOURCES OF HEAT GAIN BY THE SPACE

The total cooling load consists of three parts.

(i) Total heat gained inside the conditioned room, i.e., room cooling load.

(ii) The total heat gained by the whole building which is the summation of individual conditioned rooms. (bedroom, kitchen, office room, sitting room, etc.). This is *building cooling load.*

(iii) The air which is conditioned by the cooling coil carries the cooling load of room/ building (for recirculating case only) and the heat gained due to ventilation air, heat gained by ducts, heat generated due to air conditioning equipment itself. This is *cooling coil load.*

- **Heat gained from indoor of the conditioned room**

This heat is gained from three types of sources. One contributes only sensible heat.

1. Contribute only sensible heat as follows:

   (a) Conduction heat transfer through the building structures (walls, roof, floor, doors and windows, even if they are glazed, etc) due to temperature gradient between outdoor and indoor.

(b) Conduction heat transfer through the partition walls (for example, when there is a kitchen beside the room).

(c) Heat transmission through glazed/transparent portions of the building (glazed windows, panels, etc), mainly, due to solar radiation and also from any other heat source beside the room.

(d) Transmission through internal partition walls (for example, the other side of the air conditioned space may be a non-airconditioned kitchen).

(e) Heat gain from electrical appliances like lights, fans, TV, fridge, etc.

2. Contribute sensible and latent heat (when it is addition of water vapour), both, as follows.

(a) Occupants release metabolic heat and perspiration. Their clothing releases sensible heat.

(b) Infiltered air as it is the outdoor moist air.

(c) Tea/coffee makers, drinking water source (pots, water bottles, etc).

- **The total Heat Gained by the whole building**

This is the summation of individual conditioned rooms. (bedroom, kitchen, office room, sitting room, etc.). This is 'building cooling load.

- **The air which is conditioned by the cooling coil**

The air, to be conditioned by the cooling coil, carries the cooling load of room/building, Ventilated air and the heat gained due to supply ducts, supply duct leakage, heat generated due to air conditioning equipment itself and load to be compensated for the 'by-pass factor of the cooling coil'. This is cooling coil cooling load.

In summary: The sum of room sensible heat and room latent heat is called *room total heat* (RTH).

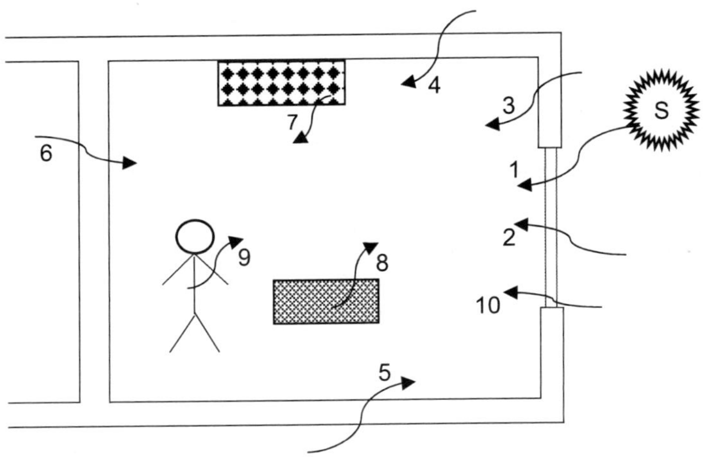

Figure 10.1: Room heat gain from different sources. (1) Direct sunrays through glazed surfaces; (2) Diffused ray from surrounding; conduction (3) through walls (includes glazed surfaces); (4) through roof; (5) through floor; (6) through partition; (7) from electrical appliances; (8) from equipment; (9) from occupants; (10) infiltration

The actual cooling load on the cooling coil is the sum of (i) RTH, (ii) sensible and latent heat gain due to the ventilation air, and (iii) leakage through ducts outside the space. This 'actual cooling load' is called grand total heat (GTH) load.

10.3 FACTORS CONTRIBUTING TO THE COOLING LOAD

Let us discuss, in detail, some factors contributing to the cooling load.

10.3.1 Solar Heat Gain

Solar heat gain is the solar radiations received by the air-conditioned space. This consists of three parts. One part is the direct beam radiation from sun with an intensity (I_D) falling upon the surface. A major part of the sun's radiation, when strikes the molecules like N_2, O_2, water vapour, dust particles, etc., get scattered in all directions. A part of this scattered radiation comes back to the earth's surface being reflected again and again. This part is also called diffused radiation with an intensity I_d. It is considered to be uniformly falling on the surfaces irrespective of its orientation. **In the cloud sky the intensity of diffused radiation is more than the direct beam radiation.** The third part is the reflected ray from the surroundings of the structure with an intensity I_r. This part is considerably high if the surroundings are covered with concrete structures and other reflecting materials. Thus the total solar radiation intensity, $I_t = I_D + I_d + I_r$.

However, the estimation for total solar radiation falling upon the surface is tedious because it involves a number of variables like solar position expressed in solar angles, season, time, the latitude and longitude of the place, etc.

Solar Angles subtended at earth's surface

It is a good knowledge with the common people that solar heat is maximum in Equatorial Zone, lesser in Tropic of Cancer and Tropic of Capricorn Zone and least as we move towards more north or more south. Basing upon this, the earth is divided into three zones. (a) Torrid Zone (very hot) 23.5 N to 23.5 S Latitude; (b) Temperate Zone (not very hot or cold) 23.5 N to 66.5 N and 23.5 S to 66.5 S Latitude and (c) Frigid Zone (very cold) 66.5 N to North Pole and 66.5 S to south Pole.The heat gain due to solar radiation depends upon the relative position of earth with respect to sun. So it is important to have knowledge on the angles subtended by sun with respect to earth.

Basic Solar Angles: (Refer Fig. 10.2) The relative position between the sun and a building on the earth, time of the day and day of the year play a vital role in estimating the solar energy gained by the building at any instant. These three factors are conveniently expressed in terms of latitude (L), hour angle (h), and sun's declination (d). In Fig. 'P' is a place in northern hemisphere whose location is to be expressed in terms of basic solar angles. 'O' is the centre of the earth.

Latitude (L): The angle subtended by OP with its projection OA on the Equatorial plane at O is the latitude (L) of the place (angle POA).

Hour Angle (h): The angle subtended by OA with the projection of OX (the line joining O and the sun) on the Equatorial plane, i.e., OB at O is the hour angle (angle AOB). As the time passes from morning to noon, OA come closer to OB. At noon hour angle becomes zero. Hour angle actually expresses the time of day with respect to solar noon.

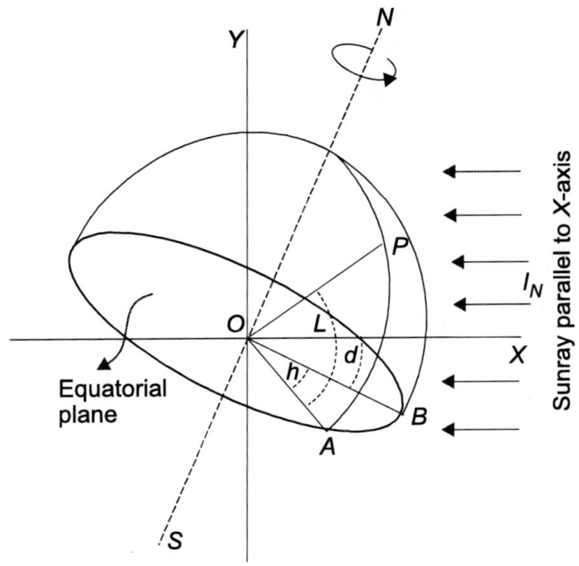

Figure 10.2: Basic solar angles

Declination (*d*): This angle is subtended by *OX* and *OB* at *O* and is denoted by *BOX*. Its maximum value is 23.5 degree on 22nd June and minimum value is –23.5 degree. For a place south to equator, its value is negative. On 22nd September and 22nd March its value is zero. This can be calculated from the correlation

$$d = 23.46 \sin [360 \times (284 + n)/365] \qquad [10.1]$$

where, *n* is the day number starting from 1st January.

However, the position of a place on the earth's surface can be expressed in terms of its latitude and longitude.

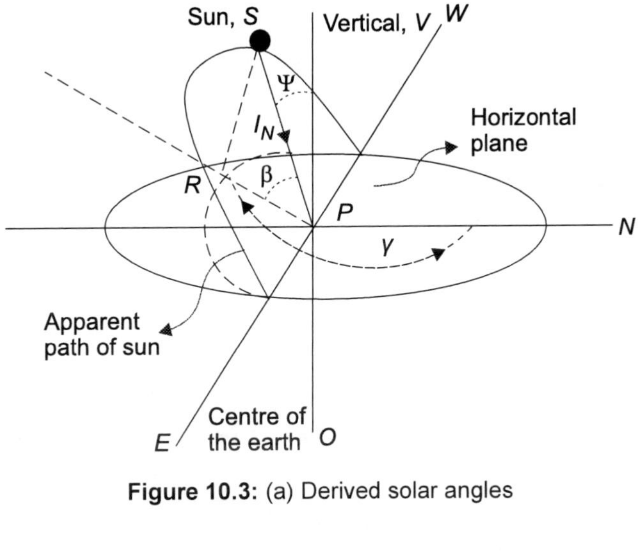

Figure 10.3: (a) Derived solar angles

Derived solar angles: The derived solar angles are shown in Fig. 10.3a. Sunray *SP*, with an intensity of I_N is directed towards the place *P* on the earth's surface.

Altitude angle (β): It is the angle in the vertical plane, subtended by sunray *SP* and its projection *PR* on the horizon, denoted by *SPR*.

Zenith angle (ψ): It is the angle in the vertical plane between *SP* and *PV* and denoted by *SPV*. It can be shown from Fig. 10.3 that $\beta + \psi = \pi/2$.

Azimuth angle (γ): It is the angle in the horizontal plane subtended by due North *PN* and *PR* at *P* and denoted by *NPR*.

Angle of incidence (θ) (Fig. 10.3b): It is the angle between the sunray and the normal to the surface.

The relationships among basic solar angles and derived angles are as follows.

Zenith angle (ψ): $\cos \psi = \cos L \cos h \cos d + \sin L \sin d$ [10.2]

Altitude angle (β): $\sin \beta = \cos L \cos h \cos d + \sin L \sin d$ [10.3]

Azimuth angle (γ): $\sin \gamma = \sec \beta \cos d \sin h$ [10.4]

Angle of incidence (θ): $\cos \theta = \cos \beta \cos \alpha \cos \varphi + \sin \beta \sin \varphi$ [10.5]

where, φ is the angle of tilt of the surface to the vertical.

If the surface is vertical ($\varphi = 0$), then $\cos \theta = \cos \beta \cos \alpha$ [10.6]

If the surface is horizontal ($\varphi = \pi/2$), then $\cos \theta = \sin \beta = \cos \psi$ [10.7]

These relationships are derived from the trigonometric principles.

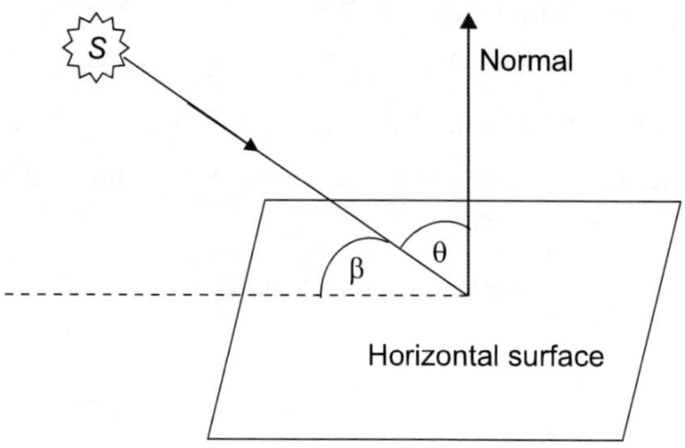

Figure 10.3: (b) Incident angle for horizontal surface

The usefulness of these solar angles in estimating the solar radiation (direct and diffused) falling upon any surface on the earth is illustrated in Example Problem 10.2.

Now calculations are made how much of these solar radiations get into the conditioned space through glazed doors/windows, how much of it is absorbed, conducted and convected into the conditioned space through the building structure.

Heat gain through transparent surfaces like glazed windows and doors, etc

Now a part of the solar radiation falling upon the transparent surface is reflected, part of it is absorbed and third part is transmitted through the glass. Again, outdoor temperature is higher than the indoor temperature. So heat is also conducted from outdoors to indoors through these surfaces.

Total heat gained by the room due to solar radiation, $Q_t = Q_{\Delta T} + Q_\zeta + Q_C$ [10.8]

where $Q_{\Delta T}$ is the heat flow due to temperature difference between outdoor and indoor, Q_ζ is the radiation directly transmitted through the glass, Q_C is the inward flow of absorbed radiation. $(Q_\zeta + Q_C)$ is called the solar heat gain (Q_{SHG}).

Taking the energy balance of a transparent surface (window glass), energy absorbed by the glass = energy lost due to conduction and convection to outdoor + energy lost due to conduction and convection to indoor + energy stored

$$= F_S I_D \alpha_D + I_d \alpha_d + I_r \alpha_r = h_i (t_{gi} - t_i) + h_O (t_{gO} - t_O) + q_{Stored} \quad [10.9]$$

where, α is the absorptivity of the glass which varies with the wavelength of the falling radiation, t_g is the temperature of the glass, subscripts i and O refer to indoor and outdoor. F_S is the fraction of the glass exposed to the direct solar radiation. For simplification the following assumptions are made.

Absorptivity of the glass is invariant with the wavelength.

Neglect the thickness of glass. So $t_g = t_{gi} = t_{gO}$ and q_{Stored} is negligible.

The whole of the window is exposed to the direct solar radiation. So $F_S = 1$

Now the above equation will be $I_t \alpha = h_i (t_g - t_i) + h_O (t_g - t_O)$ [10.10]

Total heat transferred through the glass, Q_t = Heat transmitted through the glass + Fraction of absorbed heat by the glass convected to indoor = $I_t \zeta + h_i (t_g - t_i)$ [10.11]

where, U is the overall heat transfer coefficient = $1/h_O + 1/h_i$

From Eq. (10.10); $t_g = (I_t \alpha + h_i t_i + h_O t_O) / (h_i + h_O)$,

Substituting the value of t_g in Eq. 10.11, we get

$Q_t = I_t \zeta + (U/h_O)(I_t \alpha) + U(t_O - t_i) = I_t\{\zeta + (\alpha\, U/h_O)\} + U(t_O - t_i)$ [10.12]

$I_t\{\zeta + (\alpha\, U/h_O)\}$ term is called solar heat gain factor (SHGF).

So total heat transferred through the glass, $Q_t = \text{SHGF} + U(t_O - t_i)$ [10.13]

10.3.2 Conduction Heat Gain through Walls, Roof and other Partitions

The calculation of heat transfer through walls, roof is bit complicated because the ambient temperature is not steady due to varying solar radiations, and wind speed throughout the day. The maximum intensity of radiation that reaches the earth's surface is nearly 0.95 kW/m^2, and the intensity of radiation varies throughout the day as described above. The amount of this radiation absorbed by the surface depends upon the absorptivity and the angle of incidence of the radiation. So this must be calculated on hour-to-hour basis. To simplify this, sol-air temperature, as discussed below, is considered in which the effects of all these variations are combined to get this single temperature.

As heat starts to pass inwards through the walls or roof surface, some will be absorbed in warming the outer layers and there will be a time lag before the effect reaches the inner face, depending upon the mass, conductivity and specific heat capacity of the wall materials. Some of the absorbed heat will be retained in the material and then lost to ambient at night. So we feel more hot after some time of the sunset, but not during the peak outdoor heat at around 2.00 pm. The effect of thermal lag can also be expressed mathematically.

First let us calculate the amount of heat flow (q_O) from the environment or the outdoor (at t_O) to the outer surface of the wall (at t_{WO}).

So $q_O = h_O(t_O - t_{WO}) + \alpha(I_{\text{Direct}} + I_{\text{Diffused}} + I_{\text{Reflected}})$ [10.14]

Where 1st part of RHS is the heat transfer due to convection and the second part of RHS is the amount of heat absorbed where, α is the absorptivity of the wall surface and assumed to be same for the different wavelengths.

Or this can also be expressed as $q_O = h_O(t_O - t_{WO}) + \alpha(I)$ [10.14a]

Now these two parts can be combined and conveniently expressed as $q_O = h_O(t_e - t_{WO})$, where t_e is called *sol-air temperature.*

And expressed as $t_e = t_O + \alpha\ (\text{I}/h_O)$ [10.15]

Sol-air temperature is defined as the temperature of the outdoor air that, in the absence of all radiation changes, gives the same rate of heat entry into the surface as would the combination of incident solar radiation, radiant energy exchange with the sky and other outdoor surroundings, and convective heat exchange with outdoor air.

Example 10.1

What is the sol-air temperature for a wall with the given conditions:

Total direct and diffused solar radiation 300 W/m^2

Surface absorptivity 0.9; outside surface htc 12 W/m^2-K; outside temperature 40°C

Ans: Sol-air temperature, $t_e = t_O + \alpha\ (I/h_O) = 40 + (0.9 \times 300)/12 = 62.5$°C

If the thermal capacity of the wall/roof is ignored, the instantaneous rate of heat transfer through the walls/roof will be $Q = UA(t_e - t_i)$, where U is the overall heat transfer coefficient (calculated from wall outside htc, indoor htc and wall/roof conductivity), A is the wall exposed area. t_e and t_i are the sol-air and indoor temperature, respectively.

On an average basis the mean heat flow is given by

$$Q_m = UA(t_{em} - t_i),$$

where, t_{em} is the mean sol-air temperature for the time period considered.

When the thermal capacities of the walls and roof are considered, sensible heat gained by room will be contributed by two factors. One is due to the factor as discussed above, i.e., conduction heat transfer and the second factor is due to the heat radiation and convection from the walls/roof after the lag period as shown in Fig. 10.4.

$$Q_\zeta = UA(t_{em} - t_i) + UA\lambda(t_{e,(\zeta-\varphi)} - t_{em}) \qquad [10.16]$$

where φ is the hours before the time ζ when heat transfer to be calculated, $t_{e,(\zeta\text{-}\varphi)}$ is the average sol-air temperature during the hour ζ to ϕ. Here ϕ is the time lag, i.e., the time required to transfer the absorbed heat by the wall/roof to indoor. Sometimes Q_ζ is greater or less than Q_m depending upon whether, $t_{e,(\zeta\text{-}\varphi)}$ is greater or less than t_{em}, φ hours before. This method of determining the heat transfer through the walls/roof is called decrement and time lag method. Actually it is very difficult to determine the time lag and decrement factor. Hence another method, called Equivalent Temperature Differential or Cooling Load Temperature Difference (CLTD) method is adopted. In this method an Equivalent Temperature Differential is defined from the equation

$$\begin{aligned} Q_\zeta &= UA(t_{em} - t_i) + UA\lambda(t_{e,(\zeta-\varphi)} - t_{em}) \\ &= UA\Delta t_{eE}, \text{ where, } \Delta t_{eE} = (t_{em} - t_i) + \lambda(t_{e,(\zeta-\varphi)} - t_{em}) \end{aligned} \qquad [10.17]$$

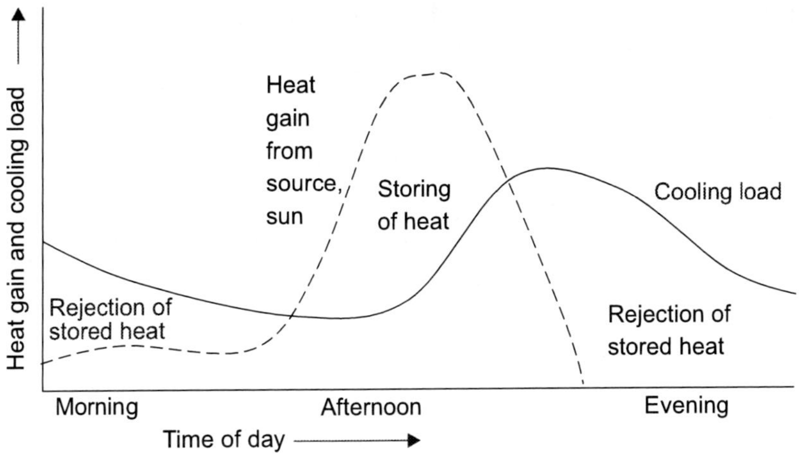

Figure 10.4: Heat gain and cooling load variation due to solar radiation during daytime

Figure 10.4 shows that solar heat intensity and the heat gained by the room is maximum in the afternoon, i.e., at around 2.00 PM. Sun is overhead at 12.00 noon. This maximum solar radiation is received by earth and it gets heated up. Then earth also starts reradiating the heat. Thus, the environment receives direct solar radiation and reradiated heat. This whole process takes around 2 hours. That is why it is estimated that the room receives maximum solar heat at around 2.00 p.m. This solar heat is absorbed and stored by the walls, roof and the materials in the room. This is shown, by dotted curve. After certain time, i.e., at around 5.0 p.m., these walls, roof, furniture, etc. start rejecting the stored heat, because by that time temperature of the walls, roof, etc., become higher than the indoor air temperature. The time required to attain that higher temperature after receiving the outdoor heat is called 'time lag'. This depends upon the thermal capacity of the structure. If the thickness or mass of the wall is very high, then the time lag will be even more. This heat is gained by the indoor air of the room. Hence, the peak cooling load in the room is considered to be at around 5.0 p.m. due to solar radiation. This is shown by the solid curve.

10.3.3 Infiltration

Infiltration is the uncontrolled leakage of outdoor air through the cracks in the buildings, around the doors and windows, joints, and most importantly through door openings while people get in or out of the room. This is more prominent in case of banks, offices and public establishments. Other factors are (a) wind pressure, (b) stack effect. (A) If the space is maintained at positive pressure, the conditioned air will leak out through the cracks and during door opening. So more make up (fresh) air is added to the return air, which increases the cooling load. If the space is maintained at negative pressure, the outdoor air will leak in through the cracks and during door opening, which increases the cooling load. Also when wind flow suddenly stops striking the wall of a building, pressure (stagnation) rises outside the building. This induces outdoor air to leak through the cracks or openings of the building. (B) In tall buildings, the pressure difference caused due to cold air coloumn inside the building and that of warmer air outside the building and the air density difference due to difference in elevation induces air flow in or out of the building. This is known as stack effect. All these factors combined, makes the estimation of cooling load due to infiltration very difficult.

However, there are two most prevalent methods of estimating the infiltration. They are: (a) air change method, and (b) crack length method.

(a) One air change means the total volume of air in the room will be replaced by fresh air in one hour. If V m^3 is the volume of the room, then one air change implies an infiltration rate of V m^3/hr. In general infiltration in m^3/s is given by Q_V = (number of air changes) (volume of the room)/3600 m^3/s. Table 10.1 gives the recommended air changes for residences and cold storages per 24 hours.

Table 10.1: Number of air changes for different room sizes

Room Volume (Residential) (m^3)	Number of Air Changes per 24 hours (Average load)	Number of Air Changes per 24 hours (heavy load)	Room Volume (Storage) (m^3)	Number of Air Changes per 24 hours
10	20.5	32.26	60	11.26
20	18.25	28.76	70	10.69
30	16.60	26.04	80	9.92
40	15.29	24.45	90	8.92
50	14.25	23.99	100	8.26
70	12.70	--------	120	7.22
90	11.67	--------	150	6.13

The cooling load (Q) due to infiltration and ventilation is calculated from the following equation.

$$Q = \rho Q_V (h_O - h_i) \quad [10.18]$$

where, h_O and h_i are the specific enthalpy of outdoor air and indoor air respectively. Sometimes number of air changes can be calculated from the equation as suggested by Cobientz and Achenbach (1963);

$$\text{Number of air changes} = a + bV + c\,(t_o - t_i) \quad [10.19]$$

where, a, b and c are 0.2, 0.015 and 0.014, respectively. V is the velocity of air (outdoor).

Crack method calculates the flow produced by pressure difference acting on each leakage path or building component. The flow is given by $Q = C(\Delta P)^n$, where, Q is the volume flow rate of air (lit/s), C is the flow coefficient, i.e., volumetric flow rate per unit length of crack or unit area at a unit pressure difference (assumed to be 1.25 for general purpose), ΔP is the pressure difference and n is the flow exponent (varies between 0.5 and 1.0 and usually taken as 0.65 for leakage opening).

Besides this, ASHRAE has developed charts in which pressure drop due to wind effect and stack effect are given. The infiltration rate through different types of widows and doors can be calculated using the different relations suggested by ASHRAE. Also direct charts, in Carrier hand book, are available to estimate the infiltration for different type of commercial buildings (like banks, hospitals, offices, etc.) with different types of windows and doors.

10.3.4 Ventilation

The process of supplying fresh air to occupied spaces to maintain the desired level of purity and odour of air is called ventilation. This may be natural or forced. Fresh air can get into the conditioned space through windows naturally due to pressure difference between outdoors and indoors or fresh air can be made to flow into the space by induced or forced fan. The amount of fresh air requirement for different places and different activities is taken from the table. However, outdoor air (due to infiltration and ventilation), being moist, add sensible and latent heat to the cooling load.The difference between infiltration and ventilation is that the former is not desirable and is taken care to minimise it whereas the later is allowed to enter the room as per the requirement.

Table 10.2: Required amount of ventilation air

Application	Smoking status	Recommended (cmm/person)
Offices	Occasional	0.3 to 0.5
Restaurants	Some	0.45
Departmental stores, shops	None	0.24
Theatres	None	0.24
Hotel rooms	Heavy	0.8
Hospital wards	None	0.82
Hospital operation theatres	None	100% fresh air

10.3.5 Internal Heat Gain from Occupants and Equipment

The metabolic heat and the perspiration released by the occupants and the heat carry with them through the clothing and other belongings as they come from outdoors are the part of the RTH gain. The amount of heat released depends upon the occupants' activities (discussed in Chapter 12). Perspiration add sensible heat as well as latent heat to the cooling load.Household equipment, walls and roof release stored heat (shown in Fig. 10.2) as the indoor air temperature is decreased. Besides, the occupants using coffee burner, toaster, etc., add heat to the RSH. Drinking water sources, tea maker, etc. add sensible and latent heat to the cooling load as the water evaporates.

10.3.6 Internal Heat Gain from Electrical Appliances

Electrical appliances like lights, fans (sometimes used for ventilation), fridge, TV, etc., release heat as they run. A fridge releases hot air from the condenser. Fluorescent lamp: Q = Total watts $\times$ 1.25; Incandescent: Q = Total rated watts. Thus, a 20 W electric bulb releases 25 W of heat at steady state. Similarly, a motor running at an efficiency of η, gives out heat of an amount $(1 - \eta) \times$ Input. Thus, the heat dissipation from the different running electrical appliances are to be added up to get an approximate value of heat added to RSH.

Example 10.2

In Bhubaneswar a classroom of size 9 m $\times$ 9 m $\times$ 3 m is to be air conditioned. Calculate the direct and diffuse solar radiations at 2.00 p.m. solar time on May 21st on the roof which is 10 cm concrete slab with 1.25 cm plaster on inside only. Also calculate the instantaneous heat transfer through the roof. Assume outdoor and indoor temperatures are 40°C and 25°C. Outdoor convective htc is 25 W/m^2–°C and absorptivity of roof is 0.85. Overall htc is 3.6 W/m^2–°C.

Ans: Data from the question are:

Latitude (L) = 20°, hour angle (h) at 2.00 p.m. = 15 × 2 = 30°, angle of tilt (Φ) = 90° (since it is horizontal)

Angle of declination, $d = 23.46 \sin [360 \times (284 + n)/365] \approx 20°$ (20° 14′ in the chart available 17.5 of CP).

Altitude angle $\sin \beta = \cos L \cos h \cos d + \sin L \sin d = 0.8817$; So $\beta \approx 62°$

Normal solar radiation can be approximated from the correlation,

$$I_n = 1082\, e^{-0.182\, M};$$

where M is the air mass and is defined as $1/\sin \beta$.

So $I_n = 1082 \times e^{-0.182/\sin(62)} = 880.5$ W/m^2

For horizontal surface, angle of incidence (θ) is expressed as $(90° - \beta) = 90 - 62 = 28°$

Direct solar radiation intensity, $I_D = I_n \times \cos \theta = 880.5 \times \cos 28 = 777.4$ W/m^2

Since this is a horizontal surface, angle factor between the surface and the ground is zero.

But the angle factor between the sky and the surface is 1. Coefficient for sky radiation, as suggested by ASHRAE, for May 21st = 0.121.

So diffuse sky radiation, $I_d = 0.121 \times 880.5 \times 1 = 106.5$ W/m^2.

So total radiation on the 9 × 9 = 81 m^2 roof surface area = (777.4 + 106.5) × 81 = 71596 W.

Now sol-air temperature is calculated from $t_e = t_O + \alpha\, (I/h_O)$

So $t_e = 40 + 0.85\,(777.4 + 106.5)/25 = 70°C$

So instantaneous heat transfer $Q = UA\,(t_e - t_i) = 3.6 \times 81\,(70–25) = 13122$ W

Example 10.3

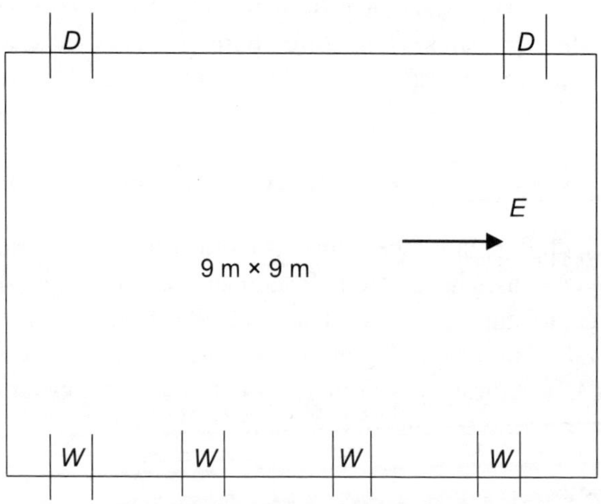

Figure 10.5: Plan of the classroom

In Bhubaneswar a classroom of size 9 m × 9 m × 3 m is to be air conditioned. The plan of the room is given in Fig. 10.5. 4 Windows each of size 1.2 m × 1.2 m are on south side, and 2 doors each of size 1.2 m × 2.4 m are on north side. Other data are as follows.

Walls: 25 cm thickness of brick, and plaster of 1.25 cm on each side; Roof: 10 cm concrete slab with 1.25 cm plaster on inside only; floor: 20 cm concrete; doors are of wooden panels and windows are fully glazed with wooden frame (15%); occupancy is 60 students; lights: 10 numbers of 20 W tube lights and 125 W OHP.

Outdoor condition: 44°C dbt and 27°C wbt; indoor condition: 26°C dbt and 50% RH. Convective heat transfer coefficients for outdoor is 30 W/m^2-K and that of indoor is 8 W/m^2-K.
Ans: Let us, first, collect data from different tables available in data books or handbooks.

Orientation of the room is so chosen to reduce the solar heat gain. As the maximum solar radiation gain is on the west side, the doors are kept on north side and two take advantage of south wind, the windows are kept on south side.

Thermal conductivities of different materials: k_{Concrete}: 10.0 W/m-K; k_{Brick}: 1.3 W/m-K; k_{Plaster}: 9.0 W/m-K; k_{Glass}: 0.8 W/m-K

Overall heat transfer coefficient of wall: $1/U = 1/30 + 0.25/1.3 + 0.025/9.0 + 1/8 = 0.353$

$U_{\text{wall}} = 1/0.353 = 2.83$ W/m^2-K

Overall heat transfer coefficient of roof: $1/U = 1/30 + 0.1/10 + 0.0125/9.0 + 1/8 = 0.17$

$U_{\text{Roof}} = 5.88$ W/m^2-K

Overall heat transfer coefficient of floor: $1/U = 0.2/10 + 1/8 = 0.145$; $U_{\text{Floor}} = 6.9$ W/m^2-K

Heat transfer/solar heat gain through walls and roof

Mass of wall/unit area $= \rho_{\text{brick}} \times \text{brick thickness} + \rho_{\text{plaster}} \times \text{plaster thickness}$

$= 2000 \times 0.25 + 1800 \times 0.025 = 545$ kg/m^2

Mass of roof/ unit area $= \rho_{\text{concrete}} \times \text{roof thickness} + \rho_{\text{plaster}} \times \text{plaster thickness}$

$= 2000 \times 0.1 + 1800 \times 0.0125 = 222.5$ kg/m^2

Equivalent temperature differentials with corrections **(From Handbook of Air Conditioning System Design by Carrier Air Conditioning Company)**

Equivalent temperature differential at 5 p.m.:

West wall 9.4°C; East wall 8.9°C; North wall 2.8°C; South wall 10.0°C and Roof 22.0°C.

For daily range (daily range is defined as the difference of maximum outdoor designed temperature and minimum outdoor air temperature (i.e., at one hour before sunrise)) of (44 –31 =) 13°C, correction for equivalent temperature differential is (13 – 11.1)/2 = 0.95°C (i.e., for each degree increase in daily range of 11.1, there will be decrease 0.5).

For $t_O - t_i = 18$°C; 18 – 8.3 = 9.7°C.

So total correction = – 0.95 + 9.7 = 8.75°C.

Hence, the resultant equivalent temperature:

West wall 18.1°C; East wall 17.65°C; North wall 11.55°C; South wall 18.75°C and roof 30.75°C.

Wall and roof area

West wall 27 m^2; East wall 27 m^2; North wall 21.24 m^2; South wall 21.24 m^2 and roof 81 m^2.

Hence, convective heat transfer through walls:

West wall, $Q_{West} = 27 \times 2.83 \times 18.1 = 1383.02$ W; East wall $Q_{East} = 27 \times 2.83 \times 17.65 = 1348.64$ W

North wall $Q_{North} = 21.24 \times 2.83 \times 11.55 = 694.26$ W; South wall $Q_{South} = 21.24 \times 2.83 \times 18.75 = 1127.05$ W

Roof $Q_{Roof} = 81 \times 5.88 \times 30.75 = 14645.61$ W

Floor $Q_{Floor} = 81 \times 2.5 \times 6.9 = 1397.25$ W (a temperature differential of 2.5°C is assumed).

Heat transfer/Solar heat gain through doors and windows

Door area = $2 \times 1.2 \times 2.4 = 5.76$ m^2; Overall heat transfer coefficient may be assumed for standard condition 2.8 W/m^2-K. Temperature difference is $t_O - t_i = 44 - 26 = 18$°C

So $Q_{Door} = 5.76 \times 2.8 \times 18 = 290.3$ W

Window area = $4 \times 1.2 \times 1.2 = 5.76$ m^2

Heat gain through ordinary glass window **(From Handbook of Air Conditioning System Design by Carrier Air Conditioning Company)** for south facing is 28 W/m^2.

So for total window area of 5.76 m^2, total heat gain, which includes direct and diffused radiation, $5.76 \times 44 = 253.44$ at 4 p.m. (44 W/m^2 is mostly appearing in the table. Hence, this value is considered which occurs at 4 p.m.)

Since 85% of windows is glass, the heat gain $Q_{Windows} = 253.44 \times 0.85 = 215.424$ W.

Infiltration: Number of air change method is adopted from the empirical relation given above (Eq. 10.19) (assuming the outdoor air velocity as 10 m/s) No of air change is 0.6/hr.

Q_V = ACH × Volume/3600 = 0.6×243 m^3/3600 = 0.0405 m^3/s

So Q due to infiltration = $Q = \rho Q_V (h_O - h_i) = 1.2 \times 0.0405 \times (85.0 - 53.5) = 1.531$ kW = 1531 W.

Out of this sensible heat = $0.0204 \times 0.0405 \times 60 \times (44 - 26) = 0.8923$ kW = 892.3 W

And latent heat = 1531 – 892.3 = 638.7 W. (By using Eq. 9.11, we get 605 W).

Electrical equipment: Lights (fluorescent) 125 W (OHP) + (10 Nos.) × 20 W × 1.25 (factor) = 375 W

Occupants: 60 students are doing mental work.

From Carrier handbook sensible heat and latent heat liberated/ student is 70 W each at 26°C indoor temperature.

So 60 students liberate 60 × 70 W = 4200 W sensible heat and 4200 W latent heat.

Total sensible heat gain = 27450.0 W (rounded) + 10% towards supply duct heat gain and safety factor, etc., = 27450 + 2745 = 30195 W; rounded to 30000 W or 30 kW.

Total latent heat gain = 4840.0 W(rounded) + 5% factor of safety = 5082 W; rounded to 5100 W or 5.1 kW

Ventilation requirement: As per the recommendation 0.3 cmm/student of fresh air is required for doing mental work. So total fresh air required is 0.3 × 60 = 18.0 cmm.

From psychrometric chart

Outdoor air condition for dbt 44°C and wbt 27°C,

Specific humidity (ω_O) is 0.0156 kg/kga and h_O is 85 kJ/kga

Indoor air condition for dbt 26°C and RH 50%

Specific humidity, ω_i is 0.0106 kg/kga and h_i is 53.5 kJ/kga.

Now the new problem statement is as follows.

A classroom is to be airconditioned for the above outdoor and indoor air condition. Sensible heat gain is 30 kW and latent heat gain is 5.1 kW. Ventilation air (fresh air) requirement is 18 m^3/min. The supply air temperature is 12°C. (Assuming the diffusion temperature, i.e., the temperature difference between the indoor air and the supply air to 14°C). Calculate the total volume of air entering the A/C space, ADP and the load on the cooling coil if the bypass factor of the cooling coil is 0.15.

Ans: Total volume of air entering the A/C space = RSH/{0.0204 × ($t_i - t_{Su}$)} (From Eq. 9.4)

= 30/ {0.0204 × (26 –12)}

= 105.04 cmm, rounded to 105.0 cmm

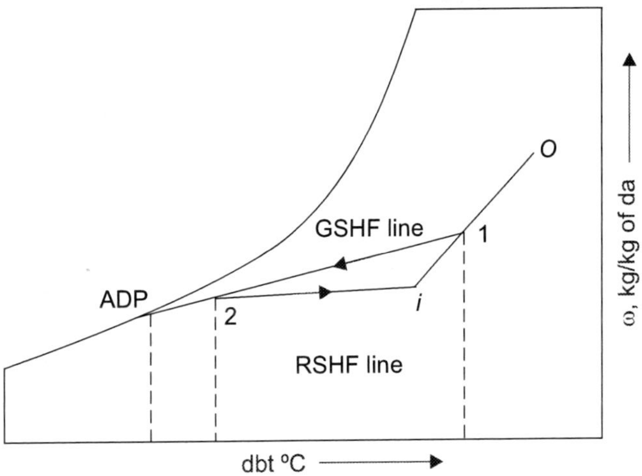

Figure 10.6: Process of Example 10.2 on the psychrometric chart.

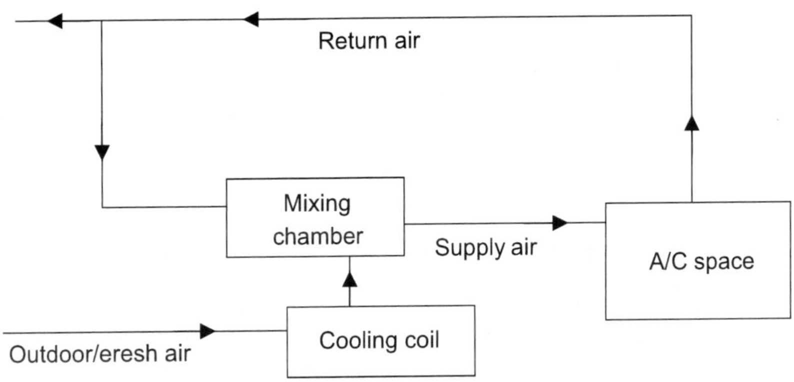

Figure 10.7: Schematic of the air flow of Example 10.2

Ventilation air requirementis 18 cmm; So return air amount is 105 – 18 = 87 cmm

For mixing state of return air and fresh air; $t_1 = (87 \times 26 + 18 \times 44)/105 = 29.1°C$

$h_1 = (87 \times 53.5 + 18 \times 85)/105 = 58.9$ kJ/kga

$\omega_1 = (87 \times 0.0106 + 18 \times 0.0156)/105$

= 0.01146 kg/kga

$BF = (t_2 - t_{ADP})/(t_1 - t_{ADP})$

$0.15 = (12 - t_{ADP})/(29.1 - t_{ADP})$

$t_{ADP} = 9°C$

Again $0.15 = (h_2 - h_{ADP})/(h_1 - h_{ADP})$

h_{ADP} (from chart) = 27.3 kJ/kga

So $h_2 = 32.04$ kJ/kga

Load on the cooling coil= 1.14 (i.e., density of air at '1') × (105/60) × $(h_1 - h_2)$ = 119.7 × (58.9 – 32.04)

= 53.586 kW = 15.22 TR

NB: (Above problem result seems to be an overestimation, like safety factors and duct leakage, and roof is without any insulated ceiling. Otherwise, this would have reduced the sensible heat gain load. The cooling load calculation is a rough estimation.)

EXERCISE

1. What is the difference between heating load and cooling load?
2. List the sources of cooling load for a room in terms of sensible heat gain and latent heat gain.
3. Sun is overhead at 12.00 noon. But it is estimated that peak cooling load received by the room is at around 2.00 p.m. Why?
4. Why the cooling load due to solar radiation is considered maximum at around 5.00 p.m. when peak heat gain from sun is at around 2.00 p.m.?
5. What is the difference between infiltration and ventilation air?
6. List the factors which contribute to the room sensible heat and room latent heat (at least 5 from each).
7. Show the different angles subtended by sunray at the earth's surface.
8. What do you mean by sol-air temperature and what is its significance in estimating the heat transfer from outdoors to indoor?
9. What will be the expression for sensible heat gain by the room, considering the heat capacity of the walls and roof?
10. Discuss the factors contributing to the internal heat gain by the room.

11. List the sources of cooling load which are avoidable, and which are not avoidable, so that total cooling load can be reduced.

12. Design an air conditioning system for a smart classroom of capacity 100 occupants in the department of mechanical engineering, OUTR, Bhubaneswar of following data. Assume $X = 0.15$.

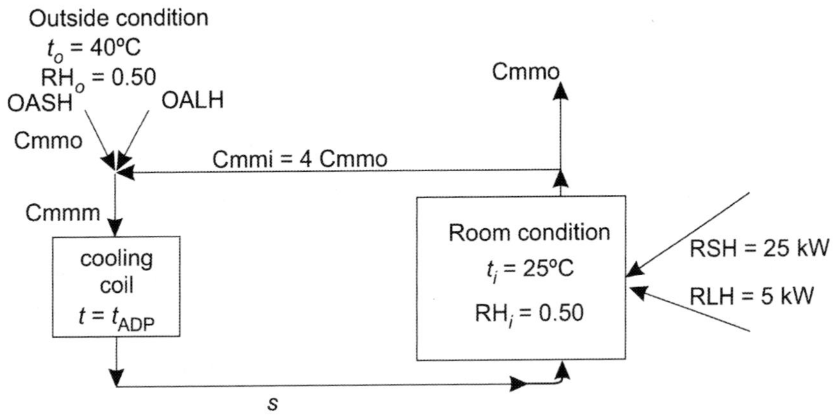

MULTIPLE–CHOICE QUESTIONS

1. Cooling load is the amount of —— to/from the air of the conditioned room.

(a) heat added (b) heat removed

(c) heat and moisture added (d) heat and moisture removed

2. Infiltration of outdoor air means outdoor air

(a) added for ventilation (b) leaked through openings

(c) filtered for cleaning (d) removed

3. Diffused part of the solar radiation is the ray —— entering the room.

(a) directly (b) getting reflected from the surrounding

(c) scattered in the sky (d) transmitted

4. Sol-air temperature is the —— temperature of ——.

(a) actual, solar radiation (b) actual, outdoor

(c) equivalent, outdoor (d) equivalent, indoor

5. Solar heat gain by the room is maximum at

(a) 12 noon (b) 1.00 p.m.

(c) 2.00 p.m. (d) 3.00 p.m.

6. Peak cooling load of the indoor air assumed to be at

(a) 2.00 p.m. (b) 3.00 p.m.

(c) 4.00 p.m. (d) 5.00 p.m.

7. Peak cooling load of the indoors is at 5.00 p.m. though peak solar heat gain is at 2.00 p.m. because

(a) west side is more heated

(b) setting sun has more intensity

(c) building walls and appliances radiates more

(d) this is just an assumption

8. Temperate climate is felt in —— latitude.

(a) 30°N (b) 15°N

(c) 70° N (d) Equatorial line

9. If V m^3 is the volume of the room, then one air change implies an infiltration rate of

(a) V m^3/Min (b) V m^3/sec

(c) V m^3/hr (d) (V m^3 × Density) of air/ hr

10. On comparison between infiltration and ventilation, it may be said that

(a) both are not desirable (b) 1 is desirable, 2 is not

(c) 2 is desirable, 1 is not (d) both are desirable

11. Consider the following factors:

1. Wind velocity
2. Type of activity
3. Indoor design conditions
4. Door openings

Occupancy load in cooling load calculations depends upon

(a) 1 and 2 (b) 1 and 3

(c) 1 and 4 (d) 2 and 3

12. An air-conditioned room has length, width and height of 20 m, 30 m and 4 m, respectively. The infiltration is assumed to be one air change. The outdoor and indoor dry bulb temperatures are 40°C and 25°C, respectively. The sensible heat load due to infiltration is

(a) 734 kW (b) 12.24 kW

(c) 0.204 kW (d) 10 kW

ANSWERS

1. (d) **2.** (b) **3.** (c) **4.** (c) **5.** (c) **6.** (d) **7.** (c) **8.** (a) **9.** (c) **10.** (c) **11.** (d) **12.** (b)

11

Human Comfort and Other Applications

11.1 INTRODUCTION

The most common application of refrigeration and air conditioning, and to the knowledge of the common people, is a household refrigerator and a room air conditioner and nonetheless the taste of ice cream. It is the common practice with the people of our society is to preserve the food and vegetables in a refrigerator and run a room air conditioner/heat pump during summer/winter to have a comfortable stay. Besides this there are innumerable applications of refrigeration and air conditioning due to which an estimated amount of more than 60% of energy is being consumed globally in this sector. Cooling process (for hot countries) consumes more power than heating process (for cold countries) which is obvious from the COP that has been calculated for a refrigeration system and a heat pump, i.e., $(COP)_{HP} = 1 + (COP)_{Ref}$. A brief discussion on some of its applications follows.

11.2 HUMAN COMFORT

Human comfort is one of the major applications of the air conditioning. The factors affecting the human comfort are (a) temperature, (b) humidity, (c) velocity or draft of air (d) purity (includes odour) (e) noise (acoustic), and (f) lighting.

It has always been tried to create a thermally comfortable environment which is normally said to be 24°C dbt and 50% RH for Indian climate. This is also called 'human thermal comfort'. But this is not justified. The feeling of comfort depends on **the physiological and psychological state** of an individual. It is not only affected by temperature and relative humidity of air but also by its velocity, body activity, gender, clothing, etc. Acclimatisation of an individual is also taken into consideration while fixing up the comfort conditions. For example, people living in hot climate, feel comfortable in warmer surroundings than those living in cold climate. Besides, mental condition and individual preferences are some of the psychological factors which decide the comfort of that individual. Hence, comfort condition is different for each individual and depends upon so many factors, as discussed.

11.2.1 Thermoregulatory System of Human Body

However, physical comfort can be established when the body is in thermal equilibrium with the environment, i.e., whatever heat is generated due to the metabolism process during different activities of a man must be dissipated as quick as possible. A human body acts just like a heat engine, as shown in Fig. 11.1. When the heat can be dissipated from the body freely and loss from the body is within certain limits, one feels comfortable. If it is more, cold is felt and when it is less, hot is felt.

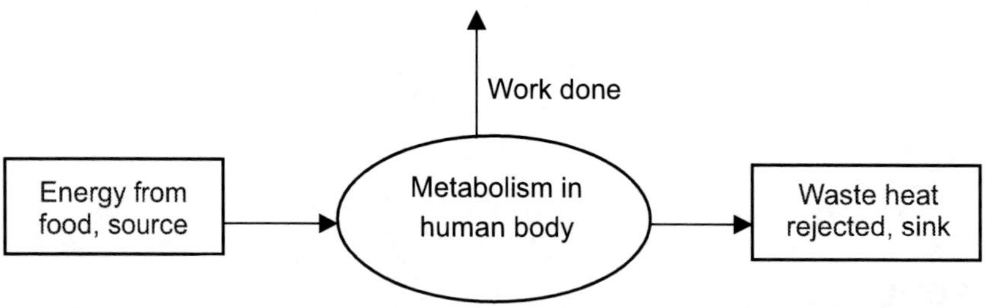

Figure 11.1: Flowchart of thermal balance of human body

Thus, the thermal balance of the body is expressed as

$$M - W = \pm Q_{Con} \pm Q_{Rad} + Q_L + Q_{Res} \pm Q_S \qquad [11.1]$$

where, M is the heat generated due to metabolic process in the body, and

W is the energy utilised by the body for a certain activity.

Q_{Con} and Q_{Rad} are the net heat exchange between the surroundings and the body due to convection and radiation respectively. It is negative in summer as the heat flows from environment (at higher temp.) to body and positive in winter.

Q_L is evaporation from the skin (in the form of sweat), heat is lost in the air that one exhales and with the urine which will be always positive because it does not depend upon the surrounding temperature. There are two modes by which body loses heat by evaporation. One is the evaporation of sweat from the skin and other one is 'diffusion' or insensible evaporation. Sweating is controlled by the thermoregulatory system whereas diffusion is a constant process.

Q_{Res} is the heat lost by the body through respiration (exhalation of air) process. It is always positive.

Q_S is the energy stored in the body. In summer the outdoor air temperature is more than the body temperature. So heat will transfer from outdoors to the body by convection and radiation. So Q_{Con} and Q_{Rad} will be negative. So the total heat of the body will be very high which cannot be dissipated by evaporation from body surface. So there is a tendency of the temperature rise of the body. That means some amount of energy is stored in the body. So Q_S will be positive in summer. On the other hand, in winter outdoor temperature is very much less than the body temperature. Heat will be lost from body surface to environment, which sometimes is more than the heat generated. In this case Q_S will be negative. When body activity and surroundings are relatively stable Q_S term will be zero.

- Here is given a rough estimation of the amount of heat lost from the human body when he is in a conditioned room.

Assume the skin temperature of the human body is 35°C and the skin surface area (A) is 1.9 m^2. The conditioned room temperature is 24°C and air speed is 0.25 m/s.

Heat lost due to convection is calculated from $h\ A\ \Delta t$. Convective htc is calculated from an empirical relation; $h = 13.5\ V^{0.6}$ [11.2]

This gives $h = 13.5 \times 0.25^{0.6} = 5.88$ W/m^2–°C.

So $Q_{Con} = 5.88 \times 0.7$ of $1.9 \times (35 - 24) = 86$ W.

Here it is assumed that 70% of the skin surface is exposed to convective heat transfer.

Heat lost due to radiation $= \varepsilon\sigma A\ ({T_{skin}}^4 - {T_{room}}^4)$

$$= 0.95 \times 5.67 \times 10^{-8} \times 0.7 \times 1.9\ (308^4 - 297^4) = 87.3 \text{ W.}$$

Heat lost due to diffusion or insensible evaporation (this is spontaneous)

$$= h_{fg}\ A\ C_{diff}\ (p_{skin} - p_{room})$$

where, C_{diff} is the diffusion coefficient and p_{skin} and p_{room} are the saturation pressures (in pascal) of water vapour at skin and room temperatures respectively.

$= (2450 \times 1000) \times 1.9 \times 1.2 \times 10^{-9} \times (5628 - 3000) = 14.68$ W

So the total heat loss is of the order of $(86 + 87.3 + 14.68) \approx 188$ W

But maximum amount of heat lost by the body through the sweating process. In any strenuous job, for an instant of time, a person sweats 0.3 gm/s. The latent heat the body loses due to this is $0.3 \times 2450 = 735$ W; (i.e., of the order of 800 W).

Metabolic heat generation depends upon the type of physical or mental activities, mental state, type of clothing and sex and its unit is met. **1 met = 58.2 W/m^2**. Heat generated varies from 120 W for sedentary habits to 1000 W for strenuous job. A young person on an average has energy capacity of 12 met. Women, in general, have a maximum capacity 30% lower than men. The metabolic heat generated during different activities is shown in Table 11.1.

Table 11.1

Activities	Metabolic heat generated (Met)
Sleeping	0.7
Seated (quiet)	1.0
Reading	1.0
Typing	1.1
Walking	1.7
Cooking	1.6–2.0
Leisure activities	2.4–4.4
Dancing	3.0–4.0
Tennis	3.6–4.0
Wrestling	7.0–8.7

Dissipation of heat from the body depends upon efficient heat transfer mechanism and environmental condition. Factors involved are discussed here.

1. Blood circulation in the body should not be so sluggish that generated heat cannot reach the peripherals of the body and other excretory organs from where heat can be dissipated.
2. More of the heat transferred to the skin surface from the internal organs of the body and get dissipated to the environment by convective method which is enhanced by increasing

exposed skin surface area, high wind velocity, low surrounding temperature, etc. When heat loss due to convection and radiation is not sufficient, sweat glands get active and evaporation of surface sweat from skin surface by absorbing the body heat makes the body cool. Some amount of heat is released through exhalation and through urine also.

Let us discuss how our body reacts to the varying surrounding temperature, specifically during summer and winter. Our body core temperature is normally around 37°C (98.6°F) and the skin temperature is around 35°C (95°F). This temperature gradient is maintained so that heat transfer is possible from the internal organs to the body surface. During summer the surroundings temperature is higher than the skin temperature. So there is a heat flow from the surrounding to the body by convection and radiation. The skin temperature becomes high. Body cannot lose heat by convection and radiation. Q_S becomes positive, i.e., heat is stored inside the body. So the body switches on to its thermoregulatory mechanism. There is more peripheral blood flow due to vasodilation. Sweat glands become more active so that there is excessive sweating. Evaporation of sweat takes away the heat from the body and the body feels cool. When the evaporation of sweat is not faster due to high relative humidity or low air velocity, body temperature rises and when it reaches around 105°F, one suffers from heat stroke. Similarly, during winter, more heat is lost from the body by convection and radiation as the surrounding temperature is less than the skin temperature. Q_S becomes negative, i.e., more heat is required for the body. So to decrease the heat loss from the body, peripheral blood flow is decreased by the thermoregulatory system of the body which is called *vasoconstriction.* More heat loss due to convection and radiation can be arrested by sitting cosily or by cuddling up to reduce the exposed surface area of the body to the surroundings and also by covering the body by woollen cloths (an insulating material). At rest body's also shivers (pseudo work is done) to generate more heat so that body's core temperature can be maintained. If the temperature decreases to 96°F, fatal situation may occur.

Velocity: Velocity of air in the space may vary with temperature and humidity in the room. For example, a higher velocity of air is tolerable for high dry bulb temperature. However, an air velocity of 8 –10 m per minute is acceptable for indoor air condition.

However, from the above discussion it is clear that 'comfort feeling' is a relative term and depends upon metabolism rate and heat transfer from/to the body which again depends upon the activity, sex, psychological state of each individual. So the comfort condition is different for each individual. Hence, a new term is defined *effective temperature* combining different variable factors to a single variable.

11.2.2 Effective Temperature

This is the most common environmental index that correlates with the combined effects of air temperature, air humidity, and air movement (speed) upon human comfort. ASHRAE (American Society of Heating, Refrigeration and Airconditioning Engineers) defines "Effective temperature is the temperature of slow moving (0.1 to 0.15 m/s) saturated air that gives the same feeling of warmth or cold as the given atmospheric condition". Some define it as the temperature at 50% RH at which one would feel exactly same feeling of comfort as at ET at 100% RH , and as felt in the actual environment. However, considering either of the definition, it is difficult to have a universal ET chart because it again depends upon type of activity, type of clothing, etc. At lower humidity, higher dbt of air gives same feeling of comfort for the same ET. Though dbt is high, more heat is lost by the body due to higher rate of evaporation from the skin as the humidity is low.

11.2.3 Purity

Purity of air is to be maintained to prevent the infection and other effects on occupants. Mechanical filters, electrostatic precipitators, etc., are used to remove the suspended particles. Activated charcoal is used to remove the odour of air due to presence of organic matters. Ozone and ultraviolet radiations are used as bactericides for indoor air. Proper ventilation takes care to maintain the desired level of CO_2 and O_2 in indoor air. CO_2 level is manageable up to 2.0% (1000 ppm) and O_2 level can also be managed up to 16% though their share in atmospheric air is 0.04% and 21%, respectively.

11.2.4 Noise

Sound is created when unitary air conditioning systems run nearby or during the discharge of air from a central air conditioning system into the conditioned space. Two sources of sound are: (a) at the outlet of air through the diffusers into the space, and (b) transmission of the system generated sound. In the first case, as the conditioned air from the AHU is delivered through the diffusers, sound is created, whose intensity depends upon the velocity. Sound is created as the air moves from high pressure (before diffusers) to low pressure (at indoors). This can be reduced by reducing this pressure difference. Second case normally arises with unitary systems like window ac, split ac, etc. The components create sounds are compressor, fan and blower of condenser and evaporator. This is the reason for which a suitable location for the installation of unitary air conditioning system is essential. (a) Air conditioners should not be installed in narrow corridors or corners to avoid reverberation of compressor and fan running sound, (b) noise blankets should be used around the compressor, (c) air conditioners should be maintained properly so that there should not be any lose part or there should not be any blockage to air flow. Of course, split a/cs don't pose this problem as its condensing unit is placed outside the room. Compressors, fans, or pumps generate low frequency vibrations and this is transmitted through the duct pipes, or the structural members of the building. This type of noise can be controlled by isolating these machines from the transmitting mediums by installing them on a resilient mount. It uses a spring to damp out the low frequency vibrations.

However, low frequency and high decibel (may be) sound is desirable. The noise generated due to air conditioners are expressed in noise criteria (NC). The permissible limit is NC 50 as suggested by ASHRAE. This can also be stated that a sound limit of 70 to 75 decibels is to be maintened to avoid sound pollution.

11.2.5 Comfort Chart

As discussed above human thermal comfort is influenced by psychological as well as physiological factors. Hence it is difficult to specify a single state of moist air, which can give thermal comfort to every individual. A typical comfort chart (Fig. 11.2) is constructed by drawing lines through the points at which majority of people equally clothed and equally active report the same feeling of comfort. These lines are the effective temperature lines.

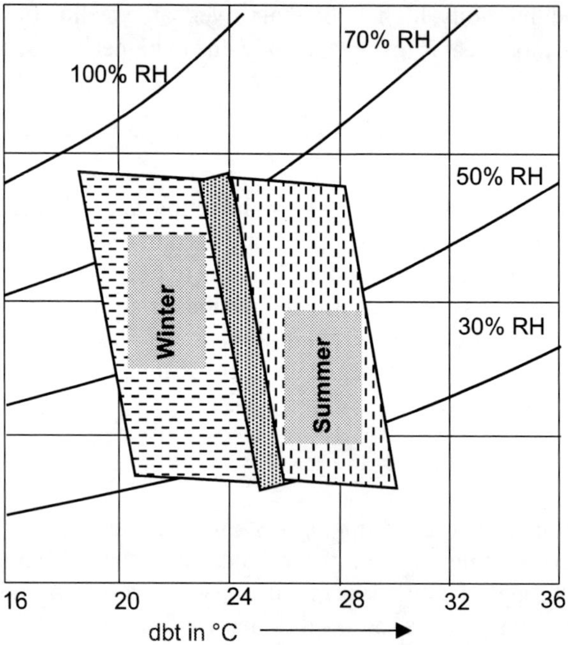

Figure 11.2: Comfort Chart for human body

11.3 INDOOR AIR QUALITY

It has no quantitative measurement. It is expressed in terms of 'acceptable or non-acceptable'. For guidance ASHRAE suggest some quantified parameters for acceptable indoor air quality. The following are some of the suggested values for inside design condition. However, selection of these quantified parameters is based upon ASHRAE Standards 62.1-2004.

(a) Indoor design condition must fall within the 'comfort zone' as shown in the comfort chart during summer and winter. The dbt should be in 20°C to 25°C and RH value should be within 40% to 60%.
(b) Average air velocity should be less than 0.15 m/s.
(c) Dew point should be more than 2°C.
(d) CO_2 content should not exceed 0.1%.
(e) There is no contaminant at harmful level concentration as determined by the competent authority.
(f) Noise level should be around 30 decibels.

11.4 DISCOMFORT AND DISEASES ASSOCIATED WITH AIR CONDITIONING

Poor indoor air quality (IAQ) has an adverse effect on human comfort, health and also activities. Normally all air conditioned buildings are made like sealed box just to prevent the infiltration of air. More the infiltered air or ventilating air more will be the cooling load and more will be power consumption. Thus, proper ventilation cannot be ensured. This leads to the increase in the concentration of pollutants like CO_2 which is contributed by the occupants and the air contaminated with fungi and bacteria laden dust particles entering the conditioned room from outdoors or carried by the occupants. This is more prevalent when there is an epidemic and also in hospitals. This causes, mainly, two types of adverse effects. One is called *sick building syndrome* (SBS) and the other one is known as *building related illness* (BRI). The symptoms of SBS are headache, dizziness, sore throat, etc., which disappear when you come out of the room. The increase in CO_2 level in the room and the unsuitable temperature and humidity level in the room for some individuals are its main causes. Water and dust particles of outdoor moist air carry allergens, the spores of fungi and the bacteria with it and they find their shelter in the poorly ventilated zone. Growth of fungi and bacteria in the moist zone of air conditioning system i.e., cooling coils and condensed water collection tanks are the causes of BRI. Amongst them the mostly found fungus is of Aspergillus species and the bacterium is Legionnaire. They cause allergic respiratory diseases like sinusitis, bronchitis, asthma, etc. Prolonged exposure to such environment may infect gastrointestinal tract, kidney and sometimes becomes the cause of facial palsy. Discomfort also includes dryness of skin due to constant exposure to low humidity environment. During any epidemic or pandemic natural ventilation should be preferred over the use of air conditioned space. It is proved that an enclosed environment or a space with limited supply of outdoor air is more susceptible for spreading of viruses. To get rid of these adverse effects, air conditioners should have their optimal use and make use of natural ventilation. Copper coils should be preferred over aluminium coils for manufacturing the air conditioning equipment because copper surface has the fungicidal effect. The a/c system should be maintained and cleaned periodically to prevent any growth of fungi and bacteria.

11.5 OTHER APPLICATIONS

It includes the fields of food and beverages preservation, cold storage (agricultural application), medicine, transport and handling of perishable goods, science and industry, Space research centres and many more. A brief discussion on these applications of refrigeration and air conditioning with typical examples follows.

Table 11.2 Average storage conditions for different types of foods (fruits, vegetables, fish, etc.)

Items	Temperature, °C	Relative humidity, %	Period of storing
Banana	13	90	15 days
Apple	2	90	6 months
Grapes	1	95	4 months
Oranges	5	85	10 weeks
Potatoes	10	90	5 weeks
Onions	1	65	7 months
Tomatoes	10	90	10 days
Cabbage	1	95	5 weeks

Mushrooms	0	90	3 days
Dried fruits	1	Low	8 months
Poultry	–1	90	1 week
Fish	2	95	15
Beer	10	65	6 months
Milk	1	—	3 months
Cream	–25	—	10 months

11.5.1 Preservation of Foodstuffs

Need of vegetables, fruits for ever growing population and their seasonal availability compel the scientists to develop the preservation methods so that each we get the required amount of these highly perishable items during the shortage of supply/production. Besides, other foodstuffs need to be preserved are meat, poultry, fish, eggs, etc. The objective of preservation is to get the foodstuffs in edible condition in the time of need while retaining its appearance, odour, taste and more importantly its vitamin contents. The agents of deterioration and spoilage of foodstuffs are: (a) the natural enzymes inherent in all organic materials, and (b) the microorganisms that grow inside and surface of the foodstuffs. The destroying actions of these two agents can be totally eliminated by heating the foodstuffs to around 75°C. Their destroying action can also be very much minimised by drying the stuffs. But heating and drying are not possible for vegetables, fruit and other stuffs, as they lose their natural appearance, taste, texture, etc. So, it is preferred to preserve it at lower temperature at which the destroying actions of these agents slow down to the minimum. (At low temperature, even at lower than –20°C, the enzymatic action goes on). Table 11.2 gives the data of different foodstuffs and their preserving temperature and relative humidity along with the period of preservation. For temporary storage of most foodstuffs for 3-4 days, storage temperature of 6 –7°C is suitable which is normally maintained in the domestic refrigerators. Some vegetables (like tomato) are not suitable to be kept in freezing condition. As the table shows most of the foodstuffs are preserved at a very high RH (around 90%). Hence, the foodstuffs, to be preserved, are to be stored being well packaged form. Otherwise they lose the moisture due to the difference in vapour pressure between the foodstuff and the storage space and get shrivelled. Also, they lose vitamins along with the moisture. So, the storage of all perishables in their natural state (unpackaged) requires close control not only of space temperature but also of space humidity and air motion. Sometimes crushed ice is spread over these items so that they can be preserved under low temperature and high humidity.

11.5.1.1 Cold storage

Cold storage made to preserve perishable and seasonal vegetables and fruit during the harvesting period and made available to the consumers throughout the year. The height of the cold storage may vary from 10 m to 30 m. The selection of rack system design and the related material handling equipment depends upon types of vegetables and fruits to be preserved. The items may be refrigerated or freezed as per the requirement which depends upon the type of stuff, size/amount of the stuff and the duration for which it is to be stored. The heating load on the cold storage comprised of sensible heat and latent heat (if freezing is required); heat of respiration of vegetables and fruits; heat gain through building; infiltration; heat of electrical equipment; heat of people and other transport equipment. Accordingly, the freezing system capacity and air flow system are decided. NH_3 is the refrigerant commonly used for cold storage refrigeration system.

Here the main problem is to avoid the formation of frost and ice. Because this creates hazard for the working people, cause damage to the material handling equipment, and hamper the performance of the cooling system. Cold storage unit are maintained at negative pressure. Hence, besides the infiltered air, outdoor air draws into the space. Again the difference in vapour pressure between the stuffs and the space of the cold storage draws out (by diffusion) some of the water vapour from the stuffs. Therefore, defrosting and cleaning the space of frost and ice in definite time interval is essential.

11.5.1.2 Freezing

Many a time a product is required to be stored in its fresh state for a longer period. Here storing is done at a temperature around –20°C or below. This includes vegetables, fruits, meats, poultry, eggs and even precooked food products. However, all variety of vegetables and fruits are not suitable for frozen storage, that is to be critically selected. Preserving the items in frozen storage, requires some pre-processing of the items. They are to be harvested at the peak of their maturity and to be kept in frozen storage as soon as possible to avoid undesirable chemical changes through enzymatic and microbial action. So the vegetables and fruit are cleaned thoroughly to remove foreign materials. Then the vegetables are put in hot water/ steam at 100°C for some time so that the natural enzymes of the vegetables are destroyed. This process is called *blanching*. The duration of it varies from 1 min to 10 min depending upon the vegetable type. Then it is immediately chilled to 10°C before packing for storage. This greatly increases the storage duration of the vegetables. (The enzymatic action of vegetables, fruits, etc., continues, though slowly, even at very low temperature.) But the fruits are not subjected to blanching though they have higher rate of enzymatic activity. Because the hot water destroys the natural freshness of the fruits. Rather they are covered with a coating of light sugar syrup or citric acid or sulphur dioxide. But for meat, poultry, or seafoods pre-processing is not required. However, it is seen that poultry items give more tenderness if freeze within 12 to 24 hours of killing.

Methods of Freezing

There are two methods of freezing process. One is 'sharp freezing' and other one is 'quick freezing'. In sharp freezing method the product is placed in a low temperature room and allowed to freeze slowly in still air within the room. It takes very long time, i.e., in the range of a day, to freeze the product. By this method a bulk amount of product, like fruit in barrel, boxed poultry, filleted and whole fish, etc., are frozen.

Quick freezing is of three types. (a) air blast freeezing, (b) indirect contact freezing, and (c) immersion freezing. In air blast freezing, low temperature and high velocity air is circulated around the product and the shelves of the freezer should be so arranged that air can circulate freely around each product. This is best suitable for freezing of non-uniform or irregular shapes of products like dressed poultry. Sometimes cooked and packaged foods can also be stored by air blast freezing. Indirect contact freezing is accomplished by keeping the products on metal plates through which evaporator/cooling coil is passed. Hence, the cooling time depends upon the thermal conductivity of the product/package. However, this method is very efficient but suitable for small products and specifically for larger commercial freezers used for consumer display. The miniature example of it is the freezer of domestic refrigerator. Immersion freezing is done by immersing the product in low temperature brine solution so that there is a good contact and heat transfer is very fast. Thus, it requires very less time to freeze the product. The only disadvantage with this method is that the brine solution (specifically, NaCl-water) infuse to the product due to osmosis. This method is particularly suitable for freezing fish and shrimp at sea.

However, quick freezing method is mostly advantageous over sharp freezing method. The principal difference between these two methods is as stated here. In sharp freezing method, there is slow cooling. So large ice crystals out of cellular fluid are formed which cause serious damage to the tissue of some products through cellular breakdown. But in quick freezing method small ice crystals which are formed almost entirely within the cell so that cellular break down is greatly reduced.

11.5.1.3 Making of ice cream

Ice cream is a product which is manufactured in three stages. First is mixing of different constituents, second is freezing to a plastic state, and third is hardening. The mix is prepared from fats (not necessarily from dairy), milk protein, sugar and additives like emulsifiers, stabilizers, etc., along with fruits, nuts, chocolate as per taste. This mix is made to liquid form, pasteurised, homogenised and cooled with chilled water. Then it is stored at 2 to 3°C for 2 to 3 hours. This is known as 'aging'. In the second stage 50% of mix along with 50% of air is made to pass through a barrel which forms the internal surface of a flooded evaporator (Fig. 11.3). The preferred refrigerant is NH_3 which is at a temperature of around –30°C in the evaporator. The aerated ice cream mix comes out of the barrel at a temperature of –5°C. The ice cream is still in plastic state as it comes out of the freezer. Then it is extruded into final sales shape (box type, cone type, etc.). In the third stage this ice cream is hardened at a temperature of –25°C or lower. Ice cream must be preserved at lower temperature at around –25°C. If it is allowed to soften, the entrained air bubbles escape out and the texture of ice cream changes. If it is refrozen, the skin or the upper surface of it becomes hard making the product inedible.

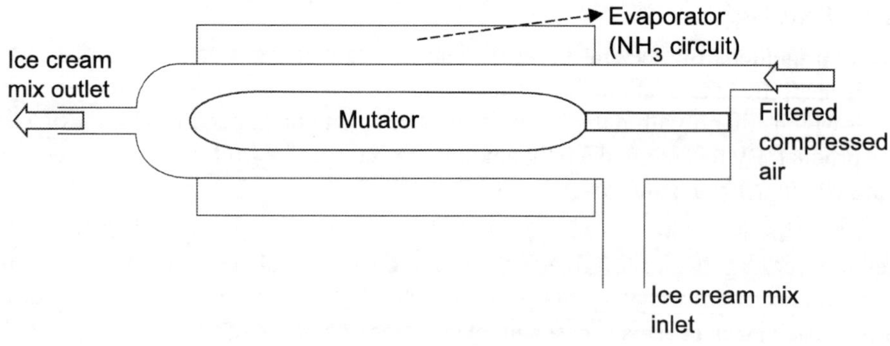

Figure 11.3: Schematic of ice cream production

11.5.1.4 Milk and milk products

This is a an important foodstuff to be preserved. This highly perishable foodstuff, in different form, is in great demand by all age group of people. Milk comes from cow at 37°C and should be cooled down to 4°C within 2 hours under hygienic conditions. Because at this temperature microorganisms have very slow deteriorating effect on the milk. At this temperature milk is transported to creameries and other dairy farms. The whole milk is pasteurized there for human consumption. It is first heated up to 73°C, first by the outgoing milk which is at 73°C and then by hot water at 78°C for a short time. Then it is re-cooled to 4°C by chilled water at 2°C. For this plate type heat exchangers are used. The advanced method of milk pasteurization by compression-absorption refrigeration system is discussed in Section 13.3. Besides, cream and skimmed milk can be obtained by centrifuge method. Churning devices are used to get butter and butter milk. Butter, after being churned out is stored at 4°C to crystalize the fat. For long-term storage, it should be preserved at –25°C. Milk is treated with rennet to get cheese and with cultured bacteria to make yoghurt. Milk powder is prepared by drying the milk. However, the whole milk and all the milk products (except milk powder) are to be stored at a temperature of 4°C.

11.5.2 Space Research (Mostly Cryogenic Application)

Liquid oxygen and liquid hydrogen are used as the rocket fuel. These are produced by gas liquification process as discussed earlier. Spacecraft cabin is also to be conditioned to maintain the temperature at 24°C and RH 50% of the cabin.

11.5.3 Medicine Sector

The most important application in medicine is the preservation of total blood and some blood components. In this case the temperatures are to be strictly maintained. Total blood (i.e., the blood we donate) is preserved at temperature 2°C to 6°C. Above 6°C, there is chance of bacterial growth and below 2°C RBC become haemolysed. Hence, this range of temperature is to be strictly monitored. Similarly, blood plasma is to be stored at –30°C to –40°C for number of days. But it can be stored at 4°C only for 24 hours. Platelets are stored at 22°C to 24°C. Besides this, anti-coagulants and certain additives are added to prevent clotting and then provide nutrients to maintain red cell viability and function throughout storage.

Operation theatres are to be maintained at 24°C and 50% RH with positive pressure of 2.5 mm of water more than the atmospheric pressure. Pre-matured baby clinics maintain the incubators, where the pre-matured baby is kept for treatment, temperature at 37°C. The reason is that the metabolism process is so slow in case of a pre-matured baby to maintain the normal body temperature of 37°C that there is a chance of baby getting collapsed. Besides this, some medicines, injections are also to be kept at lower temperature, even for retail sale.

11.5.4 Industrial Applications

Textile industries: Air conditioning is an essential for manufacturing good quality thread and minimise the wastage like thread breakage and cloth warping. The textile materials are very hygroscopic. Here humidity control is more important than any other factors. For different types of threads, on an average, the environmental temperature to be maintained is around 22°C and the RH to be maintained is 60% for most of the operations. For cotton type of thread, relative humidity to be maintained is in the range of 70% to 80% because they are very brittle.

Paper industry: Paper industries need an environment of temperature 22.5°C and relative humidity of 40 to 50%.

Concrete works: While cement, sand, chips, water get mixed up to prepare concrete heat is generated. After casting of concrete, if this heat is not removed, the casting will develop crack due to thermal stress. Hence, precooling of the constituents of the concrete by chilled water is done or chilled environment is employed during casting process.

11.5.5 Miscellaneous Applications

Library and museum: Here humidity control is more important because high humidity spoil the books, documents, etc., due to its corrosive action on paper articles. The environmental condition for such places should be maintained at an average value of 24°C and 45% RH.

Photography: The base of the photographic film is coated with gelatin emulsion which contains salt of silver. The preparation of this film requires careful control of temperature and humidity. For processing of film, the recommended temperature and relative humidity are around 27°C and 55%.

Transport, Handling and Distribution

The preserved foodstuffs, medicines, bloods, etc., are required to be transported from one part to other, globally, in view of their demands. In this context a suitable term used is 'cold chain'. This term applied to the whole process of refrigerating or freezing of any item, transportation to the place of demand with proper handling, then distributing it or selling it. This requires transport, different types of storage and display. During this entire process, storage temperatures and humidity, needed for the items, are to be maintained in the cargo planes, ships or road vehicles. The main problem in air or sea mode of transport is the waiting time for the items to be loaded or to be moved out at the air ports or harbours. During this period, they are exposed to the local climate, which may either be hot or too cold. So suitable freezing or warming arrangements are to be made at those places. Many a time solid CO_2 or liquid nitrogen is used for this type of short-term cooling process.

EXERCISE

1. Write the thermal energy balance equation for the human body.
2. What is a 'comfort chart'? Write its utility.
3. If an air conditioner is taken to a higher altitude, how does its performance is affected?
4. How 'effective temperature' is defined? What are the factors that affect the value? What is its significance in determining the human comfort condition?
5. What is the difference between comfort air conditioning and industrial air conditioning?
6. Write short notes on:
   (i) Comfort air conditioning (ii) Blanching
   (iii) Discomfort in air conditioning (iv) Thermal balance of the body

MULTIPLE–CHOICE QUESTIONS

1. A human feels hot when heat loss to the environment is — than heat generated in the body.

(a) more (b) less

(c) same (d) no effect of heat loss from the body

2. The comfortable temperature for a human in winter is — than that in summer.

(a) more (b) less

(c) same (d) no effect of season change

3. Comfort feeling for a human depends upon

(a) psychological factor (b) physiological factor

(c) both (a) and (b) (d) none of (a) and (b)

4. Unit of metabolic heat generated in the body is

(a) met (b) watt

(c) joule (d) kcal

5. Effective temperature for human comfort is — of

(a) dbt of air (b) wbt of air

(c) an estimated temperature considering the affecting variables

(d) a fixed temperature and relative humidity fixed by ASHRAE

6. The air conditioning system should be cleaned periodically, more importantly, to prevent-reduce.

(a) bad odour (b) growth of fungi and bacteria

(c) rust (d) power consumption

7. The most common disease associated with the use of air conditioner is

(a) hypertension (b) diabetes

(c) asthma (d) leucoderma

8. Fruits are preserved under — temperature and — humidity respectively.

(a) low and low (b) low and high

(c) high and low (d) high and high

9. Blanching is a process in which vegetables are exposed to — to destroy their enzymatic activity.

(a) chilled water (b) ice blocks

(c) normal water (d) steam

10. Platelets are stored at temperature —

(a) 22°C (b) 0°C

(c) –12°C (d) –22°C

11. An air-conditioned room of volume 10 m^3 has infiltration of air equivalent to 3 air changes per hour. Density of air is 1.2 kg/m^3, specific heat cp is 1 kJ/kg K and temperature difference between room and ambient air is 20 K. What is the sensible heat load due to infiltrated air?

(a) 60 kJ/hour (b) 12 kJ/hour

(c) 0.45 kW (d) 0.2 kW

12. The heat load from the occupants in air conditioning load calculation is a source of:

(a) Sensible heat only (b) Latent heat only

(c) Both sensible and latent heats (d) None of the above

13. Which one of the following statements is correct?

The optimum effective temperature for human comfort is:

(a) higher in winter than that in summer (b) lower in winter than that in summer

(c) same in winter and summer (d) not dependent on season

14. Effective temperature depends on dry bulb temperature, and

(a) wet bulb temperature only (b) relative humidity

(c) specific humidity (d) wet bulb temperature and air motion

15. What are the general comfort conditions in an air-conditioning system?

(a) 20°C dbt, 80% RH (b) 24°C dbt, 60% RH

(c) 25°C dbt, 40% RH (d) 25°C dbt, 100% RH

16. The desirable air velocity in the occupied zone for comfort for summer air conditioners is in the range of

(a) 7 - 9 m/minute (b) 4 - 5 m/minute

(c) 2 - 3 m/minute (d) 0.5 - 1.5 m/minute

ANSWERS

1. (b) **2.** (b) **3.** (c) **4.** (a) **5.** (c) **6.** (b) **7.** (c) **8.** (b) **9.** (d) **10.** (a)
11. (d) **12.** (c) **13.** (b) **14.** (d) **15.** (b) **16.** (a)

12

Heating, Ventilating, and Air Conditioning (HVAC) Systems

12.1 HVAC SYSTEMS

Till now discussion is made on principles of creating a low temperature environment (refrigeration) and principle of conditioning the air (psychrometric processes). To fulfil the need for human comfort and other applications as described in Chapter 11, there must be devices to create the desired condition of air (temperature and humidity) and efficient air distributing systems so that the conditioned air of required quantity and velocity with cleanliness can reach the end user. In this chapter, we will discuss different devices which are operated on those thermodynamic principles.

Heating, ventilating and air conditioning systems have the complete facility to heat, cool, humidify, dehumidify, clean and distribute the conditioned air into the room so as to meet the indoor air condition for human comfort or industrial applications throughout the year. In large installations like shopping complexes, academic complexes, research laboratories, hotels, etc., the desired temperature and humidity level may vary from point to point, room to room or individual to individual. In such cases an air conditioning zone is defined. **When a system is designed to provide independent control in different spaces, each space is called a zone.** It may be a room or a group of rooms in which comfortable conditions can be maintained by a single controlling device. A single zone system has one thermostat and one humidistat which control one heating or cooling unit to maintain preset indoor condition. Thus the total complex is divided into different air conditioning zones. Instead of supplying same conditioned air to different zones, desired level of conditioned air can be distributed to different zones. Thus, the energy in conditioning air for zones, where it is not required, can be saved. In some cases, the temperature can be controlled by varying the air volume flow rate or by supplying a little colder air and providing heaters. In complexes where the units are widespread or in case high rising buildings, the cold/hot waters in required amounts is generated at a central unit. They are transported to different units through underground insulated ducts and made to pass through the cooling/heating coils mounted for different zones. Air, from separate air handling units (AHUs), is blown over these coils to get the desired level of cooling/heating. The following sections have discussed details of the distribution conditioned air to our need.

12.2 DOMESTIC REFRIGERATOR

(Figure 12.1) This is the only refrigeration system that every common man has come across. The main components are same as for a vapour compression refrigeration system. The hermetic compressor is fixed at the lower most part and back side of the cabinet. The condenser is of wire and tube (or plate and tube) type which is fixed just on the back side of the cabinet. The capillary tube is also put on the back side and it is clearly visible like a small wire coil. Then comes the evaporator coil which is placed, usually, at the top inner side of the cabinet. The evaporator coil (D-X coil) is fixed around this box called freezer section. The coil temperature is maintained at around –15°C. The chilled air from the freezer section flows down due to natural convection and cools the other sections of the refrigerator below the freezer section. The temperature inside refrigerator increases as the distance from freezer section increases (normally from 2°C to 10°C). However, the average temperature of the storage space (excluding freezer) of the refrigerator comes out to be around 6°C. The food stuffs can be preserved safely for a short duration (3-4 days) with this temperature. Freezer section can be used to make small ice cubes, ice cream, etc. The domestic refrigerators are available on the basis of their gross volume of 90 litres, 165 litres, 250 litres, 380 litres, etc.

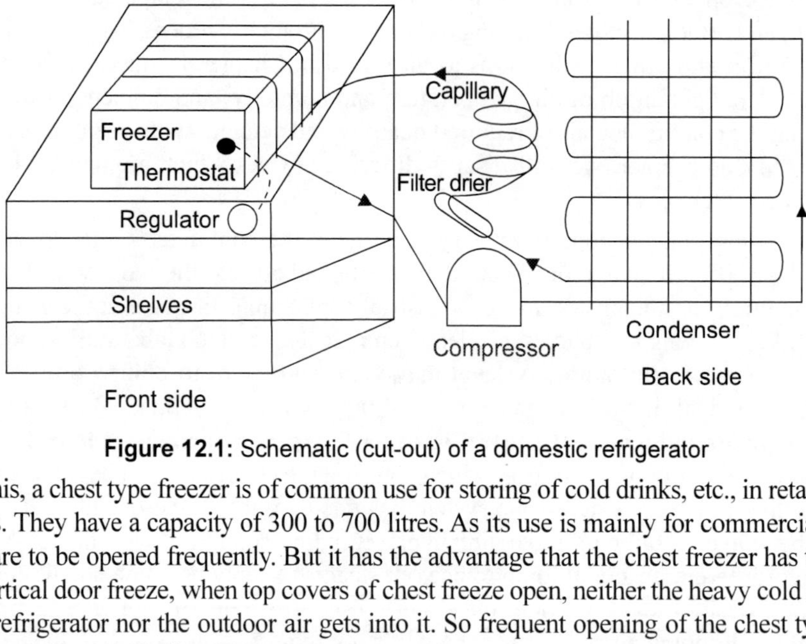

Figure 12.1: Schematic (cut-out) of a domestic refrigerator

Besides this, a chest type freezer is of common use for storing of cold drinks, etc., in retailing shops and hotels. They have a capacity of 300 to 700 litres. As its use is mainly for commercial purpose, its doors are to be opened frequently. But it has the advantage that the chest freezer has top covers. Unlike vertical door freeze, when top covers of chest freeze open, neither the heavy cold air escapes from the refrigerator nor the outdoor air gets into it. So frequent opening of the chest type freezer door has less effect on cooling load on the freezer coil.

12.3 CLASSIFICATION OF AIR CONDITIONING SYSTEMS

Air conditioning systems can be classified in a number of ways.

1. Basing upon the type of equipment arrangement:
    (a) *Unitary systems:* The unitary systems are factory assembled or packaged units and have the provision of direct cooling. It cools the air directly as it passes over the

evaporator coil. The major components, like compressor, condenser, expansion valve and evaporator are assembled to provide conditioned air for a single zone. Condenser is mostly air cooled. They are of three major types: (1) window air conditioner, (2) split air conditioner, and (3) rooftop units

   (b) Central systems

2. Depending upon the cooling or heating load demand for the conditioned space, i.e., on the basis of zones
   (a) Single zone system
   (b) Dual duct system
   (c) Multiple zone system
   (d) Variable air volume (VAV) system
3. Basing upon the type of secondary fluids used for conditioning the room air in multi-zone system:
   (a) All air system
   (b) Air-and-water system
   (c) All-water system
4. Basing upon the seasons
   (a) *Summer system:* In summer, there may be two environmental conditions. One is high temperature and high humidity (coastal belt). Second is high temperature and low humidity (desert areas). First one involves cooling and dehumidification of air. Both can be accomplished by a cooling coil and for very high humidity chemical dehumidification (before passing over the cooling) is done. Second one involves mostly evaporative cooling process where dry air is made to pass through a medium filled with moist straw-like material. Thus, the air gets cooled and humidified. Therefore, in this case, 'desert cooler' is preferred over air conditioning system.
   (b) *Winter system:* In this case air is normally cold and dry. So moist air needs to be heated and humidified. This can be achieved by injecting hot steam to the air to be conditioned. Or these two processes can be done separately, i.e., first by heating the air followed by addition of water vapour to the moist air.
   (c) *All year-round system:* These systems involve both cooling/heating and humidification/dehumidification apart from improving the purity of air. Actually, these are called the heating, ventilating and air conditioning (HVAC) systems. These have automatic controls to switch them over to heating and humidifying systems during winter and vice versa during summer.

12.3.1 Window Air Conditioner

Window air conditioners are frequently used for domestic purpose and for low cooling loads (1-3 TR). It is mounted on the space provided for window. As shown in Fig. 12.2, this unit is divided in two halves by a partition wall. The outdoor unit contains compressor, condenser and indoor unit contains evaporator and expansion valve. An axial fan draws the atmospheric air over the compressor and throws it over the condenser coil. Thus, it cools condenser rejecting heat to the environment and also cools the compressor to some extent. Another centrifugal fan draws the indoor air and made to flow it over the evaporator coil. This is how the same indoor air is re-circulated being cooled and dehumidified. A vent is provided in the partition to allow some fresh air to be inducted to the room continuously. Both the fans are mounted on a single motor shaft.

In extreme climate (very cold during winter and very hot during summer), the unitary units can also be used as heat pump during winter. This is achieved by using a four-way valve in the circuit. During summer condenser is placed outdoor and evaporator is placed indoors. During winter the refrigerant flow direction is reversed. Now the outdoor unit receives heat from outdoor and reject it to the indoors through indoor unit.

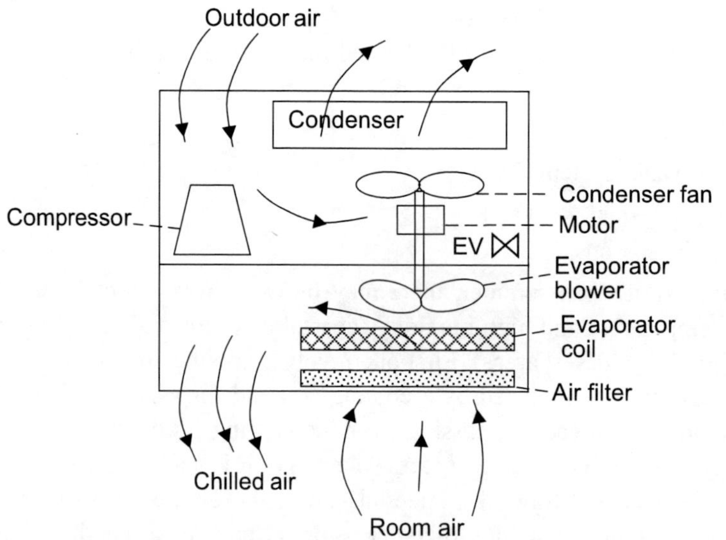

Figure 12.2: Schematic of a window air conditioner

12.3.2 Split Air Conditioners

Unlike window air conditioner, compressor and condenser, condenser fan and motor as one unit, called 'condensing unit' (outdoors) and expansion device are split from evaporator (direct expansion coil), a fan with motor which form another unit, called 'cooling unit' (indoors). (Fig. 12.3). Expansion device (mostly capillary tube) may either be placed in the outdoor unit or in indoor unit. If placed in the indoor unit, the merit is that the liquid refrigerant from condenser can flow down to capillary tube.

A comparison between window and split air conditioners is as follows.

Advantages with split air conditioner: The cooling unit can be fixed at any desired place of the room in the wall. The outdoor unit, compressor, condenser and the fan with motor (outdoor unit) can be placed at a suitable, shaded place away from the room. So the undesirable sound of running the compressor and the condenser fan is eliminated. The maintenance of the parts is easier as they are easily accessible.

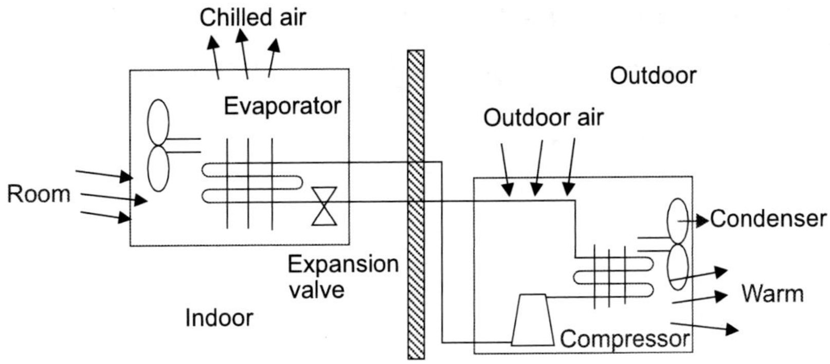

Figure 12.3: Schematic of a split air conditioner

But the disadvantages with this system are that the tubing to connect the outdoor and indoor unit is too long (sometimes as long as 10 m). So there is more pressure drop and hence there is more power consumption. Secondly, outdoor heat gain through tubing is also very large if it is not perfectly insulated. This increases the cooling load on the cooling coil. There is no scope for introducing fresh air into the room. Two separate motors are used for two fans (one for condensing unit and another for cooling unit). Therefore, power consumption is more. One more thing to be considered while installing the outdoor unit. It should be installed above or below the level of the indoor unit. If outdoor unit is above the indoor unit, the liquid refrigerant from condenser flows down to expansion device. If it is reverse, compressor has to develop the pressure to overcome the potential head between the outdoor unit and indoor unit.

12.3.3 Rooftop Unit

This is a factory assembled or packaged unit. These may be mounted on floors or on rooftops. Return air is drawn up into the base of the unit and the supply air is blown vertically down from the bottom of the unit into the space below. As an alternative, the ducts can project from the end of the unit to run across the roof before entering the room. The advantages of this unit are: (1) no working unit in the occupied space so that maintenance of the unit becomes easier, (2) no space is built inside the room for the unit. The disadvantages are: (1) regular maintenance is required in heavy rain, snow or high wind as the unit is installed on the rooftop; (2) as the mutually matching components are factory assembled, choice of performance is limited to the available set of components.

Condensing Unit

This is not an air conditioning system, but a part of it. A condensing unit comprises a compressor and a condenser mounted on a single frame. The condenser may be air cooled or water cooled. This unit has all the required tubing and electric wiring inside it. The accessories like liquid receiver, condenser fan and other control units are also housed inside it to make the set functional. This unit can directly be connected with evaporator on the site. They are available for cooling capacity ranging from one kW to more than 100 kW. They are designed in such a manner that compressor capacity matches with the performance of the condenser.

12.3.4 Central Air Conditioning System

Willis Carrier is known as the 'father of air conditioning'. In 1904, he designed a central air conditioning plant using air washer. Around 1930, the manufacture of air conditioning equipment started on a commercial scale and by 1950 the industry was firmly established.

The main components of this system are: (1) central plant, (2) air handling unit, (3) water cooling equipment, (4) other devices like pumps, dampers, etc.

This system provides chilled or hot air or water at a central location. Then it sends them to desired locations through ducting. In a central room the main components of the refrigeration system, i.e., condensing unit (compressor, condenser) and expansion device are installed which is otherwise called central plant. Normally, open or semi-open type compressors are used. Liquid refrigerant from condenser flows to expansion device and then through direct expansion coil (DX coil) or cooling coil placed in the **air handling unit (AHU)**. (Refer Fig. 12.4) In AHU return air and some percentage of fresh air (as decided previously) are mixed up in a chamber called plenum chamber. Dampers are provided to regulate the flow rate of these two streams of air. This mixed up air passes through air filter and then flows over the cooling coil (or heating coil if warm air is required). This chilled (or hot) air (or water) is sent to different zones as per requirement. In case of water cooled condensers, the hot water, if needed to be re-circulated, is sent to **water cooling equipment** where it gets cooled and returns as a coolant to the condenser.

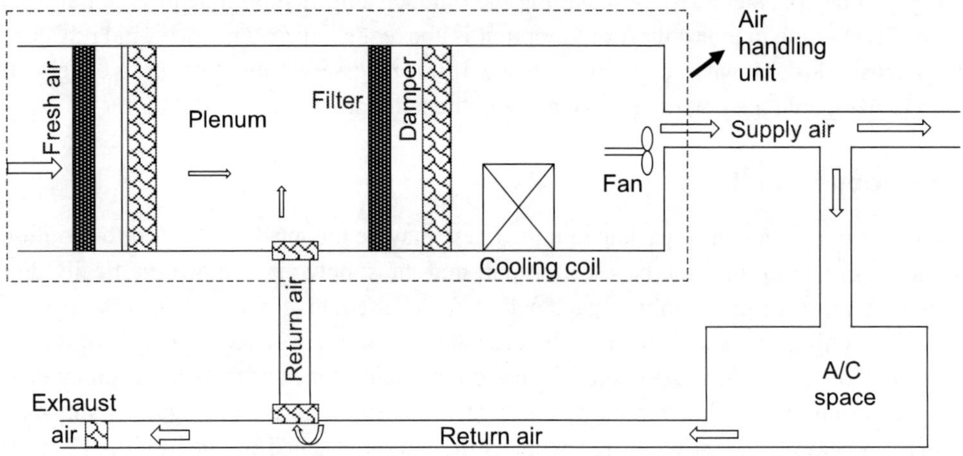

Figure 12.4: Schematic of a central air conditioning system with AHU

12.4 AIR HANDLING UNIT

This is the most important section of the central air conditioning system. Different components of this section are as follows.

1. **Air inlet and mixing part:** This is a mixing chamber, otherwise called as plenum chamber. Inlet louvers and screen are mounted at the inlet to prevent the entry of rain, snow, flying objects etc. Next to it dampers are mounted. **Function of dampers is to regulate the air flow quantity and direction.** Dampers for outdoor air and return air direct the air streams towards each other in such a manner so that there is a good mixing of two streams of

air. The dampers are adjusted in a way to allow the designed proportion of outdoor and return air to the mixing chamber. The rest amount of return air (relief air) is passed though the relief air damper. This type of simultaneous operation of three dampers can be easily regulated by a single device called 'actuator'. When ventilation air is not required, outdoor damper and the relief damper may be shut off. There will be 100% recirculation of air.

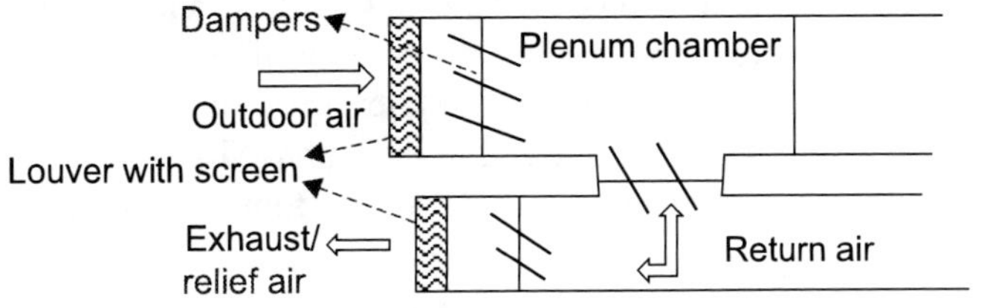

Figure 12.5: Schematic of the mixing part of an AHU

2. **Temperature sensor:** This is provided in the path of the mixed air stream. If the mixing of outdoor air and return air is not proper or the proportion of outdoor and return air is not as per the designed condition, the mixing stream temperature does not reach the estimated value. Then the control mechanism of dampers resets the position of damper blades, so that the temperature can be reached by proper mixing process or by maintaining the proper proportion between the two air streams. The control of dampers may be done manually or by solenoid valve operated auto mechanism. Solenoid valve is actuated by the thermostats or temperature sensor.
3. **Filter:** Air filter is mounted just after the plenum chamber to ensure dust free mixed air enters the air conditioning system.
4. **Cooling/heating coil:** Cooling coil may be the evaporator coil of the refrigeration system or the chilled water coil (minimum temperature can be maintained at 5°C). Its temperature is normally maintained below dew pint temperature so that dehumidification occurs. Care must be taken to see whether it is humid zone (like coastal belt) or dry zone (like deserts where humidification is needed). Heating coils may be an electrical heater coil or gas fired hot air coils.
5. **Humidifier:** This is provided to humidify the dry air with water spray so that required relative humidity of supply air can be maintained. The humidity level is indicated by a humidistat.
6. **Fan:** The fan develops the required total pressure to drive the air through the system. In some cases return fans are provided to have control over the building pressure.

12.5 WATER COOLING EQUIPMENT

Different water cooling equipment available are: (a) natural and spray ponds, (b) evaporative condenser, and (c) cooling towers

In natural and spray ponds, a natural or artificial pond or big water body is used. Hot water from condenser is directly discharged to the water body. The surface water of the water body gets cooled by the natural process, i.e., evaporative cooling. Sometimes hot water from condenser is sprayed into the air over the pond resulting in faster evaporation and cooling. This low temperature water from water body is again supplied for condenser cooling.

In case of evaporative condenser (Fig. 5.12) low temperature water is sprayed over the condenser coil. Atmospheric air is induced to flow over the condenser coil from the bottom of the condenser coil to top. Thus, there occurs heat transfer from condenser to sprayed water as well as to air. Much of the water particles get evaporated taking the heat from condenser coil. There is also heat transfer from hot water particles and the air moving up due to direct contact between the water particles and air. Thus, air takes all the heat and goes out and water gets cooled. As the air goes out, water vapour and some amount of water particles also are carried away with it. Therefore, to avoid drifting loss of water, drift eliminators are provided at the outlet of air. Low temperature water is collected in the sump. Make-up water is added to it to compensate any water loss due to drifting. Then the water from sump is recirculated.

(The minimum temperature air can reach by evaporative cooling is its wet bulb temperature theoretically. Practically, it reaches a temperature 4°C higher. Performance of a cooling tower is described below.

12.5.1 Cooling Tower

Cooling tower is provided to cool the hot fluid coming from the condenser. This is mostly a hyperbolic or cylindrical structure. The components are spray nozzles, fills, packing and a centrifugal or propeller fan. Basic principle of operation is as follows. Hot water from condenser is sprayed through a series of spray nozzles from the top of the cooling tower. At the same time atmospheric air is drawn from the bottom of the tower either by forced draft or by induced draft as shown in Fig. 5.11. Basing on this, cooling towers are classified as 'forced draft cooling tower' and 'induced draft cooling tower'. Usually, there is one or more propeller or centrifugal fans moving air vertically up or horizontally through the tower. So the flow of water droplets and air mostly of counter-flow type and sometimes of cross flow type. Heat transfer from water particles to air occurs due to intimate mixing of water particles and air. Thus, the water is sensibly cooled and the air gets heated. This is a good example of direct contact type heat exchanger. Some of the water particles also get evaporated and cools the water further. Then hot air goes out and water, being cooled, flows down and is collected at the bottom. To enhance the performance of the cooling tower 'fills' (made of wood or plastics) are arranged around the top inner periphery of the tower. Hot water particles from spray nozzles are directed towards the fills. This causes more atomization of water particles. More atomization of water particle increases the surface contact area between water particles and air. So heat transfer rate increases. To minimise the water loss due to drift, drift eliminators (this prevents the water particles to go out along with the hot air) are provided at the outlet of the air. Make-up water is provided to the tank at the bottom to compensate the drift loss.

There is also 'natural draft cooling tower'. The shape of this tower is made hyperbolic. The density difference of air at the top and the bottom of the tower made the air to flow from bottom to top.

Energy balance in cooling tower: In a cooling tower heat lost by the water particles is equal to the sum of latent heat due to evaporation of water and the sensible heat gained by the air.

$$M_W C_{PW}(T_{Wi} - T_{Wo}) = M_a(h_{ao} - h_{ai}) + W_e L \qquad [12.1]$$

where, M_W is mass of water flow rate, C_{PW} is the specific heat of water,

T_{Wi} and T_{Wo} are the inlet and outlet water temperatures, M_a is the mass of air flow rate

W_e is the amount of water evaporated and L is the latent heat of vaporisation, h_{ao} and h_{ai} are the outlet and inlet enthalpies of air, respectively.

The effectiveness of a cooling tower is defined as the ratio of temperature range through which the water is cooled to which the temperature range through which it would have been cooled down to the wet bulb temperature of the incoming air.

Mathematically, it is expressed as: $\varepsilon_{\text{Tower}} = \{(T_{Wi} - T_{Wo})/(T_{Wi} - T_{\text{Wet bulb}})\} \times 100$ [12.2]

Example 12.1

A 100 TR refrigeration plant is operating in an outdoor design condition as dbt 40°C and wbt 30°C. Condenser water at 45°C is to be cooled down to 33°C by an induced draft cooling tower. If the air leaves the tower at 35°C and 90% RH, find the approximate amount of evaporated water and amount of air to flow through the tower. Manufacturer's specification is 600 litres/hr of water per TR cooling capacity is required to cool the condenser.

Ans: From psychrometric chart specific humidity and specific enthalpy of air leaving the tower is 0.031 kg of wv/kg of da and 110.5kJ/kg of da.

And specific humidity and specific enthalpy of air entering the tower is 0.023 kg of wv/kg of da and 100.2 kJ/kg of da.

So the moisture evaporated, W_e = (0.031 – 0.023) kg of wv/kg of da = 0.008 kg of wv/kg of da

Mass of water required is (600 × 100 × 1)/3600 = 16.67 kg/s

Heat removal rate from water is 16.67 × 4.18 (45 – 33) = 836.17 kW

From the above energy balance equation, 836.17 = M_a (110.5 – 100.2) + 0.008 (2500)

Mass of air to be circulated, M_a = 79.24 kg/s

Air volume flow rate = 79.24 kg/s × 0.9 = 71.3 m^3/s

Effectiveness of the cooling tower $\varepsilon_{\text{Tower}} = \{(T_{Wi} - T_{Wo})/(T_{Wi} - T_{\text{Wet bulb}})\} \times 100 = \{(45-33)/(45-30)\} \times 100 = 80\%$

The calculation shows that a very high air volume flow rate is required.

12.6 CLASSIFICATION OF AIR CONDITIONING SYSTEMS ON THE BASIS OF 'ZONES'

There are multiplexes where different kinds of activities take place. Hence, the demand for cooling/heating load varies from one part of the building to the other. When a system is designed to provide independent control in different spaces, each space is called **a zone**. A zone may be a separate room or part of a large space.

12.6.1 Single Zone System

This system is designed to supply conditioned air to a single temperature zone and has single supply air duct. This duct is branched out to supply air to different parts of a building at the same temperature and humidity. This is mostly applicable for residential installations to small departmental stores. The best example of this system is a window air conditioner and the central air conditioning system as shown in Fig. 12.4.

12.6.2 Dual Duct System

Refer Fig. 12.6. In dual duct system, mixing boxes are placed outside of AHU and near is the zones. Separate hot and cold air ducts are run from the heating and cooling coils of AHU. Hot and cold air is directly tapped from the main ducts to the mixing boxes. The proportion of cold and hot air is maintained by the box dampers which are controlled by the room thermostats. The hot and cold air get mixed up in mixing boxes and the mixed air is sent to the respective zones.

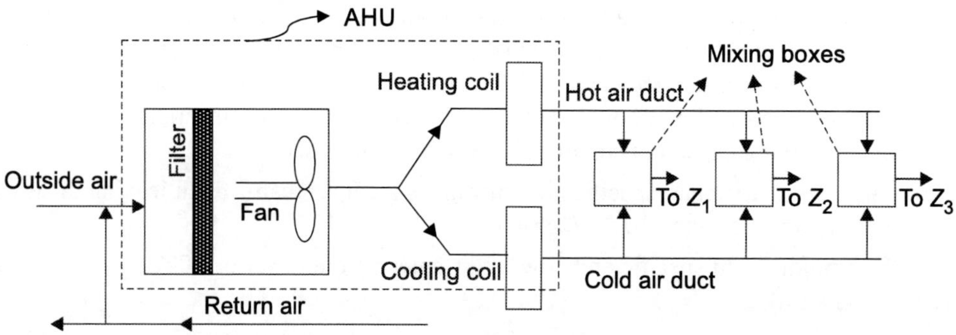

Figure 12.6: Schematic of a dual duct system

12.6.3 Multi-Zone System

In multi-zone system (Fig. 12.7) the zone dampers are provided in AHU across the hot and cold decks at the outlet of the units. Separate ducts are run from each set of dampers to each zone. Proportionate mixing of hot and cold air is done by the dampers as per the zone requirement in the AHU. Thus, the conditioned air passes through a long distance through the tubes, i.e., from AHU to the zone, whereas, In dual duct system conditioned air passes through a very short distance as the mixing boxes are placed close to their respective zones.

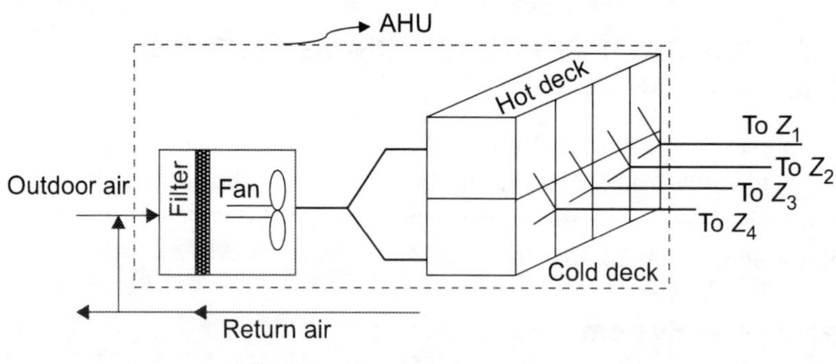

Figure 12.7: Schematic of multi-zone system

12.6.4 Variable Air Volume (VAV) System

Refer Fig. 12.7 (a). Air supplied to different zones, as discussed in the above systems, is of constant air volume amount. The indoor air temperature is maintained by varying the supply air temperature in AHU. In VAV system, indoor air temperature and humidityare maintained by varying the supply of conditioned air volume flow to the different zones (Z_1, Z_2, Z_3, etc). VAV boxes are installed for each zone. They have adjustable dampers or valves so that air volume supplied to the space can be varied. VAV boxes are supplied with conditioned air directly from AHU through a single main duct. Room thermostats located in each zone control the dampers in their respective zone VAV boxes to maintain the desired room set point temperature.

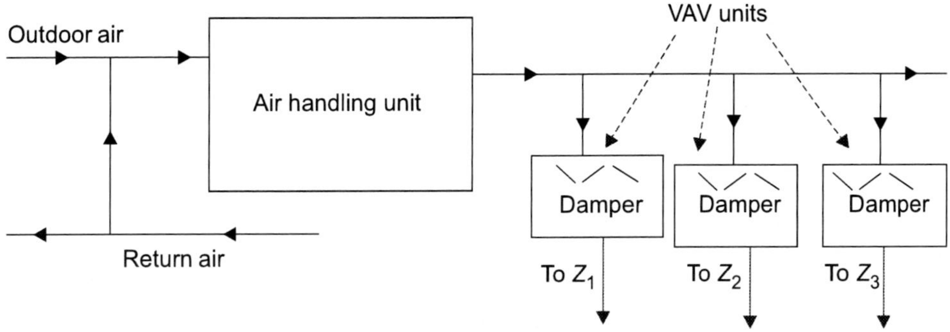

Figure 12.7(a): Schematic of a VAV system

12.6.5 All-Air System

All the air conditioning processes are achieved through the flow of air from a central unit into each zone. These systems need substantial space for ducting the air to each zone. To change the heating or cooling capacity of the air supply to one zone, the system must either alter the supply temperature or alter the flow rate to the zone. The first one is called *reheat system* and the second one is called *variable air volume (VAV) system*. In reheat system there is a heater in the duct to each zone. A zone thermostat can then control the heater to maintain the desired zone set point temperature. In VAV system, the duct to each zone is fitted with a control damper that can be throttled to reduce/increase the air flow rate to maintain the desired temperature.

12.6.6 Air-and-Water System

Also known as Hydronic system, in this system the required amount of ventilation air to each zone is supplied from the central system. The central system provides most of the humidity control with this ventilation air. The local units are supplied with hot or chilled water from another chilling plant or hot water unit. The ventilation air from central system air, when blown over it, achieves the required temperature.

12.6.7 All Water System

In this system all processes in air conditioning, other than ventilation, are provided by local equipment supplied with hot and chilled water from a central plant. The ventilation air is supplied through natural ventilation by opening windows or by some other means.

12.7 AIR DISTRIBUTION SYSTEMS

We have discussed 'how to maintain the desired temperature and humidity of the indoor air' till now. Besides, air flow pattern and the air velocity (draft) inside the room are important factors for comfort feeling. Because when the air velocity increases, the convective heat transfer from the body also increases. This causes a chilly feeling. This is known as *draft condition* or *chill factor*. The conditioned air should reach the occupied zone in such a manner that the temperature, relative humidity, velocity and purity of the air must conform to the comfort requirement. Occupied Zone is the space in the room up to 1.8 m above the floor level and 0.6 m from the walls. Again it should be distributed in such a manner that all parts of the space may attain uniform temperature, humidity and air velocity after getting mixed with the indoor air. The flow of air within a space is complex and cannot be easily predicted the following terms are used in the study of this system.

Commonly used terms in air distribution system

Refer to Fig. 12.8

Occupied zone: It is that portion of the room volume between the floor and around 1.8 m (6 ft) above the floor level. Otherwise it can be stated as it is the space above the floor level which is being occupied by the human being.

Diffusion temperature: The supplied air is introduced into the space in a manner causing rapid mixing of supply air with the room air due to temperature difference, thus reducing both the temperature difference and the velocity before the air enters the occupancy zone. This process is known as 'diffusion' and the temperature difference is termed diffusion temperature. This temperature is, normally, in order of 10°C to 12°C for cooling and as high as 28°C in case of heating. For example, in summer, the room temperature is 25°C and the supply air temperature is 13°C. So the diffusion temperature is 12°C.

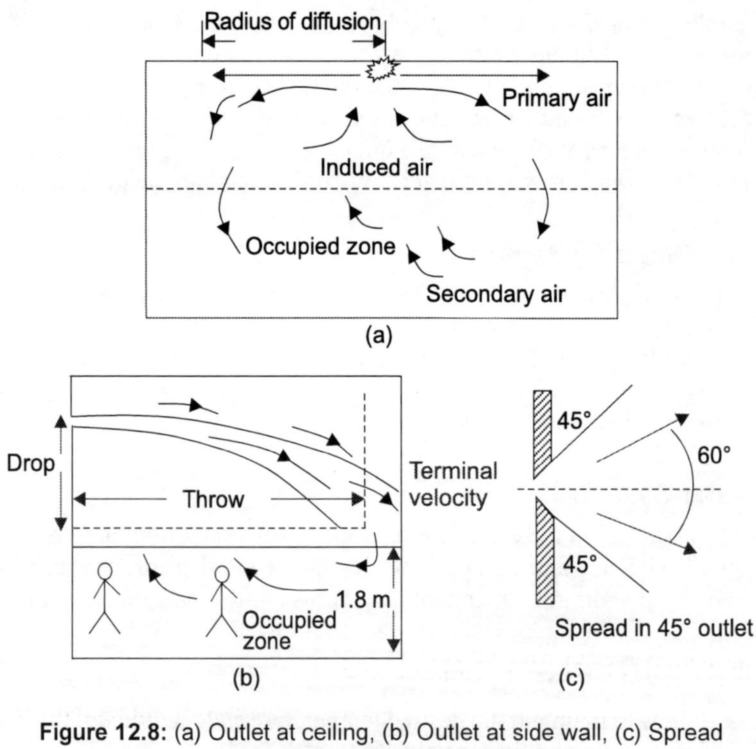

Figure 12.8: (a) Outlet at ceiling, (b) Outlet at side wall, (c) Spread

Induction/entrainment ratio: This involves the estimation of mixing of supply air with the room air quantitatively. The air leaving the outlet or inducted into the room is the primary air. As it leaves the outlet point it entrains some of the room air termed the *secondary air*. The sum of the primary air and secondary air is called the *total air*. The induction or entrainment ratio is defined as the ratio of the total air to the primary air.

Induction ratio, $$Q_{Total}/Q_{Outlet} = C_{En}V_{Outlet}/V_{Total} \quad [12.3]$$

where, Q_{Total} is the total air mixture at certain distance (x) from outlet (m^3/s), Q_{Inlet} is the discharge from outlet (m^3/s); C_{En} is the entrainment coefficient (varies from 1.4 to 2.0); V_{Outlet} is the outlet velocity; V_{Total} is the velocity of mixed air at certain distance (x).

Throw (blow) and radius of diffusion: Both the terms refer to the horizontal distance that the supply air travels before reaching a specified low velocity known as the terminal velocity. This is usually in the range of 0.4 m/s to 1.0 m/s, depending upon the application, so that the residual velocity, i.e., velocity of air in occupied zone will be around 0.15 m/s. Generally, throw or blow refers to the discharge from side wall outlet and radius of diffusion refers to the discharge from the ceiling.

Drop: It is the vertical distance that the supply air drops by the end of its throw (i.e., the point where air reaches its terminal velocity).

Spread: It is the angle of divergence of the air stream. It depends upon the type of vanes (straight/converging/diverging) used at the outlet diffuser/grille.

Draft: Draft is the localised feeling of coolness or warmth of any portion of the body due to both air movement and air temperature, with humidity and radiation considered constant.

Air velocity: The preferred indoor air velocity in the occupied zone ranges between 0.1 and 0.25 m/s.

Air distribution performance index (ADPI): A measure of the effective temperature difference between any point in the occupied space and the controlled conditions is called the effective draft temperature and it is defined as

$$t_{draft} = (t_{local} - t_{room}) - 8\,(V_{centre} - 0.15) \quad [12.4]$$

where, t_{local} is the local air stream dbt, t_{room} is the average room dbt, V_{centre} is the local air stream centre line velocity (m/s).

Research shows that in many cases the comfortable effective draft temperature is between –0.15°C and 1°C and the air velocity is less than 0.35 m/s. Thus, ADPI is defined as the percentage of location amongst many other locations in the occupied zone that meets the above effective draft temperature criteria. ADPI helps in selecting and locating the air diffusers so that an ADPI of 100% can be achieved which means all the occupied locations in the room are within the comfort zone.

12.8 AIR DUCT DESIGN METHODS

Ducts carrying the conditioned air to different zones are to be designed in a manner such that (a) they can carry and deliver the required amount air to the specified location, (b) power consumption, and the occupied space by the ducts should be minimum, (c) objectionable noise should not be generated. Power consumption can only be minimized by reducing the pressure drop throughout the flow length.

12.8.1 Pressure Drop in the Air Ducts

Hence, it is important to know about the pressure of the fluid while it flows through the pipe or duct. The total pressure of a flowing fluid consists of two parts. The pressure exerted by the fluid at rest is *static pressure* and the pressure developed by virtue of its movement (velocity) is *dynamic pressure*. The summation of these two is called the *total pressure*. In Bernoulli's equation $p/\rho + V^2/2 + gz = C$; p is the static pressure which contributes to overcome the wall and fluid friction. Dynamic pressure developed due to fluid velocity maintains the flow velocity in the duct. They are interchangeable. As the flow velocity decreases, as in case of diffuser or gradual increasing flow area, the dynamic pressure converted to static pressure. Static pressure is converted to dynamic pressure or velocity in case of flow through nozzle. Following diagrams show the measurement of these pressures by manometer. Figure 12.9 (a) gives the static (gauge) pressure (h_S) of the fluid and Fig. 12.9 (b) gives the total (gauge) pressure (h_T). The open end of the manometer tube is placed in the middle of the tube. The difference of these two gives the dynamic pressure. A manometer, combining the above two, can be prepared which can directly measure the dynamic pressure. The principle of measurement has been discussed in Chapter 1.

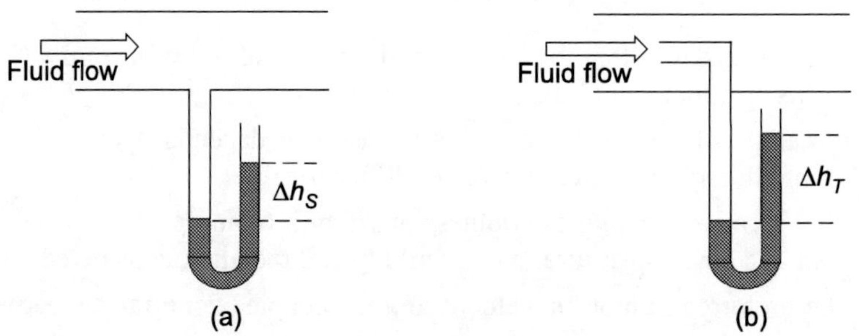

Figure 12.9: Measuring (a) the static (gauge) pressure, and (b) the total (gauge) pressure

Air duct design is very important for the following reasons.

(a) Air should have the velocity such that it reaches the different points of air conditioning system with recommended velocity and acceptable sound level.

(b) The pressure drop should be minimum, so that fan power consumption is less.

A suitable velocity of air should be given so that there will be minimum pressure drop (static and dynamic) as well as acceptable noise level. The volume flow rate of air is estimated from the cooling load, occupancy use, etc. The velocity of air, as recommended by ASHRAE or any other reliable institution for different applications, is considered. The cross-sectional area of the duct is calculated from

$$Q_V = A_C X V$$

where, A_C is the cross-sectional area of the duct [12.5]

Larger the velocity, smaller is the size of the duct. This causes higher frictional pressure drop, which is evident from the following equation and hence the motor power. Larger the duct size, velocity is low and frictional pressure drop will be low. But the larger size of duct causes more material cost and occupies more space. So there should be trade-off between the size of the duct and the velocity of air. While designing the ducting system, the shortest path between the source of conditioned air and the conditioning space should be chosen. Number of sharp bends, joints where there will be change of flow direction, should be avoided to minimize the dynamic loss. Dampers should be provided in all branches to balance the flow in the total zone.

Table 12.1

Applications	Noise level point of view		Friction factor point of view			
	Velocity (mpm)	Sound level (dB)	Main ducts		Branch ducts	
			Supply	Return	Supply	Return
Residences	180	40	300	240	180	180
Hotel and hospital	300	40	450	390	360	300
Offices	360	60	600	450	480	360
Theatres	240	30	390	330	300	240
Studios	120	30	240	180	120	120

The duct materials used are either GI sheets or aluminium sheets. For moisture resistance and light duct work aluminium sheets are preferred.

The total pressure drop is contributed by, mainly, two factors:

(a) Friction loss (friction between pipe wall and the fluid, the viscosity and the turbulence).

(b) Minor losses (pressure drop due to fittings, sudden expansion and contraction, bends, elbows and change in direction of flow, at inlet and discharge point of the duct, etc).

Pressure drop due to friction is calculated from the following relation.

$$\Delta P = \frac{fL\rho v^2}{2d_h} \qquad 12.6]$$

where, f is friction factor, L is length of the tube/duct, ρ is density of the fluid, v is average velocity, and d_h is hydraulic diameter of the tube/duct and expressed as $4A_C / P_W$;

where, A_C is area of cross section, P_W is wetted perimeter.

Friction factor is expressed in two forms. One is 'fanning's friction factor', (f_f) and other one is 'Darcy-Weisbach' friction factor, (f) and they are related by $f = 4f_f$. in Moody's chart the friction factor is the Darcy-Weisbach friction factor.

For laminar and fully developed flow friction factor f = 64/Re or f_f = 16/Re; Re is the Reynolds number. For different types of fluid flow many more correlations are available to calculate the friction factor. The correlation is selected meticulously to fit our flow condition.

Friction factor charts are also available which are separately prepared by ASHRAE and many other reliable institutions. This is a chart of flow rate (m^3/s) vs friction loss (Pa/m). Flow velocity and the duct diameters are the parameters. Static loss is only determined from this chart.

This frictional loss during the straight flow through a duct/tube is called the *static loss*. The additional loss, in excess of the straight duct friction loss, is the *dynamic loss*. This loss is usually contributed by minor losses as mentioned above.

The dynamic losses are mainly contributed by elbows (90°), branching off, sudden enlargement and sudden contraction.

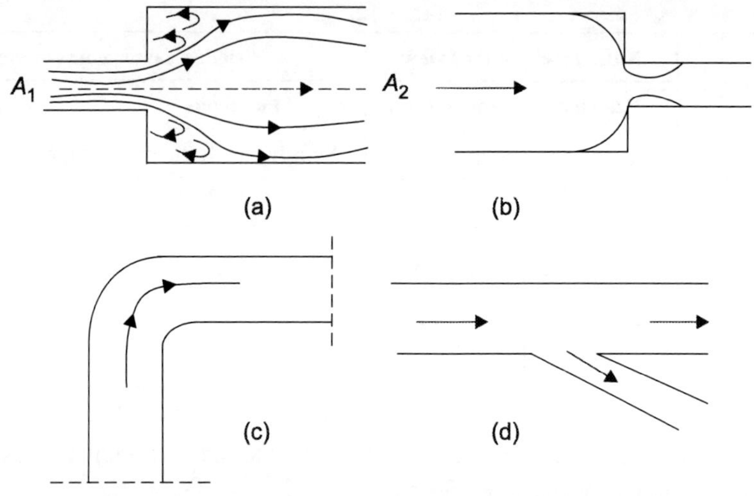

Figure 12.10: Flow through (a) sudden enlargement, (b) sudden contraction, (c) bend, and (d) branching

12.8.2 Design of Air Duct System

Different methods are adopted for the design of air duct. These design methods, mainly, aim at determining the duct sizes that are reasonable with respect to the space requirements and velocities.

(a) Equal friction method
(b) Static regain method
(c) Velocity reduction method

Equal eriction method is commonly used for duct designing when the flow velocity through the duct is low. In this method the pressure loss due to friction per metre length of duct throughout the piping network is kept constant (mainly the static loss). This is achieved by selecting a suitable duct size (diameter). Initially, the maximum velocity of air flow is fixed for the total required amount of air flow rate so that at each end of the branch, the required velocity and discharge of air are achieved.

The method of solving the problem to determine the total frictional pressure drop for the above ducting arrangement is discussed here.

The limiting velocity is fixed at 10 m/s.

Referring to the 'duct friction chart' (Refer Fig. 12.11), for total volume flow rate of 5 m^3/s, and velocity 10 m/s, the frictional drop is 1.3 Pa/m.

Now this value (1.3 Pa/m) is kept constant for all other branches of duct.

For branch BP: 1.3 Pa/m line and 1.5 m^3/s volume flow rate line intersect at a point on the chart. The corresponding velocity and duct diameter are found to be 7.8 m/s and 490 mm, respectively.

Similarly, for branch CQ: The velocity and duct diameter corresponding to 1.3 Pa/m and volume flow rate of 0.5 m^3/s are 6.1 m/s and 325 mm, respectively.

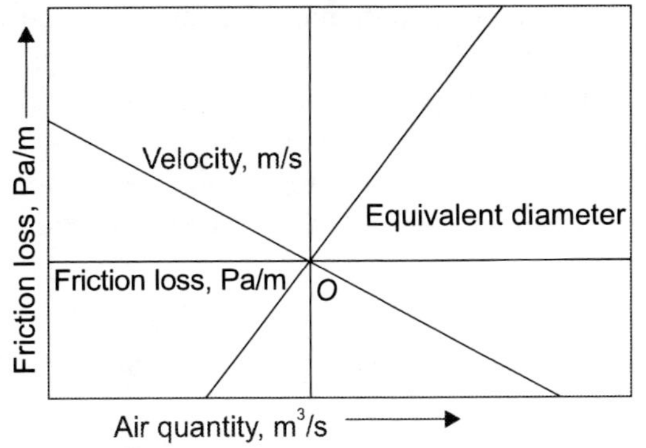

Figure 12.11: Duct friction chart

Similarly, for branch DR: The velocity and duct diameter corresponding to 1.3 Pa/m and volume flow rate of 1.0 m^3/s are 7.0 m/s and 410 mm, respectively.

Similarly, for branch ES: The velocity and duct diameter corresponding to 1.3 Pa/m and volume flow rate of 2.0 m^3/s are 8.4 m/s and 560 mm, respectively.

Total pressure loss will be same as that of the circuit having the maximum pressure loss which is obviously the longest one, otherwise called the '*index circuit*'.

(The duct run having the largest pressure drop is called the *index run*. This is usually the longest duct run, but not so all the time)

So the total (static) pressure loss in the longest circuit (*A-B-C-D-E-S*)

$$= (1 + 0.5 + 1.5 + 1.5 + 1.5) \text{ m} \times 1.3 \text{ Pa/m} = 6 \times 1.3 = 7.8 \text{ Pa}$$

Pressure losses at fittings, dynamic losses, (joints, bends, grilles and dampers) may also be added. These pressure loss values can be had from ASHRAE handbook or any other reliable sources.

Thus, the total friction loss can be calculated and the fan power can be determined.

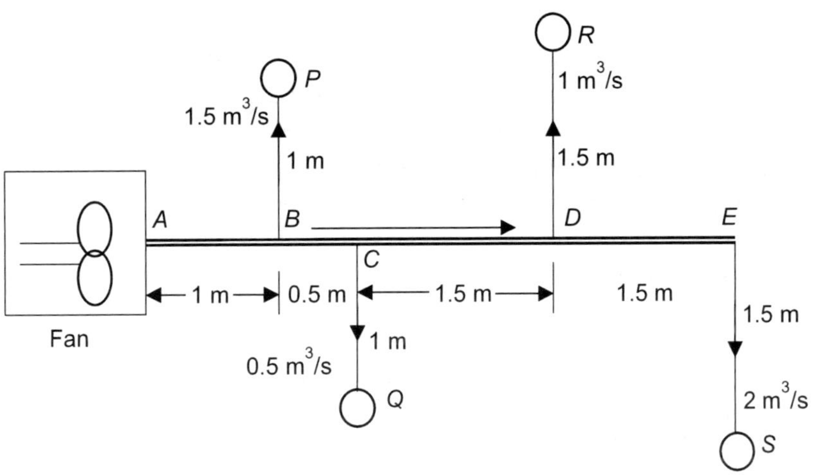

Figure 12.12: Layout of a ducting system

Static regain method: This method of designing the duct is practised when the air flow velocity is medium or high. In an air flow duct system, as the flow proceeds the volume flow rate decreases. If the duct diameter is kept same, the flow velocity (or kinetic energy) decreases and some of the kinetic energy is converted to pressure energy. This is called *regain of pressure*. This increased static pressure will be available for the air flow in the succeeding section to compensate for the different losses.Thus the static pressure remains approximately same at the entrance to each branch.

The steps followed in this method are noted below. (Refer Fig. 12.12)

(a) Assume a velocity for the main duct section *AB*, V_{AB} (i.e., after fan). (normal assumed value is 8 to 10 m/s).

(b) Find the pressure loss ($\Delta P/L$) from equal friction method. Also calculate the dynamic pressure loss due to bend or elbow.

(c) For the next section (*B-C*), assume a velocity, V_{BC} (must be less than the velocity of the main branch). The flow rate will be (total flow – the flow through branch *B-P*). Pressure loss is found out from the duct friction chart. Total friction loss will be (rate of pressure loss × length of section *B-C*). Again the static regain is expressed by $\Delta P_{SR} = SR\,(V_{AB}^2 - V_{BC}^2)$; where, *SR* is the static regain factor and the value may be assumed in the range of 0.6 to 0.9 (mostly 0.75). Compare the two values; total friction loss and the static regain value. If they do not match, assume another value of velocity in section *B-C*. Proceed in the same manner till the two values match. Then find the corresponding duct diameter from the duct friction chart.

(d) Same procedure, i.e., (c) is to be followed for succeeding sections.

Velocity reduction method: Like the above two methods, suitable air flow velocity through the main duct, i.e., just after the fan, is assumed (8 to 10 m/s). This assumption of flow velocity is limited by noise level and frictional resistance point of view. After the first branch of duct, the volume flow rate will decrease. For the next branch, *B-C*, if velocity is to be kept constant, flow area (*A*) is to be reduced. This will increase the pressure loss. If duct size is kept constant, the velocity will decrease and there will be regain of pressure and the duct material cost will increase. Hence, there must be a balance between these two. From experience and terminal velocity of air flow as recommended by the manufacturer, the velocity of air flow is to be so chosen that the velocity and pressure are sufficient to meet the requirement of the succeeding section.

After getting the diameter of the circular duct from the chart (as discussed in the above examples), the same can also be converted to equivalent rectangular duct dimension. This is done assuming the same surface roughness, same volume flow rate and same rate of pressure drop ($\Delta P/L$). The rectangular ducts having **aspect ratio (width: depth)** not exceeding 8 will have the same head loss for same length and mean flow velocity as a circular duct of same hydraulic diameter. However, the preferred aspect ratio is 4:1.

Mathematically, it can be shown that $d = 1.3\left[\dfrac{(ab)^{0.625}}{(a+b)^{0.25}}\right]$ [12.7]

Here *a* and *b* are the dimensions of rectangular duct. One of these two dimensions is fixed depending upon the availability of ceiling height.

If the surface roughness values are not same, the equation to be used is

$$d = 1.265 \sqrt[5]{\frac{f_C (ab)^3}{f_R (a+b)}} \quad [12.8]$$

where f_C and f_R are roughness values of circular duct and rectangular duct, respectively.

The circular duct uses less amount of material and the friction loss is less. But it occupies more space (blank side space when fitted of a circular duct) when fitted to a ceiling. It also looks odd. That's why in many cases rectangular or square duct is preferred.

Finally, fan power is calculated from the following expression.

Fan power = Total pressure drop in the air duct × Volume flow rate of air

$$= \Delta P_{\text{Total}} \times Q_V \text{ watt} \quad [12.9]$$

12.9 AIR SUPPLY DEVICES

Room air distribution is very important from comfort point of view. Variation of temperature throughout the occupied zone should be within 1°C. The air velocity should be within 8 to 10 m/min for rest and may be higher for other activities. It is also desirable to place the room air supply devices at higher level, i.e., at high wall or ceiling for cold air, because it comes down to mix with the room air and at lower level, i.e., at floor or lower wall for warm air, because it moves up to mix with the room air. The supply air is called *primary air* and the room air is called *secondary air*. It should be seen that as the primary air comes down/up, it should touch the glass surfaces because the temperature difference is more with glass surfaces than the walls.

Devices used for the purpose are: (a) grilles and registers, (b) ceiling diffusers, (c) slot diffusers, and (d) plenum ceilings. (Students may confuse with dampers which are provided, somewhere, in the ducting work or at the entry point to a zone to regulate the air flow quantity.)

Grilles consist of frame and parallel bars which are either fixed or movable. They are used to change the air direction and adjust the throw and spread of air. If the damper is placed behind the grilles, the combination is called *registers*. Ceiling diffusers consist of separate concentric rings with a collar or neck to connect to the duct. They can be designed to distribute the air uniformly in any desired direction. It has an aesthetic look. *Slot diffuser* is a long strip shaped outlet with one or more narrow openings. Sometimes a combination of fluorescent lamp and slot diffuser is used. Cooling of fluorescent lamp increases the efficiency of the lamp and this also gives an aesthetic look. *Plenum ceiling* is a suspended or hung ceiling with slots or perforations throughout the ceiling which serve as the air outlets. The space above the ceiling is used as the large plenum through which the supply air is delivered.

12.10 FAN

Fan plays an important role in air transmission and distribution system. The performance of fan includes the pressure rise, its efficiency and power requirement as a function of air volume flow rate. The rotation of the impeller imparts kinetic energy to the air and thereby increases the static pressure as it passes through the diverging passage of the blades. Normally, two types of fans are used for the air conditioning purpose. One is centrifugal fan and another is axial fan. Centrifugal fan draws air along its axis and leaves it radially outward, but axial fan draws air along its axis and also leaves it in the same direction. Centrifugal fans are mostly preferred due to its ability to develop large pressure and quiet performance. The quantity of flow rate does not matter much. So this fan is used where ducting is done for air distribution as in central air conditioning system. Even in package system like window air conditioner centrifugal fan is used. Axial fan has the ability of high volume flow rate but with low pressure development. Its operation is noisy. So it is preferred for exhaust purpose. In many cases where noise is not a criterion, as in industries, axial fans are used in ducting as well as for exhaust.

Total pressure rise across the fan is expressed, $\Delta P_{Total} = (P_2 - P_1) + \rho_{air}(V_2^2 - V_1^2)/2$ [12.10]

Of these RHS two terms, first term is the static pressure rise and the second term is the pressure needed to provide the kinetic energy setting air in motion.

So the power requirement, Fan power = $\Delta P_{Total} \times Q_V/60$ watt [12.11]

where, Q_V is the air volume flow rate in cmm.

Fan efficiency is expressed as: Power consumed (ideal)/Power consumed (actual)

However, there are some fundamental relations among fan capacity, speed, pressure, and power. These are called *fan laws*.

1. Air volume flow rate (Q_V) is directly proportional to fan speed, N. If we combine it with the area of the impeller, it can be expressed as $Q_V \alpha\, D^3N$;
   where, D is the diameter of the impeller (m), N is rpm. So $Q_{V1}/Q_{V2} = N_1/N_2$ [12.12]
2. The total fan pressure is directly proportional to the square of the fan speed.
   $P_{Total} \alpha \rho_{air} D^2N^2$. So, $P_{Total\ 1}/P_{Total\ 2} = (N_1/N_2)^2$ [12.13]
3. The fan power is directly proportional to the cube of the fan speed, $P_{Fan} \alpha \rho_{air} D^5N^3$.
   So (fan power)$_1$/(fan power$_2$) = $(N_1/N_2)^3$ [12.14]

Operating Condition of Fan

Figure 12.13 shows the graphical representation of fan characteristic. This performance data is supplied by the fan manufacturer in tabular form as shown in Table 12.2. For example, a fan rotating at 730 rpm delivers 5.5 m^3/s of air with an outlet velocity of 11 m/s raises the pressure by 900 Pa and consumes 8.1 kW.

Pressure drop in the system, as the air flows through ducts, can be calculated from the methods discussed in Section 12.8 (constant friction method or static regain method). This is also shown in Fig. 12.13. These two curves intersect each other at point A which represents the design point of the fan. That means at point A, the pressure rise across the fan is equal to the pressure drop in the system and the air volume flow rate through the duct is same as that through the fan. Now if the pressure drop in the duct is reduced due to cleaning of air filter or any other reason, volume flow rate is increased as shown at point B. Pressure loss due to frictional resistance in the duct system varies as the air volume flow rate

as $$P_{Loss}\ \text{(in duct)}\ \alpha\ (Q_V)^2 \quad [12.15]$$

Hence for a change in flow rate, it can be expressed

$$(P_{Loss}2/P_{Loss}1) = (Q_V2/Q_V1)^2 \quad [12.16]$$

Increase in fan speed increases the total fan pressure. Velocity pressure which is a part of the total fan pressure (refer Eq. 12.13) depends upon air density. So if the air density increases total pressure across the fan (FTP) increases, so also the fan power requirement.

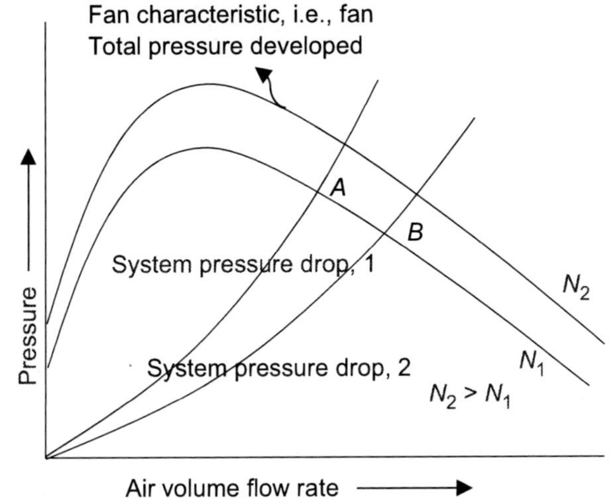

Figure 12.13: Fan operating conditions

Table 12.2: Performance data as provided by a fan manufacturer: (650 mm wheel diameter and 0.5 m^2 outlet area)

Air flow rate Q_V (m^3/s)	Outlet velocity m/s	800 Pa		900 Pa		1000 Pa	
		rpm	kW	rpm	kW	rpm	kW
5.0	10.0	715	6.5	750	7.5	820	8.5
5.5	11.0	700	6.75	730	8.1	810	9.00
6.0	12.0	692	7.12	720	8.5	790	9.36
6.5	13.0	708	7.8	710	8.7	780	9.7

Example 12.2

A fan, used for ventilation, delivers 5.0 m^3/s of air at 750 rpm. It requires 7.5 kW. It is required to increase the air delivery rate to 5.5 m^3/s. What should be the speed of the fan. What should be verified before making such a change?

Ans: Using the fan laws: $N_2 = (Q_V2/Q_V1) \times N_1 = (5.5/5.0) \times 750 = 825$ rpm

(Fan power$_1$)/(Fan power$_2$) = $(N_1/N_2)^3$; (Fan power$_2$) = 7.5 kW/$(750/825)^3$ = 9.9825 kW

Thus, the required motor power should be 10 kW.

But the original motor rating must not have more than 8.0 kW because it is designed for delivery of 5.0 m^3/s of air. So the same fan motor will be overloaded if it is used for delivery of 5.5 m^3/s of air.

Example 12.3

A HVAC system is required to deliver 6.0 m^3/s at a total pressure rise of 1.0 kPa. Determine the speed, the fan power, and the fan efficiency. Fan has the wheel diameter 650 mm. The fan performance data is given in Table 12.2.

Ans: Table gives the static pressure rise.

Velocity pressure rise required = $\rho V^2/2 = 1.2 \times 12^2/2 = 86.4$ Pa

So total fan pressure rise = 1000 + 86.4 = 1086.4 Pa.

For 6.0 m^3/s, from the table the fan speed should be 790 rpm and fan power requirement at shaft is 9.36 kW.

But the ideal power consumption is

Total fan power rise = $\Delta P_{Total} \times Q_V$ watt = 1086.4 Pa × 6.0 = 6518.4 W = 6.5184 kW

So the fan efficiency = 6.5184/9.36 = 0.6964 = 69.64 %

EXERCISE

1. What is 'zone' in an air conditioning system and how this zone system helps in reducing the power consumption?
2. Give examples of single zone and multi-zone systems.
3. In a household refrigerator how temperature gradient is created, i.e., at freezer section it is around –15°C and gradually reduced to around 8°C at vegetable tray cabinet.
4. Give the detailed classification of air conditioners basing upon their applications.
5. Describe the function of air handling unit.
6. Why dampers are used in air supply system?
7. Match the following:

1. A large indoor stadium	(b) All water systems
2. Individual rooms of a large hotel	(b) All air systems
3. An existing building with good ventilation	(c) Unitary systems
4. An existing, small office building	(d) Air-water system

8. Describe the different types of noise control element used in air conditioning system.
9. Why ventilation is required and also explain why different ventilation standards for different purposes are recommended?
10. Give the classification of fans and explain the working principles on which they work.
11. Discuss the working principle with suitable diagrams of unitary refrigeration systems and also describe its advantages, disadvantages and applications.
12. Write short notes on:
    (a) Comfort air conditioning
    (b) Air duct design
    (c) Acoustic elements uses in air conditioning systems
    (d) Cooling tower

MULTIPLE–CHOICE QUESTIONS

1. An a/c system is designed to provide independent control in different spaces, each space is called a

(a) block (b) zone
(c) unit (d) space

2. Frequent opening of the chest type freezer door has — effect on cooling load of the freezer coil.

(a) more (b) less
(c) negligible (d) equal

3. Condensing unit of split a/c contains

(a) condenser, throttle valve (b) throttle valve, evaporator
(c) evaporator, compressor (d) compressor, condenser

4. There is no mixing of outdoor air in case of

(a) window a/c (b) split a/c
(c) central a/c (d) rooftop a/c

5. Air handling unit (AHU) is a part of

(a) window a/c (b) split a/c
(c) central a/c (d) rooftop a/c

6. Cooling coil of central a/c is placed

(a) before AHU (b) after AHU
(c) inside AHU (d) compressor room

7. Dampers function is to

(a) regulate the air flow quantity and direction
(b) block the air flow
(c) regulate the direction of air flow
(d) regulate the quantity of air flow

8. Cooling tower is provided to cool the

(a) the refrigerant from condenser
(b) the coolant from condenser
(c) the compressor lubricant
(d) the indoor hot air

9. The space above the floor level of a conditioned room which is occupied by the human being is called

(a) utility zone (b) occupied zone
(c) sitting area (d) conditioned space

10. The difference between the supply air temperature and indoor air temperature is called

(a) temperature lag (b) temperature lift

(c) diffusion temperature (d) coil temperature

11. The popular method used to design the duct size for air supply in an a/c system is

(a) equal friction method (b) static regain method

(c) velocity reduction method (d) none

12. Consider the following statements pertaining to duct design:

1. Aspect ratio of ducts should be high.
2. In the equal friction, method of design, use of dampers cannot be eliminated by any means.
3. The static regain method is not suitable for long ducts.
4. The velocity reduction method is employed only in simple systems.

Which of the statements given above are correct?

(a) 1 and 2 (b) 3 and 4

(c) 1 and 3 (d) 2 and 4

13. Which of the following method(s) is/are adopted in the design of air duct system?

1. Velocity reduction method
2. Equal friction method
3. Static regain method

Select the correct answer.

(a) 1 alone (b) 1 and 2

(c) 2 and 3 (d) 1, 2 and 3

14. Which one of the following statements is true for air conditioning duct design?

(a) Static regain method is used, when the duct work is extensive, total pressure drop is low and flow is balanced.

(b) Static regain method is used, when the duct work is extensive, total pressure drop is high and flow is unbalanced.

(c) Equal friction method is used, when the duct work is extensive, total pressure drop is low and flow is balanced.

(d) Equal friction method is used, when duct work is extensive, total pressure drop is low and flow is unbalanced.

15. Consider the following statements related to all-air air conditioning system:

1. All-air system uses air as heating or cooling fluid.
2. When hot air is circulated through rooms, dehumidification is necessary to control relative humidity.
3. Return air ducts are required for recirculation.

Which of the statements given above are correct?

(a) 1, 2 and 3

(b) 1 and 2 only

(c) 1 and 3 only

ANSWERS

1. (b) **2.** (b) **3.** (d) **4.** (b) **5.** (c) **6.** (c) **7.** (a) **8.** (b) **9.** (b) **10.** (c)
11. (a) **12.** (d) **13.** (d) **14.** (b, **d)** **15.** **(c)**

13

Sustainability in Refrigeration and Air Conditioning

13.1 INTRODUCTION

Nowadays most of the research activities are focused on the **sustainable development**. Sustainability is defined as the preservation of productive capacity for a foreseeable future. Productive capacity needs resources of material and energy, technology and conducive environment to lead a healthy life. When a society enjoys the developmental benefits preserving the productive capacity for the foreseeable future it is said it undergoes a sustainable development. A world-wide estimate shows that around 40% of energy is consumed by refrigeration and air conditioning sectors. Preservation of goods and human comfort are the two most energy consuming applications of refrigeration cycle. In view of the sharp declining of the fossil fuel and sudden rise in the environmental pollution, scientists and researchers are striving to develop new technologies so that energy consumption can be reduced or renewable energy can be used and as a result environment pollution can be reduced. In this context, it is essential to discuss the existing technology for refrigeration systems and their use. In this regard **a scientific approach towards adopting advanced and efficient technology for refrigeration and air conditioning and conservative approach for the use of these systems** will definitely lead us towards a sustainable development. After successful implementation of Montreal Protocol to phase out ozone depleting substances like CFCs since 1987, Kigali amendment in 2016 came into fore for phasing out of HFCs, particularly high GWP HFCs. This amendment was specifically meant for climate system protection by reducing the use of high GWP HFCs and promoting the use of energy efficient technology. In short it can be said that this amendment provides an opportunity for search of 'sustainable technologies'. Basically, 'sustainability' is a concept. There is nothing like sustainable energy. All types of energy sources, that is, fossil fuel, nuclear energy, non-conventional/renewable energy, etc., should be made sustainable by utilizing them economcally and judiciously. It is normally perceived that sustainable system means (a) use of renewable energy, (b) saving of energy using developed technology, and (c) saving of environment by using environment-friendly refrigerants/working fluids for refrigeration systems. Most of the research has been carried out and feasibility studies have been reported on vapour absorption, thermoelectric refrigeration systems using solar energy. It has already been discussed about replacing environment-friendly refrigerant in Chapter 2. Let us now discuss how to save energy in refrigeration and air conditioning.

13.2 ENERGY SAVING TECHNIQUES

It is estimated that world consumes around 40% of total energy consumption in refrigeration and air conditioning industry. Therefore, it is essential to find ways and means for optimal utilization of these systems so that energy can be saved. Here some energy saving methods are discussed which should be followed religiously to save energy.

Energy saving is only possible only when users' aptitude, commitment and knowledge about saving of energy goes side by side with the technological development. Thus, all the factors that help saving energy can be categorized into **two groups.** First group is the user's awareness and second group is the technological advancement.

13.2.1 User's Awareness

- It is said that **the best way of conservation of energy is by energy preservation.** So turn off all the energy consuming devices when they are not in use.
- **Reducing the cooling load by having green building concept:** A green building is one which uses less water, optimizes energy efficiency, conserves natural resources, generates less waste and provides healthier space for occupants, as compared to a conventional building. Cooling load is reduced by adopting passive cooling techniques. These techniques bring down the cooling load and have natural cooling in the conditioned space. The most effective way of cooling the building is to shade the windows, walls and roof of the building from direct solar radiation. Other methods are evaporative cooling, utilizing night cool air and vegetation (by planting trees around the building to check the direct solar radiation). It is emphasized to have natural ventilation system and to utilize sunlight (for natural lighting of building) by having proper orientation of the building and proper positioning of the windows (for cross ventilation).
- **Ventilation principles:** Ventilation principle shows how natural ventilation will function based on the design of the building, internal thermal loads and the position of the openings (typically windows). All natural ventilation principles are based on ensuring healthy and comfortable indoor climate through minimal energy consumption and at minimum cost. Three fundamental principles are: (a) single sided ventilation,in which Fresh air goes in and warm air goes out of the indoor through the single window; (b) cross ventilation, fresh air goes in through one opening and warm air goes out through another; (c) displacement ventilation, fresh air goes in through the opening at the floor level slowly and steadily and drives out the warm air of indoor through the opening at the top. (Fig. 13.1). Good ventilation system also reduces the chance of disease infection caused due to air conditioning as discussed in Section 11.4.

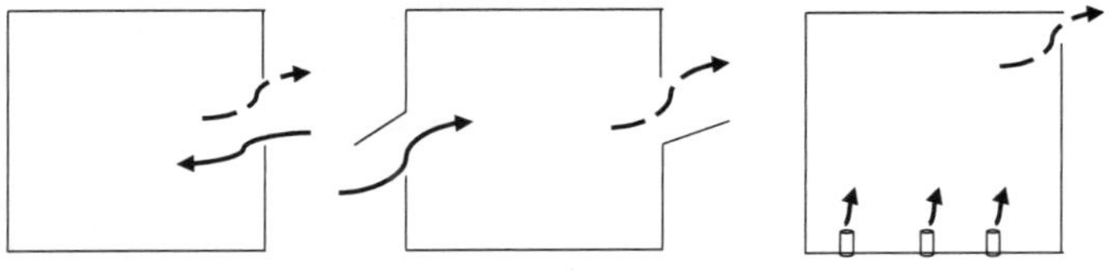

(a) Single side (b) Cross (c) Displacement

Broken arrow – Warm air from indoor; Solid arrow – Fresh air from outdoor

Figure 13.1. Type of ventilation

Using green building materials like hollow bricks is also useful in reducing the cooling load.

Understanding effective temperature for comfort (Refer Section 11.2.2) so that evaporator temperature can be set at the required value instead of setting unnecessarily at 17°C or 18°C. A simple calculation on vapour compression refrigeration system working on S-S-S cycle shows that one degree centigrade increase in evaporator temperature reduces 4% to 5% of power consumption.

Example 13.1

A vapour compression refrigeration system of 10 TR capacity works between a temperature range of 5°C and 40°C uses R22 as refrigerant. Find the required mass flow rate, compressor swept volume and work done, and heat rejection from condenser and COP.

Also calculate the above parameters if evaporator temperature is increased by 5°C.

Ans: Refer Fig. 4.4 and Example 4.1.

Here assumption is made that the vapour compression refrigeration system works on S-S-S cycle.

Mass flow rate, $m = 0.2234$ kg/s

Compressor work, $W_c = 5.375$ kW

COP = 35.167 / 5.375 = 6.543

(a) If evaporator temperature increases by 5°C, $t_e = 10$°C.

At state point 1, $P_1 = 0.6807$ MPa, $h_1 = 408.76$ kJ/kg, $s_1 = 1.7377$ kJ/kg–K

State point 2, from superheat table of 40°C, $h_2 = 427.9$ kJ/kg, $t_2 = 53$°C

Mass flow rate, $m = 10 \times 3.5167/ (408.76 - 249.08) = 0.22$ kg/s

Work done by compressor = $0.22 \times (427.9 - 408.76) = 4.2108$ kW

COP = 35.167/4.2108 = 8.35

So compressor work (power consumption) is decreased by 21.65% and COP is increased by 27.6%.

- Selecting a suitable type of a/c (window/split/rooftop) and the required capacity of air conditioning system. In low humid climate a desert cooler (which works on evaporative cooling principle) should be used instead of an air conditioner. In high humid climate, when dbt is less (as in rainy season), desiccant wheels /dehumidifiers should be used which consumes comparatively less power. Cooling coil having coil surface temperature less than the dew point temperature of the outside air for hot and humid climate (for cooling and dehumidification).
- Good maintenance of the system (correct positioning of installation of a/c so that there is good air circulation throughout the room and air throw should be such that most of the space is covered. During installation of a/c condensing unit must be placed in a shady or relatively cool place so that heat rejection from the condenser is more efficient and the COP increases and there saving of power. Cleaning of air filters is essential on regular interval of time. The air filter is clogged by dust and dirt. So the conditioned air cannot reach us in total and there is a feeling of discomfort. Then we have tendency to reduce the coil temperature which increases the power consumption.

Bureau of Energy Efficiency (BEE) is an organization of Government of India. This organization tests all the electrical appliances and gives a certificate of its efficiency which is known as *Energy Efficiency Ratio* (ERR). EER is defined as the ratio of cooling effect produced to the Power input and this is same as COP of a refrigeration system. This EER is expressed in number of stars as mentioned. 1-star has EER 2.7, 2-Star 2.9, 3-star 3.1, 4-star 3.3 and 5-star 3.5. Higher the star rating, lower the power consumption. The COP or EER of a cooling system is increased by increasing heat transfer characteristics, compressor based upon inverter technology, good refrigerants, etc. This will have initial cost of the system. But in the long run it pays off by saving more power.

13.2.2 Technological Advancement

Improvising the system technology by (a) improving the refrigeration system, and (b) using energy efficient working fluids or refrigerants.

13.2.2.1 Inverter technology (variable speed motors)

At present the refrigeration systems follow off/on cycle process. It means when the system reaches the temperature as set by the thermostat the compressor stops. After the compressor stops, the liquid refrigerant in the evaporator continues to absorb the heat from the surroundings, thus over-cooling the system. After certain time when the thermostat senses a temperature higher than the preset temperature, compressor starts again taking up the starting current which is 3-4 times higher than the running current and runs at its full capacity. Thus, the compressor consumes the full load energy all the time whether there is a demand or not for the same and the other disadvantage is that at each on-off cycle, a/c space gets overcooled for a while and also gets warmer for a while. (Refer Fig. 10.4). In inverter technology the compressor has a variable speed motor. The speed of the a/c motor is controlled by variable frequency drive, i.e., the speed of the motor is varied by varying the supply power frequency. It changes the flow rate of refrigerant, by varying the speed, based on the heat of the room. When heat is less, the flow rate is less and when heat is more flow rate is more. Because mass flow rate is expressed as m = (Swept volume × rpm/60)/(Specific volume of vapour).

It does not switch of the compressor ever. The compressor continues to run at the required speed drawing optimum current for the actual load. It makes sure that if temperature setting is 25°C, it is maintained at that level. By this it saves 30-50% of power.

In Fig. 13.2 it is shown the air conditioner with inverter technology maintains the preset temperature as long as the system runs. In non-inverter a/c system compressor runs with its full capacity, overcools and then it stops. When thermostat senses a higher temperature, compressor stars again and runs at its full capacity.

Inverter technology is a solid-state power electronics conversion system consisting of three sub-systems. (i) a rectifier bridge converter which is a three-phase, six-pulse, full-wave diode bridge, (ii) a direct current link consists of a capacitor which smoothens the converter's DC output ripple and provides a stiff input to the inverter, (iii) this filtered DC voltage is converted to quasi sinusoidal AC voltage output using the inverter's active switching elements. An embedded microprocessor governs the overall operation of the VFD drive. Most drives are AC-AC drives in that it converts AC line input to AC inverter output. In some applications

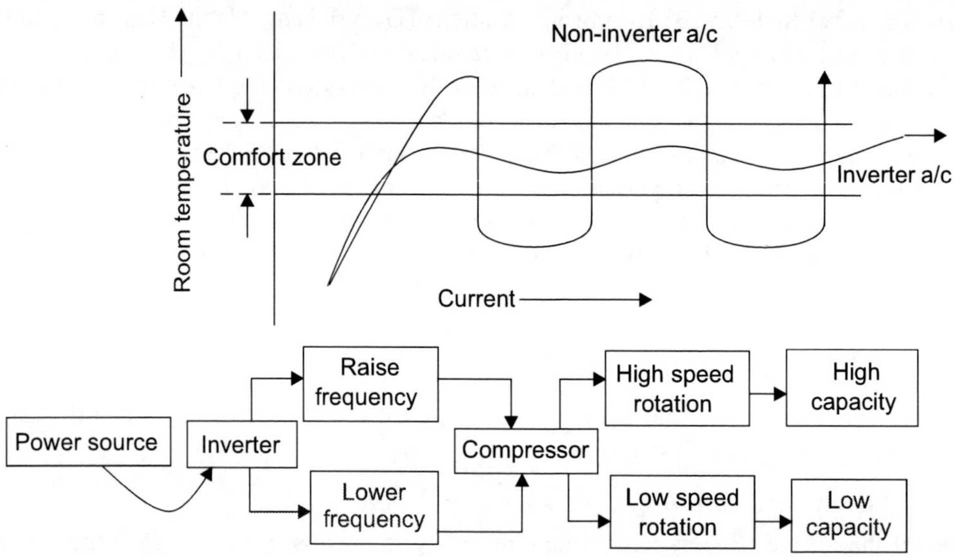

Figure 13.2: Performance of air conditioner with Inverter and without inverter

such as common DC bus bar or solar applications drives are configured as DC-AC drives. Now let us discuss how does it save energy. In starting a motor, a VFD initially applies a low frequency and voltage, thus avoiding high initial current flow as in case of direct on-line motors. A regular motor needs 3-4 times more current during start up as compared to running current. But in inverter technology, it needs much less current at start up as it starts gradually. For example, for 1.5 TR AC which requires 10 amp running current may need 30 amp starting current for which a 5 kVA generator is needed. In contrast inverter technology needs 6-7 amp current and not more during start up. Thus, needs a 2.0 kVA generator. In non-inverter air conditioner or refrigerator, as it follows a off/on cycle, it starts many times in a day and hence consumes much amount of power. Regular motor has a lower power factor, but inverter motor has a power factor approaching one. So power consumption is less. Inverter technologies are also best suited for air conditioning systems using solar PV panels. This also gives a soundless operation as frequent starting and stopping of compressor makes undesirable sound in non-inverter air conditioners.

13.2.2.2 Use of CO_2 sensor

Use of CO_2 sensor for ventilation control (so that required amount of ventilating air is added to the return air instead of excess of that) also reduces cooling load on the a/c system. Carbon dioxide sensor helps in reduced ventilation when people density is known to be lower. This is more prevalent in cinema halls, hotels, etc., where the occupancy may vary from 20% to 100%. In such cases, a CO_2 sensor is installed in the conditioned space which measures the CO_2 level and connects it to a controller to open the outside air dampers to maintain the CO_2 level at no higher than 1000 ppm. This is very often called *demand controlled ventilation* (DCV). As the name implies DCV is a tool for estimating the fresh air intake as per the demand. The demand for fresh air is predicted by CO_2 concentration in the indoor air. Let us consider one example. A classroom has a capacity of 50 students. During classroom activity the metabolic rate is about

1.0 met or 55 W/m^2. By this process the average CO_2 exhaled is 0.005 L/s. To this is added the CO_2 from outdoor air which is around 350 ppm (for non-polluted city). It is to be calculated how much should be the ventilated air so that 1000 ppm of CO_2 level is maintained. This process is expressed by $VC_{\text{classroom}} = v + VC_{\text{outdoor}}$ where V is the volume of outdoor air entering the space, C is the concentration of CO_2 and v is the volume CO_2 produced by the student. So $V \times 0.001\ \text{m}^3/\text{m}^3 = 0.000005\ \text{m}^3/\text{m}^3 + V \times 0.00035\ \text{m}^3/\text{m}^3$; this gives $V = 0.0077\ \text{m}^3/\text{s}$, i.e., 7.7 L/s. So required ventilation rate should be 7.7 L/s. This result is the same regardless of how many students are in the classroom since everything is proportional. Setting the ventilator above this value will consume more power for treating the extra amount of ventilated air. CO_2 sensor used should be highly reliable from safety point of view so that it maintains good indoor air quality and also fires alarms when reaches the preset value. CO_2 is comparatively easy to measure with various methods, chiefly chemical, electrochemical and infrared methods.

13.2.2.3 Utilisation of non-conventional energy

Utilisation of non-conventional energy, mostly solar energy, and waste heat of an industry to produce refrigerating effect also helps in reducing the conventional power consumption and reducing the environmental pollution. Efficient solar collectors can be used to trap enormous solar energy which can be utilized to heat the generator section of a vapour absorption refrigeration system so also the waste heat from an industry. This can also be utilized to generate electricity from photo-voltaic cells and run a thermoelectric refrigeration system. Here is given a feasible system descrption to utilize the solar energy for producing low temperature by thermoelectric refrigeration principle.

Thermoelectric generator and refrigerator

Thermoelectric generator works on the principle of Seebeck effect and its efficiency increases with the increase in temperature difference between the hot end and the cold end. But thermoelectric refrigerator works on the principle of Peltier effect and its performance increases with the decrease in that temperature difference.

Maximum efficiency of a TE generator and the maximum COP of a TE refrigerator can be expressed by Eqs. (13.1) and (13.2), respectively.

$$\eta_{\max TE} = \frac{T_H - T_C}{T_H} \left[\frac{\sqrt{\left(1 + ZT_{\text{avg}}\right)} - 1}{\sqrt{\left(1 + ZT_{\text{avg}}\right)} + \frac{T_C}{T_H}} \right] \qquad (13.1)$$

and

$$\text{COP}_{\max TR} = \frac{T_C}{T_H - T_C} \left[\frac{\sqrt{\left(1 + ZT_{\text{avg}}\right)} - \frac{T_H}{T_C}}{\sqrt{\left(1 + ZT_{\text{avg}}\right)} + 1} \right] \qquad (13.2)$$

The overall performance can be enhanced by having maximum utilization of the energy source and reducing the irreversibility (resistance to heat and electricity flow).The heat sources may be radioisotopes, waste heat from industries or motor vehicles, solar energy, etc.

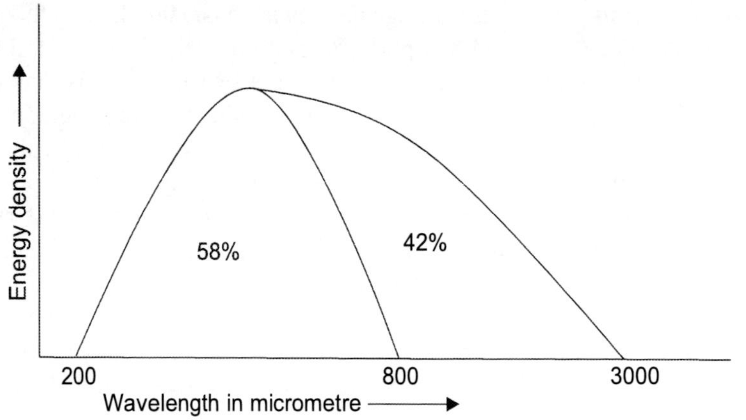

Figure 13.3: Represents the solar spectrum for PV cell and TEG

Figure (13.3) shows the range of wavelength which carries enough energy with it to energize the electrons to that is useful for photo voltaic cell (PV cell) is 58%. The rest 42% of solar energy carried by longer wavelength (specifically infrared range) cannot be utilized by PV cells. Thus, a PV module converts around 12% of total incident solar energy to electrical energy and rest of the radiations become available as thermal energy. Hence, this increases the temperature of the PV module and increase in temperature of the PV module decreases its conversion efficiency. That is why a PV module needs cooling to enhance its efficiency. Again this thermal energy can be effectively utilized by thermoelectric generators to generate electricity as shown in Fig. (13.4). Cooling of cold end of the TEG due to flow of water (coolant) causes a large temperature difference between hot end and cold end that causes increase in TEG efficiency. Hence, it is beneficial to use integrated systems, which combines PV cells (solar panels) with thermoelectric generators. It is reported that by using water in place of air as the PV module cooling fluid, 18% increase in conversion efficiency is obtained for a photovoltaic thermal (PVT) integrated thermoelectric generator. A wavelength segregator through which the whole spectrum of the solar energy can be made available for electricity conversion by PV cell and TEG. The power generated by the PV cell is used to drive a fan which cools the PV cell, TEG as well. The efficiency of a PV cell is calculated from

$$\eta_{PV} = \frac{P_{PV}}{A_{PV}\ G} \tag{13.3}$$

where, P_{PV} is the power generated by the PV cell, A is its surface area of exposure and G is the solar radiation. Efficiency of TEG is expressed by Eq. (13.3). The overall of efficiency of the hybrid system is expressed as

$$\eta_{\text{Overall}} = \eta_{\max TE} + \eta_{PV} \tag{13.4}$$

It is reported PV cell efficiency is 16.69% and TEG efficiency is 2.96% and an overall efficiency of 19.65%. The TE generator obtained an electrical power output of 2.94 W whereas PV module obtained an output of 1.93 W.

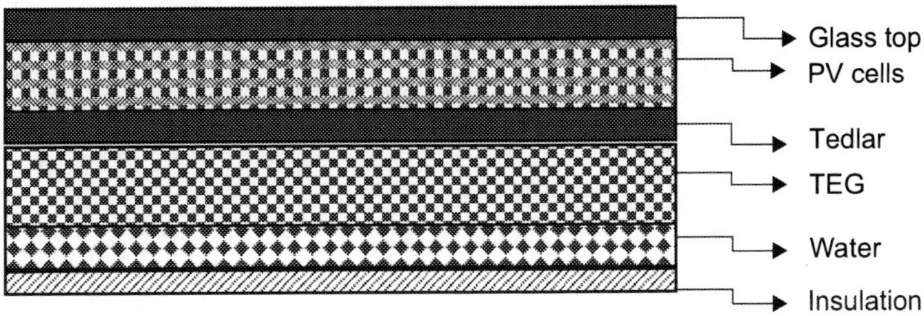

Figure 13.4: Schematic of a hybrid thermoelectric power generation

Portable thermoelectric cooler

Researches have been carried out on self-driven TEG-TEC systems, i.e., thermoelectric cooler driven by a thermoelectric generator as shown in Fig. (13.5 a). Sometimes referred to as Thermoelectric self-cooling (TSC) system. In this case TEG may be driven by waste heat from industries or motor vehicles. For space applications radioisotopes are used as the heat source for TEG. For solar systems, combination of PV cell and the TEG modules (hybrid systems) are more beneficial. They are sometimes called portable TECs. A portable TEC developed by Vijayarengan, could cool 250 m of water by 10°C in 2 hours when the ambient was 31°C. This shows this system has very poor performance (COP is around 0.2). Cooling of electronic devices can be done by placing the hot end of the TEG close to the device to be cooled as shown in Fig. (13.5 b). TEG takes up the heat from the device which is normally at 55°C to 75°C. Current generated from the TEG is sufficient to drive a fan which will act as a heat sink. Thus, an electronic device can be self-cooled. This is in contrast to the present method of thermoelectric active cooling. These self-driven TEG-TEC systems are more useful for remote areas where power grids are not available or places like aircraft, submarines, space shuttles commodities like helmets, etc.

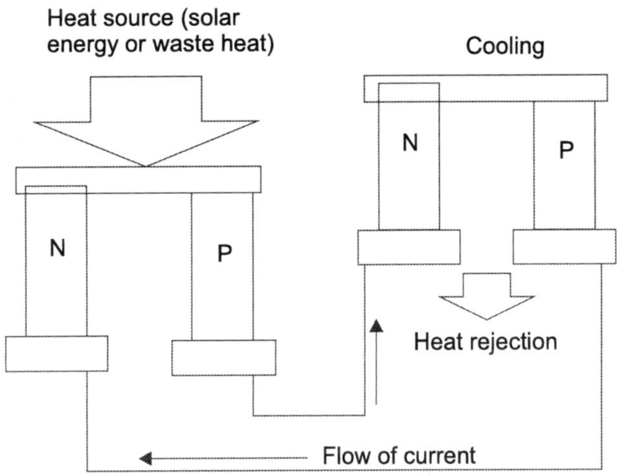

Figure 13.5: (a) Schematic of a TEG operated TEC

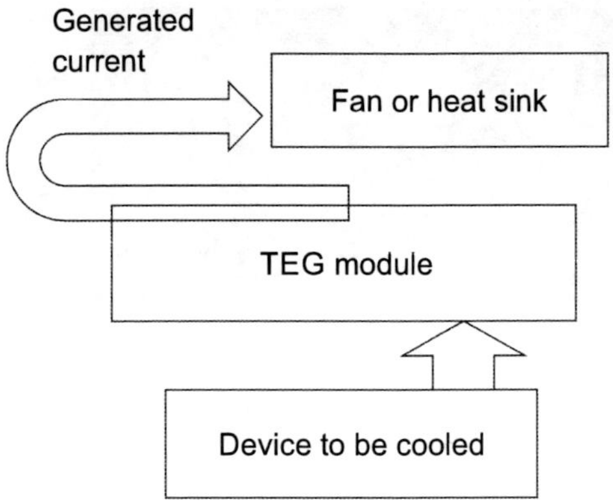

Figure 13.5: (b) Schematic of a thermoelectric self-cooling device

Integrating the phase change material with hybrid system

Performance of thermoelectric system can also be made more reliable, specifically for solar TE systems by integrating phase change material (PCM) with TEG or TEC. Heat is rejected by PCM during cooling process and gets solidified. Total heat rejection by PCM will be summation of its sensible heat and the latent heat of fusion. During 'off period' PCM absorbs latent heat and sensible heat from the refrigerated/cooling cabinet and keeps it at the desired temperature. Thermal diode is used so that heat is not leaked to the PCM when there will be power off. Studies reveal that a salt, mixture of sodium nitrate and potassium nitrate, may be used with thermoelectric generator to act as the heat source as its melting temperature is nearly 270°C.

13.2.2.4 Heat recovery

Some energy can be recovered by utilising the heat rejected by the condenser and the compressor. Heat rejected by the condenser is normally at a temperature of 55°C. It can be recovered by the water used as coolant in water cooled condenser and the hot water may be used as a preheating source for further heating process. Of course, this is possible where natural water body is available for water cooled condenser.

13.2.2.5 Availing a low temperature sink

Rejecting heat by the condenser to an environment whose temperature is less than the atmospheric air (relatively cold groundwater available from wells) saves much amount of compressor work. The following example explains 'how'.

Example 13.2

Let us consider a 100 TR capacity R22 refrigeration system working within a temperature limit of 0°C and 40°C. The condenser is a water-cooled condenser and rejects heat to the normal water which is at 30°C (10°C of temperature gradient for efficient heat rejection). Assuming S-S-S vapour compression refrigeration cycle compressor work and COP is calculated as follows. (Refer Fig. 4.4.)

Ans: At 0°C, $P_1 = 0.4976$ MPa, $h_1 = 405.36$ kJ/kg, $s_1 = 1.7518$ kJ/kg-K

At 40°C, $P_2 = 1.5335$ MPa, $h_3 = h_4 = h_f = 249.08$ kJ/kg. So the compression ratio is 3.082. State point 2 properties are determined from 40°C superheated table at $s_2 = s_1$ = 1.7518 kJ/kg-K, i.e., h_2 = 432.8 kJ/kg, t_2 = 58°C. As per the previous example mass flow rate of refrigerant is calculated to be 2.25 kg/s, and work done by the compressor is 61.74 kW. Hence, COP is 5.7. Now let us consider coolant water is available at 15°C from a 9.0 metre deep well and this water temperature remains constant throughout the year. (Of course, this temperature may vary a little from place to place for the same climatic zone). Hence, condenser temperature can be limited to 25°C instead of 40°C. Now the new temperature limit for the refrigeration system will be 0°C and 25°C. So at 25°C, P_2 = 1.045 MPa, $h_3 = h_4 = h_f$=230.05 kJ/kg. So the compression ratio is 2.1. State point 2 properties are: h_2 = 423.5 kJ/kg, and t_2 = 38°C. Then the mass flow rate is calculated to be 2.006 kg/s, and work done by compressor is 36.4 kW. Now amount of cooling water required for the condenser is calculated from the assumption that water gets heated by 15°C (assuming counterflow heat exchange). In first case heat rejected by condenser is 2.25 kg/s × (432.8 – 249.08) kJ/kg = 413.37 kW = m_W × 4.19 kJ/kg-K × 15°C; so m_W = 6.577 kg/s (m_W is the mass of water). Similarly, for the second case heat rejected by the condenser is 2.006 kg/s × (423.5 – 230.05) kJ/kg = 388.06 kW = m_W × 4.19 kJ/kg-k × 15°C; m_W = 6.17 kg/s. So to lift 6.17 kg/s of water from a 9.0 m deep well the pump work required will be $m_W gh$ = 6.17 × 9.81 × 9 = 544.75 W = 0.544 kW. So the total work input for the second case will be 36.4 kW + 0.544 kW = 36.944 kW ≈ 37.0 kW. So the COP is 9.50. To summarize the benefits (i) compression ratio is decreased from 3.082 to 2.1; (ii) work input decreased from 61.74 kW to 37 kW; (iii) peak temperature decreased from 58°C to 38°C, COP increased from 5.7 to 9.5.

13.2.2.6 Changing the position of freeze doors

Vertical doors in refrigeration devices are inherently inefficient. As soon as the door of a vertical fridge is opened, the cold air escapes. Simply because it is heavier than the warmer air present in the room. When the door of a chest freezer (it has the door on its top) is opened, the cool air stays inside. Just because it is heavy. Any leak or wear in vertical door seal causes significant loss of refrigerator efficiency. In contrast, even if the door of a chest freezer remains wide open, the heavy cool air will still remain inside. Due to the outdoor air gets into a vertical door fridge, its performance drastically decreases. The main difference between a freezer and a fridge is that sub-zero temperature is maintained in the former whereas 4 to10°C is maintained in the later. But a freezer can be changed to fridge just by changing the temperature control. It is estimated that each household vertical door fridge waste 1.5 kWh of energy each day. So with little design change, a chest fridge can replace a vertical door fridge. Thus much power can be saved.

13.2.2.7 Desiccant cooling

Power saving by using desiccant cooling system: Desiccant cooling systems are used to dehumidify and cool the air for space cooling in buildings. The property of the desiccant material is to absorb moisture from air by creating an area of low vapour pressure at the surface of the desiccant. The common desiccant materials (called adsorbent materials) are LiCl, SiO_2 (Silica Gel), Al_2O_3 (activated alumina), LiBr, Zeolite, etc. These materials are used in the form of wheels in desiccant cooling system and called '*desiccant wheels*'. A solid DCS consists of a desiccant wheel, a heat exchanger, and direct evaporative cooler. The principle of operation of the system is as follows. Refer Fig. 13.6 (a and b). High temperature and high humid atmospheric air (state point 1) is made to flow through one half of the desiccant wheel. The desiccant material absorbs the water vapour from the moist air and relatively dry and high temperature air comes out of it at state point 2, as it happens in chemical dehumidification process. Then this low humid and high temperature air is sensibly cooled by the heat exchanger to state point 3. This sensibly cooled air is again cooled by an indirect evaporative cooler (IEC) to state point 4 and then sent to the space to be air conditioned. The process is shown on the psychrometric chart. Regeneration of desiccant is made by supplying hot air from any waste heat source, solar energy or by internal heat exchange (i.e., by passing the hot air coming out of the heat exchanger) through the other half of the desiccant wheel. In Fig. 13.6, it is shown outdoor (o) air is first cooled by direct evaporative coil to state point 5. This air gets heated by taking heat from the hot air from desiccant wheel at state point 2 to state point 6. Then this is heated to the required reactivation temperature of the desiccant material by a heater (state points 6 to 7).This air, as it passes through the saturated half of the desiccant wheel, that half gets free from the adsorbed moisture and the wheel is said to be regenerated and the air at state point 8 leaves the system. Now that used half of the desiccant wheel is ready for the next cycle of operation.

In conventional air conditioning, both cooling and dehumidification of air is carried out by evaporator of the refrigeration system. So the cooling coil temperature has to be maintained at a temperature below the dew point temperature. But in desiccant based hybrid cooling system, as discussed, the dehumidification is done by the desiccant wheel. So the evaporator coil need not be maintained lower than the dew point temperature. This results in downsizing the refrigeration system and consequent energy saving by the compressor. This type of cooling is beneficial in the climate where the temperature as well as humidity is high (mainly in coastal belt).

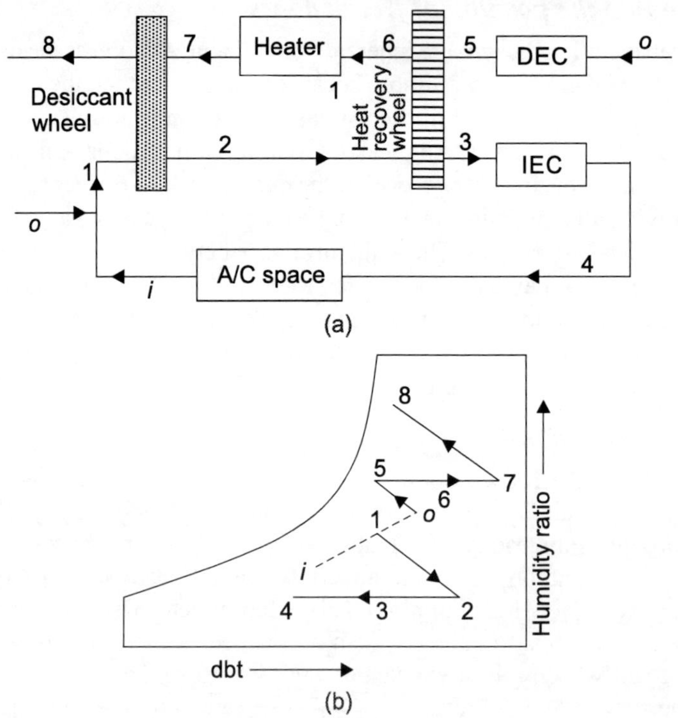

Figure 13.6: (a) Schematic of an a/c system with desiccant wheel, (b) processes on psychrometric chart

Let us discuss the following example how energy can be saved by recirculating the return air.

Example 13.3

Take the climatic condition of Bhubaneswar during summer. Temperature is 42°C and relative humidity is 90%. Desired indoor condition is 25°C and 60% RH. Calculate power requirement to maintain the desired indoor condition. The air conditioning system uses desiccant wheel and the air flow process is shown in Fig. 13.6.

Ans: From psychrometric calculation, $\phi = P_V/P_S$, from steam tables $P_S = 8.208$ kPa at 42°C. So $P_V = 7.3872$ kPa. So specific humidity, $\omega = 0.622\,[P_V / (P - P_V)] = 0.04968$ kg / kga.

Enthalpy is calculated from $h = (C_{P,a} + \omega\, C_{P,V})t + \omega\, h_{fg,0} = 168.886$ kJ/kga.

When the moist air dehumidified by the desiccant wheel the process will be along the constant enthalpy line. o –1 (Neglecting the heat of adsorption) and it should end at point 1 [as shown in Fig. 13.6(b)]. The specific enthalpy of air will be same as that at state point o. The specific humidity at state point 1 will be same as specific humidity at point 2 and 3 (i.e., 25°C and 60% RH). Calculating as above, $\omega = 0.0119345$ kg/kga at state point 3. Specific enthalpy at point 3 is calculated to be 55.534 kJ/kga.

Temperature of air coming out of the desiccant wheel is calculated from

$$h = (C_{P,a} + \omega\, C_{P,V})t + \omega\, h_{fg,0}.\text{ It gives } t_1 = 135°\text{C}.$$

Now it is cooled by the heat exchanger to $t_2 = 75$°C where the coolant air is the return air from the a/c space. It is designed on the basis that the reactivation temperature of the silica gel is 80°C to 90°C. So the temperature t_4 should be equal to 85°C. Then the air is sensibly cooled by Indirect evaporative coil (IEC).

For easy calculation supply air state is assumed same as the state of the indoor air.

The cooling load on IEC will be $C_{P,\text{humid}}(t_2 - t_3) = 1.0216\,(75–25) = 51.08$ kJ/kga.

Without desiccant wheel, the total cooling load to bring the moist air from state point 1 to state point 4 is (168.886 kJ/kg –55.534 kJ/kg) = 113.352 kJ/kga. Now for occupancy of 100 persons, if the fresh air requirement is 13 m^3/hr/person doing office work, the fresh air requirement (with 100% ventilation) will be 0.26 kg/s. With desiccant wheel the power requirement will be sensibly cooling load (51.08 kJ/kg × 0.26) = 13.28 kW plus the amount of power required by the blower to drive the air (0.26 kg/s) through the desiccant wheel and the heat exchanger which amounts to maximum 0.4 kW. (This can be calculated by taking the total friction head loss for the air flow through pipes and using the formula: Power = Volume flow rate × Pressure drop due to friction. A double tube heat exchanger may be considered.) The total power requirement will be 13.28 kW + 0.4 kW = 13.68 kW.

The power consumed by the air conditioning system without desiccant wheel will be 113.352 × 0.26 = 29.47 kW. Almost 100% power saving! (Mass of water vapour removed is negligible in comparison to mass of dry air. Hence, mass of air is taken same before and after the dehumidification.)

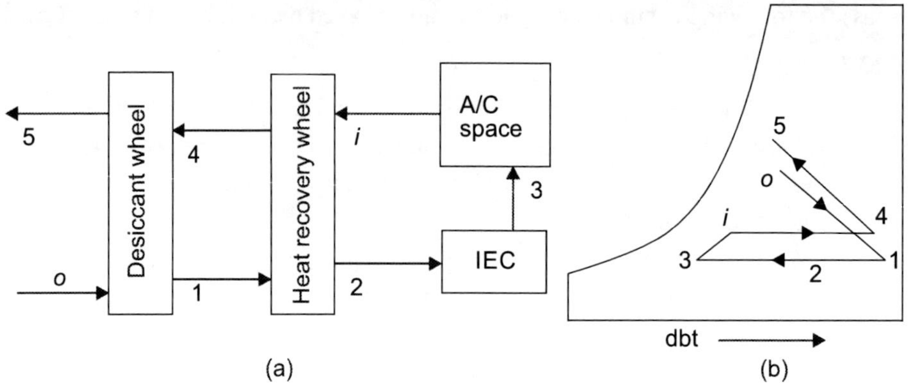

Figure 13.7: (a) Schematic (b) process on psychrometric chart of Example 13.1.

13.2.2.8 Use of heat pipes

Heat pipe is a passive heat transfer device (heat exchanger) that can transfer large quantities of heat (latent heat) through small surface areas with small temperature difference.

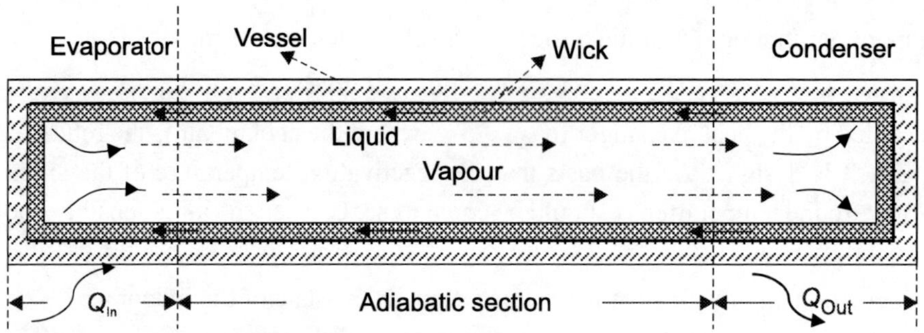

Figure 13.8: Schematic of a heat pipe.

Construction: Refer Fig. 13.8. It is a slender tubular-like structure closed at both the ends. The fully sealed tube is the vessel of the heat pipe made up of copper, nickel or stainless steel (any highly conducting material). A layer of wicking structure/material covers the inner surface of the tube. Commercially available heat pipe wick structures are grooves, wire mesh, screen mesh, sintered powder and fibre/spring, etc. The middle part of the tube is perfectly insulated and is called adiabatic zone. The two ends are made highly conducting for high heat transfer rate because one end acts as the evaporator and the other one acts as the condenser. (Adiabatic zone may be constructed from any other low conducting material and be joined to two ends.) To enhance the heat transfer rate aluminum fins are attached to evaporator and condenser sections.

Working fluids are selected depending upon the temperature range of application.Viscosity and surface tension of the fluid are most important physical properties to be considered. Ammonia, methanol, ethanol, etc., are used for refrigeration purpose whereas mercury, sodium etc., for high temperature applications.Even for temperature as low as 6.0 K, liquid helium is used as working fluid.

Working principle: Complete vacuum is created in the heat pipe. Sometimes, without creating a vacuum, the working fluid inside it may be boiled so that any non-condensable gas is removed (purging method) and then the pipe is sealed and it is partially filled with working fluid. Evaporator section of the heat pipe (hot end) receives heat from the space to be cooled. Working fluid boils off taking the latent heat and the vapour moves through the core section of the pipe towards the condenser (cold section). The vapour rejecting heat to the cooling medium, becomes liquid. This condensed liquid moves back along the periphery of the pipe due to the capillary action of the wicking material/structure to the evaporator section.This capillary action depends upon the surface tension of the working fluid and the type of wick material/structure. Heat pipe normally operates in horizontal position. In some case, the pipe is put slightly in a slanting manner so that the condensed liquid flows back to evaporator section due to gravity.

For a heat pipe to operate, the maximum capillary pumping head (ΔP_{Max}) must be able to overcome the total pressure drop in the heat pipe. Total pressure drop consists of pressure drop: (a) vapour moving from evaporator to condenser (ΔP_v), (b) liquid moving from condenser to evaporator (ΔP_c), and (c) potential head difference between evaporator and condenser ΔP_g (if any).

$$\Delta P_{\text{Max}} \geq \Delta P_v + \Delta P_c + \Delta P_g \qquad [13.5]$$

Otherwise liquid cannot move back to evaporator by capillary action and the evaporator goes dry.

Working principle shows the device is very simple and the operation is also very simple. There is no moving part. But there are many limitations in working a heat pipe. First limitation is the capillary limitation. As it is shown in the above equation, capillary pumping head developed must be greater than the total pressure drop in the pipe. Different wick-working fluid combination gives different capillary pumping head. Second is the entrainment limitation. In this case, the amount of vapour leaving the evaporator and condenser, cannot get back to evaporator by capillary action of the wick structure. So after certain time evaporator dries up. Thus, the amount of heat supply to the evaporator is limited. Third is the sonic limitation. If the vapour coming out of the evaporator reaches sonic velocity, the vapour flow to the condenser is choked. Similarly, there is a limitation to boiling of fluid in the evaporator section otherwise the evaporator gets dried up. With all these limitations, heat pipe heat carrying capacity gets limited.

This device is suitable for electronic equipment cooling (spot cooling), to be used in aerospace activities. It is light and simple in construction. Working fluids used are environment-friendly easily and cheaply available.

There is only a very little structural difference between thermosyphon and heat pipe. But the principle of transferring the heat is same. Thermosyphon is a vertical tubular structure. Vapour from evaporator section rises up and reaches the condenser section. There it condenses rejecting heat to the cooling medium and the liquid comes down to evaporator due to gravity. But in heat pipe the liquid from condenser returns to evaporator due to capillary action of the wick. So the heat pipe alignment can be horizontal, slanting. When the heat pipe is placed vertically, the condensate flows due to capillary action and gravity both. In short, it can be said that a thermosyphon is a plane vertical tubular structure, whereas heat pipe is a tubular structure with wick material for liquid transport.

13.2.2.9 Maintaining differential pressure across indoor

There is a significant effect on energy consumption and comfort due to pressure maintained inside the conditioned space. The pressure in the space may be maintained at positive or at negative to the atmospheric pressure which amounts to around 2.5 Pa. Refer Fig. 13.9 (a) and (b). Positive pressure means the space is maintained at a pressure slightly above that of the atmosphere so that indoor air will have tendency to come out of the space and vice versa. This can be measured by placing a water column under the door. In hot and humid climate, like the coastal belt of India, the space should be maintained at a slightly positive pressure so that infiltration through leakage or cracks can be prevented. But there is chance that as the door is opened the indoor air may flow out of the room. Hence, the door should be opened to inside of the room so that immediate flow out of the indoor air can be restricted. In a cold climate the space should be kept close to the outside pressure or slightly negative to prevent warm and moist air from inside the space from entering the wall where it could cause condensation or ice.

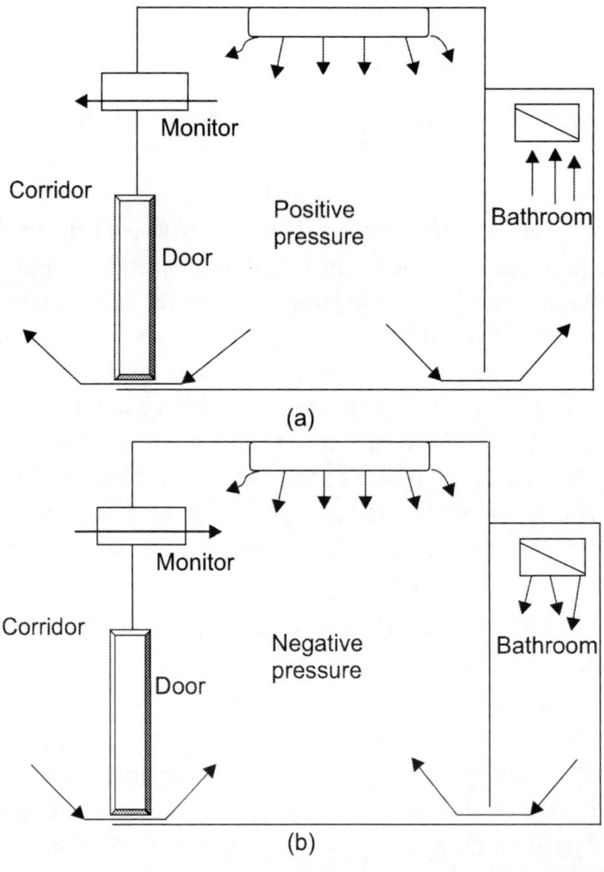

Figure 13.9: Schematic of (a) +ve pressure and (b) –ve pressure maintained indoor

This concept is also significant in maintaining the hospital indoor air quality (IAQ). The air of the space, where patients with infectious diseases are accommodated, should not come out as the air is contaminated. So the pressure of the space should be maintained negative. This is mostly recommended for tuberculosis patients. But in an operation theatre or room, the space should be very much dust free and germ free. Hence, more ventilation is required. Maintaining positive pressure inside the space helps venting out the contaminated air of the operation space. However, neither positive pressure nor negative pressure in the conditioned space is desirable as it causes discomfort, energy waste and deterioration of the building fabrics. Over and above, the conditioned space should be properly sealed to check leakage of air from or into the space.

13.2.2.10 Zoned air conditioning system

Significant amount of energy can also be saved by 'zoned air conditioning system'. Depending upon the complexity of the requirements, the heating, ventilating and air conditioning (HVAC) designers consider many more issues than simply maintaining temperature and humidity in comfortable range. There are multiplexes where different kinds of activities take place. Hence, the demand for cooling/heating load varies from one part of the building to the other. When a system is designed to provide independent control in different spaces, each space is called a zone. A zone may be a separate room or may also be part of a large space. For example, a theatre stage may be a zone, while the audience seating area is a second zone within the same big space. Each has a different requirement for heating/cooling. For this purpose, two important factors are to be considered. One is the different mediums to/from which heat can be transferred for heating/cooling of different zones. Second is the means by which the supply of the working fluids to different zones can be regulated as per the requirement. Let us discuss one application. It is discussed in Section 12.7: Air Distribution System).

1. First the main air system is chilled to the minimum required temperature to satisfy all possible cooling loads. Then heaters are provided in the duct to each zone with a zone thermostat. The zone thermostat can then control the heater to maintain the desired set point temperature. This is called a 'reheat system'. In some cases, the duct to each zone is fitted with a control damper that can be throttled to reduce the air flow to maintain the desired indoor condition. This is called *variable air volume* (VAV) system.
2. System control can be done by decoupling each zone hydraulically and thermally. A secondary pump is provided for each zone where the flow rates are regulated by microprocessor control bridges, automatic flow control devices, check valves, etc. All these controlling devices are activated by the sensors which sense the differential pressure between the zone and the main supply system as the demand for that zone increases or decreases.

13.3 ADVANCED REFRIGERATION SYSTEMS

Scientists and researchers are striving to develop refrigeration systems which will consume less amount of energy and will use environment-friendly working fluids. Some of the non-conventional refrigeration techniques have been discussed in Chapter 7 like Thermoelectric, Vortex Tube, Magnetic Refrigeration systems. These are considered to be sustainable refrigeration systems and environment friendly. But they have very low COP. Here, a few of the developments done in the area of vapour compression refrigeration system are discussed for the knowledge of the readers. It is claimed that their COP is at par or more than that of the conventional vapour compression refrigeration system.

13.3.1 Compression-Absorption System (CAS)

This is an advanced refrigeration system and falls under the category of 'hybrid refrigeration system'.

Working fluids used are the zeotropic mixture of R22 and DMETEG, R22 and NMP, R134a and DMETEG, etc. These are the solutions of refrigerant (solute) and other fluid (solvent).

Working principle

Figure 13.10 shows the schematic diagrams of a single-stage vapour compression (13.10a), single stage vapour absorption (13.10b) and vapour compression-absorption system (13.10c). It has alredy been discussed about vapour compression and vapour absorption systems. As shown in the Fig.13.10c, the system consists of a desorber, an absorber, a compressor, a solution pump, an expansion valve and a heat exchanger. A mixture of refrigerant and absorbent solution circulates through the desorber, absorber, solution pump, heat exchanger and the

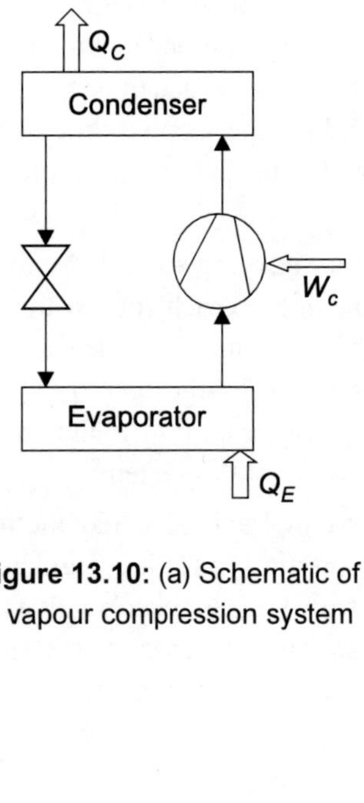

Figure 13.10: (a) Schematic of a vapour compression system

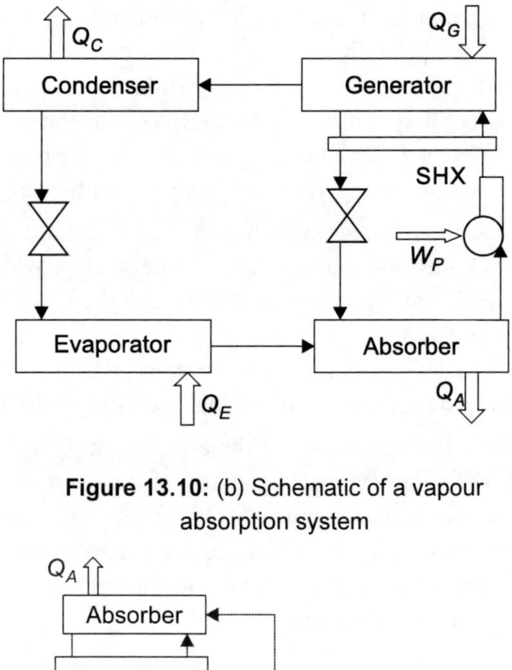

Figure 13.10: (b) Schematic of a vapour absorption system

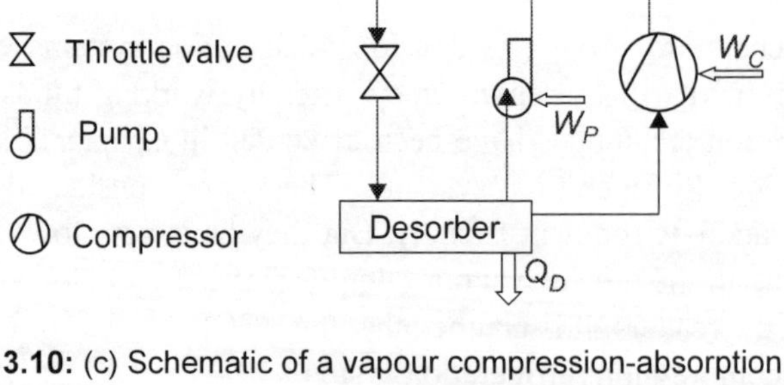

Figure 13.10: (c) Schematic of a vapour compression-absorption system

expansion valve, while the refrigerant vapour flows through the compressor. Cooling effect (Q_D) is obtained at the desorber as heat flows into the desorber and a part of the refrigerant from the solution vaporizes. The refrigerant vapour is compressed to the absorber pressure by the compressor utilizing external energy (W_c). The remaining solution in the desorber (which is weaker in refrigerant) is pumped by the solution pump utilizing external energy (Wp) to the absorber pressure. Here both the compressor and the pump use mechanical energy. The high pressure refrigerant from the compressor and the weak solution from the solution pump are recombined in the absorber. Since this is an exothermic process, useful heat is released at a high temperature in the absorber (Q_A). The resulting solution from the absorber (which is stronger in refrigerant) flows through the expansion valve where its pressure drops to that in the desorber. The cycle is completed as the low pressure strong solution re-enters the desorber. The solution heat exchanger (SHX) is employed between the weak and strong solutions to improve the cycle efficiency.

COP is calculated from the equation $(Q_D + Q_A)/(W_c + W_P)$.

This can also be designated as combined COP because the desired effect is both Q_D and Q_A where simultaneous heating and cooling effect is required.

Advantages

1. This system uses refrigerant and absorbent mixture which boils or condenses non-isothermally over a range of temperature. Thus, the temperature profiles on internal and external sides can be matched. The glide matching with the external fluid reduces the irreversibility during heat transfer resulting in higher COP and lower heat exchanger area.
2. The system operates at a lower peak pressure and the pressure ratio is lower in comparison to vapour compression system for same temperature lift. Temperature lift is defined as the difference in condenser temperature (T_C) and evaporator temperature (T_e). Hence, compressor discharge temperature is also lower, which improves its performance.
3. Capacity of the system can easily be controlled by varying the working fluid composition only.

Disadvantages

1. Leakage from the system will change the solution composition.
2. Liquid absorbent may enter the compressor causing damage to it.

Higher COP and higher temperature lift can also be obtained with multi-stage and multi-effect vapour compression-absorption system. Some of these advanced vapour compression-absorption systems are of two stage, double lift, double effect, and desorber-absorber heat exchange. All these possible configurations of compression-absorption systems are described by Alefeld and Radermacher.

Application: This system is more suitable for simultaneous heating and cooling applications, like for pasteurization of milk, paper manufacturing, etc. For example, in milk pasteurization process, milk is first cooled to 4°C in desorber and then heated to 72°C in absorber.

Performance comparison between S-S-S cycle and single-stage CAS for 1.0 TR (3.5167 kW) cooling effect is given in the following table.

Performance parameter	**S-S-S cycle (VCS)**	**Single-stage CAS**
P_{Max} (Bar)	33.2	29.74
Compression ratio	6.054	5.745
Compressor displacement (m^3/s)	0.0018	0.002
Compressor exit temperature (K)	389	368
Compressor work, W_c (kW)	1.933	1.21
Pump work (W_P) (kW)	---------	0.0228
Heat rejection at absorber (Q_A) (kW)	5.45	4.71
Combined COP	4.64	6.673

13.3.2 Auto Cascade Refrigeration System

Working Principle

Refrigerants used: R507A (an azeotropic mixture of R125, 50%, and R143a, 50%, and NBP – 46.74°C), for high temperature side; and R23 for low temperature side (NBP - 82.0°C).

A mixture of R507A and R23 in the proportion of 60% and 40% respectively is taken. (This proportion may be varied depending upon the refrigeration load, the system is to be designed for.)

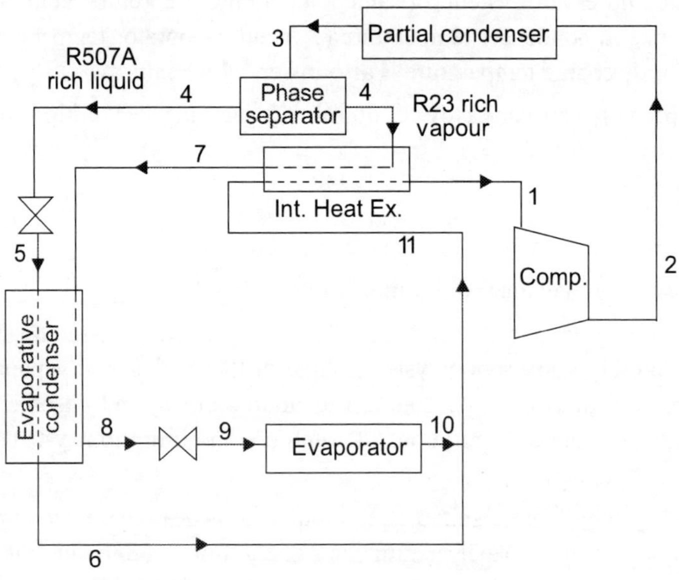

Figure 13.11: Schematic of an auto cascade refrigeration system

Working principle: Refer Fig. 13.11.

The vapour of this refrigerant mixture enters the compressor at 1. This is compressed to condenser pressure and comes out as high temperature and high pressure vapour mixture at 2. This vapour enters the condenser and rejects heat to the cooling medium. The condenser is so designed that R507A condenses to saturated liquid and R23 remains as vapour. (That is why it is called partial condenser.) It comes out at 3 and enters the phase separator where liquid phase (rich in R507A and little amount of R23) and vapour phase (rich in R23 vapour and little amount of R507A) are separated and comes out at 4. Both the vapour and liquid are at equilibrium. Hence, their state points are shown same, i.e., 4.

For easy explanation, let us assume the liquid at 4 is pure saturate R507A liquid and vapour at 4 is pure saturated R23 vapour.

R507A, at 4, is throttled to 5 and the low pressure and low temperature two-phase refrigerant enters the evaporative condenser. There it gains heat from the high pressure and high temperature R23 vapour and becomes saturated vapour. This low pressure and low temperature vapour comes out of the evaporative condenser at 6.

R23 vapour, at 4, enters the internal heat exchanger, rejects heat to the low temperature vapour mixture entering the heat exchanger at 11. This comes out of it at being cooled and enters the evaporative condenser at 7. There it rejecsts heat to low temperature, two-phase R507A and becomes saturated liquid. Then it is throttled to low pressure and low temperature, two-phase fluid at 9. Here it enters the evaporator, absorbs the heat from the cooling space and comes out as saturated vapour at 10.

Then the R507A vapour at 6 and R23 vapour at 10 get mixed at 11 and enters the internal heat exchanger. It absorbs heat from R23 vapour, gets superheated to 1. Thus, the thermodynamic cycle is completed.

If auto cascade system is compared with cascade system, the following similarities are observed.

1. The function of evaporative condenser and internal heat exchanger of auto system is similar to the cascade condenser of cascade system where the HT side refrigerant fluid receives heat from LT side refrigerant.
2. Instead of two compressors one compressor receives both the HT side and LT side vapour.
3. The refrigerant pair is chosen in such a manner that the HT side and LT condenser pressure and evaporator pressure can be maintained equal.

The disadvantage of this system lies in properly separating the R507A and R23 in phase separator.

This system is used for production of very low temperature (around –80°C, though it depends upon the LT side refrigerant boiling temperature.)

Industry 4.0 on energy saving in HVAC systems

Fourth generation of industrial development is introduced after industrial automation which is known as third generation of industrial development. Most of the researches have been

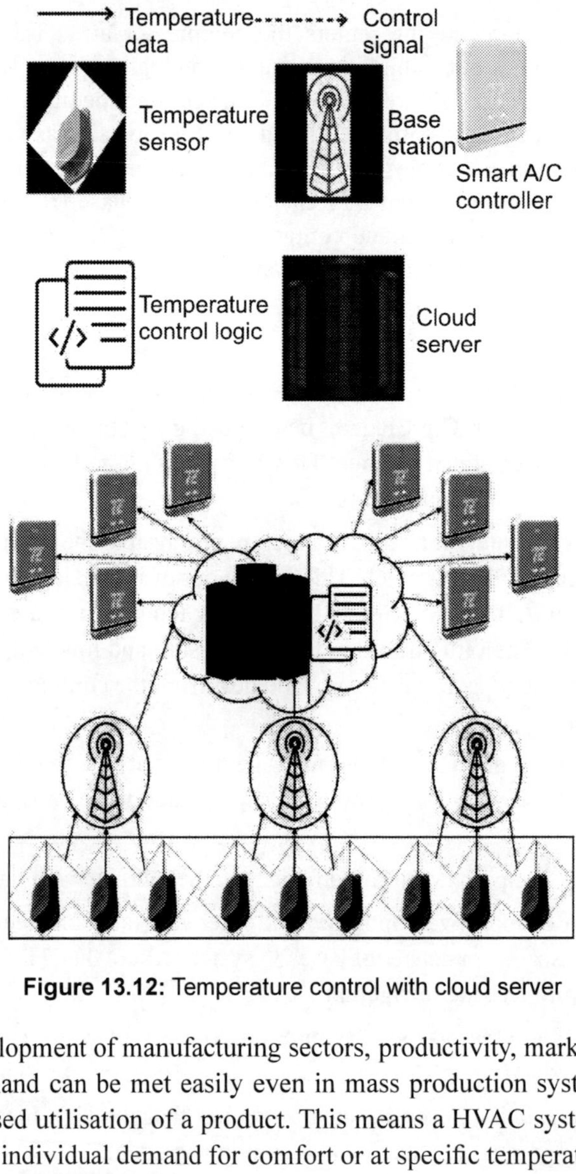

Figure 13.12: Temperature control with cloud server

focused on the development of manufacturing sectors, productivity, marketability, etc., so that each consumer demand can be met easily even in mass production systems. Nowadays it is focused on need based utilisation of a product. This means a HVAC system can be controlled automatically at the individual demand for comfort or at specific temperature and humidity for different product preservation. This reduces the wastage of conventional energy. Here is given a simple example of the working principle of the system. (Fig. 13.12)

- Cloud server is responsible for storage and maintenance of temperature data forwarded by the base station. Thus if the control logic is provided with suitable program and sufficient data for the humidity, CO_2, etc. from cloud server, a smart a/c can save energy during season change or occupancy and activity variation, etc. The temperature, humidity, and ventilated air can be very well regulated to our need, may be for comfort or for preservation, so that a considerable amount of energy can be saved by following the energy saving techniques discussed in Section 13.2. By adopting the concept of industry 4.0, a timely action can be taken with the help of well prepared data base for individual comfort or for food preserving conditions

EXERCISE

1. Define sustainability. How sustainable development can be achieved for refrigeration and air conditioning systems?
2. What are the ventilation principles? Give a brief account of these principles.
3. How air conditioner installation and maintenance help in saving energy consumption?
4. What do you mean by star rating on refrigerators and air conditioners?
5. Explain the inverter technology of an air conditioning system.
6. How the CO_2 sensor helps in saving the energy consumption in a/c systems?
7. Explain, how the solar energy can be utilized to produce refrigerating effect.
8. What is a desiccant wheel? How does it reduce the cooling load on a cooling coil?
9. Discuss the principle of operation of a heat pipe.
10. What do you mean by positive pressure and negative pressure in an air conditioned room? Explain with examples.
11. Define zone for air conditioning systems. Briefly explain how does it save energy in air conditioning a multiplex building?
12. What is Industry 4.0? Give a brief note on it.

MULTIPLE–CHOICE QUESTIONS

1. Sustainable development means:
   (a) use the resources economically
   (b) save the resources for future generation
   (c) both (a) and (b)
2. If a/c temperature is set at 24°C instead of 18°C, the power saving will be around
   (a) 10% (b) 15%,
   (c) 25% (d) 10% more power will be consumed
3. Air conditioner with inverter technology means the compressor runs
   (a) if power goes out
   (b) continuously consuming more power
   (c) continuously according to cooling load consuming less power
   (d) intermittently consuming less power
4. The maximum CO_2 level maintained in a air conditioned space should be
   (a) 100 ppm (b) 500 ppm
   (c) 1000 ppm (d) 2000 ppm

5. Dessicant wheels are used in a/c systems to reduce the
(a) sensible heat load
(b) latent heat load
(c) both (a) and (b)
(d) dust particles

6. Heat pipe is used to take away the
(a) sensible heat
(b) latent heat
(c) both (a) and (b)
(d) carry the heat source

7. Driving force for a heat pipe is
(a) gravity
(b) affinity for absorption
(c) capillary action
(d) buoyancy

8. The pressure of the indoor space where the infectious diseases are treated with respect to outdoor should be maintained
(a) positive
(b) negative
(c) equal

ANSWERS

1. (c) **2.** (c) **3.** (c) **4.** (c) **5.** (b) **6.** (b) **7.** (a) **8.** (b)

Appendices

Appendix 1

Thermodynamic Properties of Saturated Water (Liquid and Vapour)

Temp (t) (°C)	Sat Pr (P) (kPa)	Sp. Vol. Liquid (v_f) × 10^3 m^3/kg	Sp. Vol. Vapour (v_g) m^3/kg	Liquid Sp. Enthalpy (h_f) kJ/kg	Vapour Sp. Enthalpy (h_g) kJ/kg	Liquid Sp. Entropy (s_f) kJ/kg-K	Vapour Sp. Entropy (s_g) kJ/kg-K
0.01	0.6113	1.000	206.14	0.01	2501.4	0.000	9.1562
5	0.8721	1.000	147.12	20.98	2510.6	0.0761	9.0257
10	1.2276	1.000	106.38	42.01	2519.8	0.1510	8.9008
15	1.7051	1.001	77.93	62.99	2528.9	0.2245	8.7814
20	2.339	1.002	57.79	83.96	2538.1	0.2966	8.6672
25	3.169	1.003	43.36	104.89	2547.2	0.3674	8.5580
30	4.246	1.004	32.89	125.79	2556.3	0.4369	8.4533
35	5.628	1.006	25.22	146.68	2565.3	0.5053	8.3531
40	7.384	1.008	19.52	167.57	2574.3	0.5725	8.2570
45	9.593	1.010	15.26	188.45	2583.2	0.6387	8.1648
50	12.349	1.012	12.03	209.33	2592.1	0.7038	8.0763
60	19.940	1.017	7.671	251.13	2609.6	0.8312	7.9096
70	31.19	1.023	5.042	292.98	2626.8	0.9549	7.7553
80	47.39	1.029	3.407	334.91	2643.7	1.0753	7.6122
90	70.14	1.036	2.361	376.92	2660.1	1.1925	7.4791
100	101.35	1.044	1.6729	419.04	2676.1	1.3069	7.3549
120	198.53	1.060	0.8919	503.71	2706.3	1.5276	7.1296

Appendix 2

Physical Properties of some Common Refrigerants

Refrigerant	Chemical Formula	Molecular Weight	NBP (°C)	Freezing Point(°C)	Critical Temp(°C)	Critical Pressure (Bar)	ODP	GWP	
								100 yr	500 yr
R11	CCl_3F	137.39	23.7	–111	197.78	44.06	1.0	3500	1500
R12	CCl_2F_2	120.92	–29.8	–158	112.04	41.13	0.86	7300	4500
R22	$CHClF_2$	86.48	–40.8	–160	96	49.74	0.055	1500	510
R23	CHF_3	70.02	–82.1	–155	25.83	48.33	Negli	–	12000
R32	CH_2F_2	52.042	–51.75	–	78.41	58.3	Negli	–	220
R134a	CF_3CH_2F	102.03	–26.15	–96.6	101.06	40.56	Negli	1200	420
R717	NH_3	17.031	–33.35	–77.7	132.4	112.97	0	0	0
R744	CO_2	44.01	–78.52	–56.6	31.0	73.8	0	1	1
R600a	$(CH_3)_3CH$	58.13	–11.73	–160	135.0	36.45	0	3	3
R290	C_3H_8	44.1	–42.07	–187.7	96.8	42.54	0	3	3
R404A, NARM	R125 44% R143a 52% R134a 4%	97.6	–46.0		72.1	37.3	0	–	3860
R407C, Zeotropic	R32 23%, R125 25% R134a 52%	86.2	–43.6		86.7	46.2	0		1530
R502, Azeotrope	R22, 48.8% + R115, 51.2%	111.63	–45.4		82.2	40.72	0.34	–	5490
R507A, Azeotrope	R125, 50% + R143a, 50%	98.9	–46.9		70.9	37.9	0		3985

Specific Heat at Constant Pressure of Saturated Liquid (kJ/kg-K)

Refrigerants	–40°C	–20°C	0°C	20°C	40°C
R22	1.10	1.13	1.17	1.24	1.34
R134a	1.255	1.290	1.341	1.368	1.408
R717	4.414	4.510	4.617	4.755	4.932

Specific Heat at Constant Pressure of Saturated Vapour (kJ/kg-K)

Refrigerants	–40°C	–20°C	0°C	20°C	40°C
R22	0.608	0.665	0.739	0.840	0.997
R134a	0.749	0.816	0.887	1.001	1.145
R717	2.244	2.425	2.680	3.028	3.510

Thermal Conductivity of Saturated Liquid (W/m-K)

Refrigerants	–40°C	–20°C	0°C	20°C	40°C
R22	0.12	0.11	0.1	0.09	0.08
R134a	0.111	0.101	0.092	0.0835	0.0750
R717	0.690	0.622	0.560	0.410	0.425

Thermal Conductivity of Saturated Vapour

Refrigerants	–40°C	–20°C	0°C	20°C	40°C
R22	0.0071	0.0082	0.0094	0.011	0.013
R134a	0.0082	0.0098	0.0115	0.0133	0.0154
R717	0.0207	0.0218	0.0234	0.0264	0.0284

Dynamic Viscosity of Saturated Liquid (centipoise, i.e., gm/m-s)

Refrigerants	–40°C	–20°C	0°C	20°C	40°C
R22	0.343	0.272	0.218	0.175	0.140
R134a	0.472	0.353	0.271	0.211	0.163
R717	0.281	0.214	0.171	0.138	0.114

Dynamic Viscosity of Saturated Vapour (centipoise, i.e., gm/m-s)

Refrigerants	–40°C	–20°C	0°C	20°C	40°C
R22	0.0101	0.0107	0.0116	0.0127	0.0135
R134a	0.00912	0.01	0.011	0.0116	0.0126
R717	0.00786	0.00842	0.0091	0.0097	0.0103

Appendix 3

Saturated Properties of R-717, (Ammonia), Reference –40°C

Temp (t) (°C)	Sat Pr (P) (kPa)	Sp. Vol. Liquid (v_f) × 10^3 m^3/kg	Sp. Vol. Vapour (v_g) m^3/kg	Liquid Sp. Enthalpy (h_f) kJ/kg	Vapour Sp. Enthalpy (h_g) kJ/kg	Enthalpy of vaporisation (h_{fg}) kJ/kg	Liquid Sp. Entropy (s_f) kJ/kg-K	Vapour Sp. Entropy (s_g) kJ/kg-K
–50	40.8	1.425	2.623	-43.96	1370.41	1414.37	–0.1729	6.1478
–40	71.7	1.449	1.550	0.00	1387.15	1387.15	0.00	5.9518
–30	119.50	1.476	0.963	44.47	1403.11	1358.639	0.1872	5.7760
–25	151.5	1.489	0.771	66.91	1410.68	1343.776	0.2785	5.6948
–20	190.1	1.504	0.624	89.48	1417.97	1328.493	0.3685	5.6169
–15	226.3	1.519	0.509	112.17	1424.919	1312.750	0.4564	5.5423
–10	290.8	1.534	0.418	134.95	1431.409	1296.462	0.5435	5.4712
–5	354.8	1.550	0.347	157.85	1437.56	1279.71	0.6293	5.4020
0	429.4	1.566	0.290	180.88	1443.34	1262.46	0.7139	5.3368
5	515.8	1.583	0.244	204.07	1448.70	1244.63	0.7976	5.2731
10	615.0	1.601	0.206	227.40	1453.52	1226.12	0.8801	5.2112
15	728.5	1.619	0.175	250.89	1457.87	1206.98	0.9618	5.1513
20	857.4	1.639	0.149	274.58	1461.81	1187.23	1.0426	5.0931
25	1003.0	1.659	0.128	298.49	1465.24	1166.75	1.1226	5.0366
30	1167.0	1.680	0.111	322.57	1468.09	1145.52	1.2017	4.9809
35	1350.0	1.702	0.960	346.89	1470.35	1123.46	1.2800	4.9264
40	1555.0	1.726	0.0833	371.47	1472.02	1100.55	1.3579	4.8728
45	1782.0	1.750	0.0726	396.22	1473.03	1076.81	1.4349	4.8201
50	2033.0	1.778	0.0635	421.13	1473.49	1052.36	1.5115	4.7682
60	2614.0	1.838	0.0489	472.29	1472.99	1000.69	1.6685	4.6685

Superheated Table of R717 (NH_3) Ref: –40°C

Sat. Temp (°C) –10, Saturation Pressure (kPa) 290.8

Degree of superheat→	10	20	30	40	50	60
v (m^3/kg)	0.438	0.468	0.476	0.494	0.513	0.531
h (kJ/kg)	1456.3	1480.6	1504.3	1527.6	1550.8	1573.7
s (kJ/kg-K)	5.5641	5.6513	5.7291	5.8120	5.8870	5.9592

Sat. Temp (°C) 0, Saturation Pressure (kPa) 429.4

Degree of superheat→	10	20	30	40	50	60
v (m^3/kg)	0.304	0.318	0.331	0.344	0.356	0.369
h (kJ/kg)	1469.5	1494.7	1519.3	1543.5	1567.3	1590.9
s (kJ/kg-K)	5.4307	5.5189	5.6011	5.6794	5.7543	5.8201

Sat. Temp (°C) 10, Saturation Pressure (kPa) 615.0

Degree of superheat→	10	20	30	40	50	60
v (m^3/kg)	0.216	0.226	0.236	0.245	0.254	0.263
h (kJ/kg)	1481.0	1507.5	1533.1	1558.1	1582.2	1607.0
s (kJ/kg-K)	5.3067	5.3954	5.4785	5.5571	5.6632	5.7040

Sat. Temp (°C) 40, Saturation Pressure (kPa)

Degree of superheat→	10	20	30	40	50	60
v (m^3/kg)	0.088	0.093	0.097	0.101	0.105	0.108
h (kJ/kg)	1504.9	1535.7	1565.2	1593.5	1621.0	1647.9
s (kJ/kg-K)	4.9761	5.0702	5.1572	5.2385	5.3153	5.3883

Sat. Temp (°C) 45, Saturation Pressure (kPa) 1782.0

Degree of superheat→	10	20	30	40	50	60
v (m^3/kg)	0.077	0.081	0.085	0.089	0.092	0.096
h (kJ/kg)	1507.0	1538.8	1569.0	1597.9	1626.0	1653.4
s (kJ/kg-K)	4.9251	5.0205	5.1084	5.1905	5.2678	5.3412

Sat. Temp (°C) 50, Saturation Pressure (kPa) 2033

Degree of superheat→	10	20	30	40	50	60
v (m^3/kg)	0.067	0.071	0.075	0.078	0.081	0.084
h (kJ/kg)	1508.7	1541.4	1572.4	1602.1	1630.8	1658.7
s (kJ/kg-K)	4.8754	4.9723	5.0613	5.1447	5.2221	5.2960

Appendix 4

R 22: Thermodynamic Properties Saturated Refrigerant (Reference –40°C)

Temp (t) (°C)	Sat Pr (P) (kPa)	Sp. Vol. Liquid (v_f) × 10^3 m^3/kg	Sp. Vol. Vapour (v_g) m^3/kg	Liquid Sp. Enthalpy (h_f) kJ/kg	Vapour Sp. Enthalpy (h_g) kJ/kg	Enthalpy of vaporisation (h_{fg}) kJ/kg	Liquid Sp. Entropy (s_f) kJ/kg-K	Vapour Sp. Entropy (s_g) kJ/kg-K
–50	64.50	0.695	0.325	–10.69	229.29	239.98	–0.0467	1.0287
–45	82.90	0.702	0.257	–5.4	231.75	237.15	–0.0233	1.0161
–40	105.2	0.709	0.206	0.0	234.17	234.17	0.0	1.0044
–35	132.0	0.717	0.167	5.50	236.53	231.03	0.0233	0.9934
–30	164.0	0.724	0.136	11.09	238.84	227.75	0.0464	0.9831
–25	201.5	0.732	0.112	16.77	241.09	224.32	0.0694	0.9734
–20	245.4	0.740	0.093	22.53	243.26	220.73	0.0923	0.9642
–15	296.3	0.749	0.078	28.36	245.36	217.00	0.1149	0.9555
–10	355.0	0.758	0.0654	34.25	247.37	213.12	0.1374	0.9473
–5	421.8	0.768	0.0554	40.20	249.29	209.09	0.1596	0.9393
0	498.0	0.778	0.0472	46.19	251.12	204.93	0.1815	0.9317
5	583.8	0.788	0.0404	52.23	252.84	200.61	0.2032	0.9244
10	680.2	0.799	0.0348	58.30	254.46	196.16	0.2245	0.9173
15	788.1	0.811	0.0300	64.40	255.96	191.56	0.2456	0.9104
20	908.1	0.824	0.0260	70.53	257.33	186.80	0.2663	0.9036
25	1041.1	0.837	0.0227	76.69	258.58	181.29	0.2868	0.8968
30	1188.8	0.852	0.0198	82.88	259.67	176.79	0.3070	0.8902
35	1349.5	0.867	0.0173	89.12	260.61	171.49	0.3269	0.8834
40	1526.6	0.884	0.0152	95.40	261.38	165.98	0.3466	0.8767
45	1720.8	0.903	0.0133	101.76	261.95	160.19	0.3662	0.8697
50	1932.6	0.923	0.0117	108.22	262.30	154.08	0.3858	0.8626

Superheated Refrigerant R 22 (Ref: –40°C) (Degree of superheat is given). It means (for –10°C superheat table) if the state point lies at 7°C degree of superheat, temperature of superheated vapour is (–10 + 7) = –3°C

Sat. Temp (°C) –10, Pressure (kPa) 355

Degree of superheat→	5	10	15	20	30	40
v (m^3/kg)	0.0675	0.0686	0.0703	0.0719	0.0751	0.0782
h (kJ/kg)	250.79	254.25	257.66	261.12	268.65	274.96
s (kJ/kg-K)	0.9610	0.9729	0.9851	0.9976	1.0206	1.0448

Sat. Temp (°C) 0, Pressure (kPa) 498

Degree of superheat→	5	10	15	20	30	40
v (m³/kg)	0.0484	0.0496	0.0508	0.0520	0.0543	0.0566
h (kJ/kg)	254.73	258.32	261.921	265.51	272.69	279.89
s (kJ/kg-K)	0.9447	0.9578	0.9698	0.9818	1.0068	1.0298

Sat. Temp (°C) 10, Pressure (kPa) 680

Degree of superheat→	5	10	15	20	30	40
v (m³/kg)	0.0357	0.0366	0.0376	0.0385	0.0403	0.0420
h (kJ/kg)	258.18	261.97	265.74	269.50	277.73	284.48
s (kJ/kg-K)	0.9305	0.9435	0.9505	0.9685	0.9935	1.0165

Sat. Temp (°C) 35, Pressure (kPa) 1350

Degree of superheat→	5	10	15	20	30	40
v (m³/kg)	0.0179	0.0184	0.0191	0.0195	0.0205	0.0215
h (kJ/kg)	265.13	269.55	273.85	278.33	286.61	295.32
s (kJ/kg-K)	0.8981	0.9114	0.9251	0.9382	0.9641	0.9879

Sat. Temp (°C) 40, Pressure (kPa) 1526.7

Degree of superheat→	5	10	15	20	30	40
v (m³/kg)	0.0157	0.0162	0.0167	0.0172	0.0182	0.0191
h (kJ/kg)	266.09	270.69	275.19	279.61	288.31	296.84
s (kJ/kg-K)	0.8917	0.9062	0.9202	0.9332	0.9592	0.9832

Sat. Temp (°C) 50, Pressure (kPa) 1932.7

Degree of superheat→	5	10	15	20	30	40
v (m³/kg)	0.0122	0.0126	0.0131	0.0135	0.0143	0.0151
h (kJ/kg)	267.49	272.49	277.34	282.07	291.29	300.25
s (kJ/kg-K)	0.8781	0.8931	0.9081	0.9221	0.9481	0.9731

Appendix 5

R 22 Saturated Refrigerant Thermodynamic Properties (Reference 0°C)

Temp (t) (°C)	Sat Pr (P) (kPa)	Sp. Vol. Liquid (v_f) × 10^3 m^3/kg	Sp. Vol. Vapour (v_g) m^3/kg	Liquid Sp. Enthalpy (h_f) kJ/kg	Vapour Sp. Enthalpy (h_g) kJ/kg	Enthalpy of vaporisation (h_{fg}) kJ/kg	Liquid Sp. Entropy (s_f) kJ/kg-K	Vapour Sp. Entropy (s_g) kJ/kg-K
–50	64.50	0.695	0.325	145.05	384.01	238.96	0.7792	1.8500
–45	82.90	0.702	0.257	150.24	386.37	236.13	0.8021	1.8371
–40	105.2	0.709	0.206	155.51	388.70	233.19	0.8249	1.8250
–35	132.0	0.717	0.167	160.83	390.98	230.15	0.8384	1.8138
–30	164.0	0.724	0.136	166.22	393.22	227.00	0.8698	1.8033
–25	201.5	0.732	0.112	171.68	395.41	223.73	0.8919	1.7934
–20	245.4	0.740	0.093	177.21	397.53	220.32	0.9139	1.7841
–15	296.3	0.749	0.078	182.81	399.60	216.79	0.9356	1.7754
–10	355.0	0.758	0.0654	188.47	401.60	213.13	0.9139	1.7841
–5	421.8	0.768	0.0554	194.21	403.51	209.30	0.9787	1.7592
0	498.0	0.778	0.0472	200.00	405.36	205.36	1.0000	1.7518
5	583.8	0.788	0.0404	205.87	407.11	201.24	1.0211	1.7446
10	680.2	0.799	0.0348	211.81	408.76	196.95	1.0422	1.7377
15	788.1	0.811	0.0300	217.82	410.31	192.49	1.0631	1.7311
20	908.1	0.824	0.0260	223.90	411.73	187.83	1.0839	1.7246
25	1041.1	0.837	0.0227	230.07	413.03	182.96	1.1047	1.7183
30	1188.8	0.852	0.0198	236.31	414.18	177.86	1.1253	1.7120
35	1349.5	0.867	0.0173	242.65	415.16	172.51	1.1460	1.7058
40	1526.6	0.884	0.0152	249.08	415.95	166.87	1.1666	1.6995
45	1720.8	0.903	0.0133	255.62	416.52	160.90	1.1873	1.6930
50	1932.6	0.923	0.0117	262.27	416.84	154.57	1.2081	1.6864

Appendix 6

Thermodynamic Properties of R 134a (Reference 0°C)

Temp (t) (°C)	Sat Pr (P) (kPa)	Sp. Vol. Liquid (v_f) × 10^3 m^3/kg	Sp. Vol. Vapour (v_g) m^3/kg	Liquid Sp. Enthalpy (h_f) kJ/kg	Vapour Sp. Enthalpy (h_g) kJ/kg	Enthalpy of vaporisation (h_{fg}) kJ/kg	Liquid Sp. Entropy (s_f) kJ/kg-K	Vapour Sp. Entropy (s_g) kJ/kg-K
–50	29.45	0.6914	0.6062	135.67	367.65	231.98	0.7410	1.7806
–40	51.21	0.7054	0.3611	148.14	374.00	225.86	0.7956	1.7643
–30	84.38	0.7202	0.2260	160.79	380.32	219.53	0.8486	1.7515
–25	106.50	0.7281	0.1818	165.69	383.45	219.76	0.8746	1.7461
–20	132.70	0.7362	0.1474	173.64	386.55	212.91	0.9002	1.7413
–15	164.10	0.7447	0.1208	180.13	389.63	209.5	0.9255	1.7371
–10	200.60	0.7535	0.0996	186.70	392.66	205.96	0.9506	1.7334
–5	243.48	0.7627	0.0829	193.32	395.65	202.33	0.9754	1.7300
0	292.80	0.7723	0.0693	200.00	398.6	198.6	1.0000	1.7271
5	349.82	0.7824	0.0584	206.75	401.49	194.74	1.0243	1.7245
10	414.61	0.7930	0.0494	213.58	404.32	190.74	1.0485	1.7221
15	488.56	0.8043	0.0421	220.48	407.07	186.59	1.0725	1.7200
20	571.71	0.8161	0.0360	227.47	409.75	182.28	1.0962	1.7180
25	665.61	0.8287	0.0309	234.55	412.33	177.78	1.1199	1.7163
30	770.20	0.8420	0.0266	241.72	414.82	173.10	1.1435	1.7145
35	887.24	0.8565	0.0230	249.01	417.18	168.17	1.1670	1.7128
40	1016.60	0.8721	0.0200	256.41	419.43	163.02	1.1905	1.7111
45	1160.20	0.8888	0.0174	263.95	421.52	157.57	1.2139	1.7093
50	1317.9	0.9072	0.0151	271.62	423.44	151.82	1.2375	1.7072
60	1681.8	0.9500	0.0114	287.50	426.63	139.13	1.2848	1.7024

Properties of Superheated vapour (Temperature of Superheated vapour is given)

Pressure 0.2 MPa, Sat. Temp. –10.07°C

Temp °C	–10	0	10	20	30
v (m^3/kg)	0.1	0.105	0.109	0.114	0.119
h (kJ/kg)	392.77	401.21	409.73	418.35	427.07
s (kJ/kg-K)	1.7339	1.7654	1.7961	1.8260	1.8552

Pressure 0.4 MPa, Sat. Temp. 8.94°C

Temp °C	10	20	30	40	50
v (m³/kg)	0.0515	0.0542	0.0568	0.0593	0.0617
h (kJ/kg)	404.78	414.00	423.21	432.46	441.76
s (kJ/kg-K)	1.7263	1.7583	1.7892	1.8192	1.8485

Pressure 1.0 MPa, Sat. Temp. 39.4°C

Temp °C	40	50	60	70	80
v (m³/kg)	0.0204	0.0218	0.0231	0.0244	0.0254
h (kJ/kg)	419.99	430.91	441.56	452.05	462.47
s (kJ/kg-K)	1.7139	1.7482	1.7807	1.8117	1.8416

Pressure 1.2 MPa, Sat. Temp. 46.32°C

Temp °C	50	60	70	80	90
v (m³/kg)	0.0172	0.0184	0.0195	0.0205	0.0215
h (kJ/kg)	426.51	437.83	448.81	459.61	470.30
s (kJ/kg-K)	1.7226	1.7571	1.7896	1.8206	1.8504

Pressure 1.4 MPa, Sat. Temp. 52.43°C

Temp °C	60	70	80	90	100
v (m³/kg)	0.0150	0.0160	0.0170	0.0179	0.0188
h (kJ/kg)	433.69	445.31	456.56	467.60	478.53
s (kJ/kg-K)	1.7347	1.7691	1.8014	1.8322	1.8619

Pressure 1.6 MPa, Sat. Temp. 57.91°C

Temp °C	60	70	80	90	100
v (m³/kg)	0.0124	0.0134	0.0144	0.0152	0.0160
h (kJ/kg)	428.99	441.47	453.30	464.76	476.01
s (kJ/kg-K)	1.7124	1.7493	1.7833	1.8153	1.8458

Appendix 7

R22 Properties in Equation Form

Thermophysical and thermodynamic properties of R-22 can also be calculated using the following correlations. This helps in calculating the refrigeration system performance by developing your own computer program. Such correlations for many other refrigerants have also been developed by our researchers and also available on different educational internet sites.

The following correlations may be used to calculate the properties of R-22.

R-22 saturated liquid viscosity (Pa.s):

$$\mu_{r,L} \times 10^{-6} = 23.58553740779 - 0.169516424218T + 1.5511829541 \times 10^{-3}T^2 - 6.73801758861 \times 10^{-6}T^3 - 9.97968293679 \times 10^{-8}T^4$$

T is in °C, Temperature range of application: –20°C to 80°C

R-22 saturated vapour viscosity (Pa.s):

$$\mu_{r,V} \times 10^{-6} = 1.19791093598 + 4.94126334057 \times 10^{-3}T + 1.46248817918 \times 10^{-5}T^2 + 2.11874239565 \times 10^{-7}T^3$$

T is in °C, Temperature range of application: –20°C to 80°C

R-22 saturated liquid thermal conductivity (W/m.K): [101]

$$k_{r,L} \times 10^{-3} = 100.007451491 - 0.497061299686T + 2.72636611269 \times 10^{-4}T^2 + 1.35354765507 \times 10^{-6}T^3 - 9.90480031995 \times 10^{-8}T^4$$

T is in °C, Temperature range of application: –20°C to 80°C

R-22 saturated vapour conductivity (W/m.K):

$$k_{r,V} \times 10^{-3} = 9.3837854793 + 6.40858703196 \times 10^{-2}T + 9.6047166812 \times 10^{-5}T^2 - 4.47171748634 \times 10^{-6}T^3 - 5.67779208879 \times 10^{-8}T^4 + 1.85558049327 \times 10^{-9}T^5$$

T is in °C, Temperature range of application: –20°C to 80°C

R-22 saturated liquid specific heat (kJ/kg.K):

$$C_{P,r,L} = 1.171117595803 + 2.41357321228 \times 10^{-3}T + 1.93601117333 \times 10^{-5}T^2 + 3.04180050348 \times 10^{-7}T^3 + 3.04891000697 \times 10^{-9}T^4$$

T is in °C, Temperature range of application: –20°C to 80°C

R-22 saturated vapour specific heat (kJ/kg.K):

$$C_{P,r,V} = 0.72131116101 + 4.32348768858 \times 10^{-3}T + 3.83804797782 \times 10^{-5}T^2 - 3.73646207394 \times 10^{-7}T^3 - 2.1617922793 \times 10^{-9}T^4 + 2.66288723045 \times 10^{-10}T^5$$

T is in °C, Temperature range of application: –20°C to 80°C

R-22 saturated liquid density (kg/m³):

$$\rho_{r,L} = -789.60336903 + 41.1153860801T - 0.255410446126T^2 + 6.68592007073 \times 10^{-4}T^3 - 6.69907086411 \times 10^{-7}T^4$$

T is in K, Temperature range of application: –20°C to 80°C

R-22 saturated vapour density (kg/m³):

$$\rho_{r,V} = (148.313575941 - 2.40625540379T + 0.015631154964T^2 - 5.07631754315 \times 10^{-5}T^3 + 8.23528834013 \times 10^{-8}T^4 - 5.33626468465 \times 10^{-11}T^5)^{-1}$$

T is in K, Temperature range of application: –20°C to 80°C

R-22 superheatead vapour density (kg/m³):

$$\rho_{r,shV} = \begin{bmatrix} \exp(-11.82344 + 2390.321/T) \times (1.01859 + 5.09433 \times 10^{-4}T - 14.8464 \times 10^{-6}T^2 \\ -2.49547 \times 10^{-7}T^3) \times (1 + 5.23275 \times 10^{-3}(\Delta T) - 5.59394 \times 10^{-6}(\Delta T)^2 \\ + 3.45555 \times 10^{-5}(\Delta T)T - 2.31649 \times 10^{-7}T(\Delta T)^2 \\ + 5.81303 \times 10^{-7}(\Delta T)T^2 - 3.20189 \times 10^{-9}(\Delta T)^2T^2) \end{bmatrix}^{-1}$$

ΔT is the degree of superheat, T is the saturation temperature in K.

R-22 enthalpy of vaporisation (kJ/kg):

$$T_{brr} = t_{br}/T_{cr}$$

$$T_{rr} = T/T_{cr}$$

$$\alpha = -\log(P_{cr}) - 5.985303 + 6.09648/T_{brr} + 1.28862 \times \log(T_{brr}) - 0.169347 \times T^6{}_{brr}$$

$$\beta = 15.2518 - 15.6875/T_{brr} - 13.4721 \times \log(T_{brr}) + 0.43577T^6{}_{brr}$$

$$w = \alpha/\delta$$

$$R = 8.314/86.46$$

$$H_{fg} = RT_{cr}\,[7.08(1 - T_{rr})^{0.354} + 10.9w(1 - T_{rr})^{0.456}]$$

Saturation vapour pressure of R-22:

$\ln P_r(T) = E_0 + E_1/T + E_2T + E_3 \ln T$; T is in K

Correlation coefficients of equation are as follows.

$E_0 = 0.20972706E + 03$

$E_1 = -0.77382052E + 04$

$E_2 = 0.05627680E + 00$

$E_3 = -0.3479000\,E + 02$

Equations for R-22 saturated liquid and saturated vapour (T is in K)

$h_{sl}(T) = F_0 + F_1T + F_2T^2$

$h_{sv}(T) = F_0 + F_1T + F_2T^2$

Table: Correlation coefficients of equations for (h_{sl}, and h_{sv})

Coefficient	Saturated Liquid		Saturated Vapour	
F_0	–68.14840	1966.645	37.22971	–2720.182
F_1	0.06313	–12.09784	1.59952	18.11784
F_2	0.00202	0.02018352	–0.002260308	–0.02699451
TR	253.15–323.15	323.15–368.15	253.15–323.15	323.15–368.15

TR is Temperature range

Equation for enthalpy of superheated R-22 vapour:

$$h_{\text{sup},r}(P, T) = G_0 + G_1T + G_2T^2 + G_3P + G_4PT + G_5PT^2$$

P is in bar and *T* is in K

The correlation coefficients are

Constants	Value
G_0	179.0624
G_1	0.4129627
G_2	0.0003690196
G_3	–11.30278
G_4	0.0449785
G_5	–0.00004677046
ΔT_{sup}	Up to 100°C

R-22 enthalpy value is calculated with the reference enthalpy of 100 kJ/kg at 0°C.

Appendix 8

Physical Properties of some Organic Absorbents

Name	Abbreviation	Chemical Formula	Molecular weight (gm/mole)	NBP (0°C)
N, N-dimethylformamide	DMF	$CHON(CH_3)_2$	73.1	150
N, N-dimethylacetamide	DMA	$CH_3\ CON(CH_3)_2$	87.1	165
Dimethyl ether Diethylene glycol	DMEDEG	$CH_3(OC_2H_4)_2OCH_3$	134.2	162.2
Dimethyl ether Triethylene glycol	DMETrEG	$CH_3(OC_2H_4)_3OCH_3$	178.2	216.1
Dimethyl ether Tetraethylene glycol	DMETEG (E181)	$CH_3(OC_2H_4)_4OCH_3$	222.3	275.6
N-Methyl-2-pyrrolidone	NMP	C_5H_9NO	99.1	201.1
Xylene		C_8H_{10}	106.17	144.5
Cyclohexanone		$C_6H_{10}O$	98.145	155.7
Aniline		C_6H_7N	93.129	184.35
Diethylene glycol	DEG	$C_4H_{10}O_3$	106.122	245.85

Appendix 9

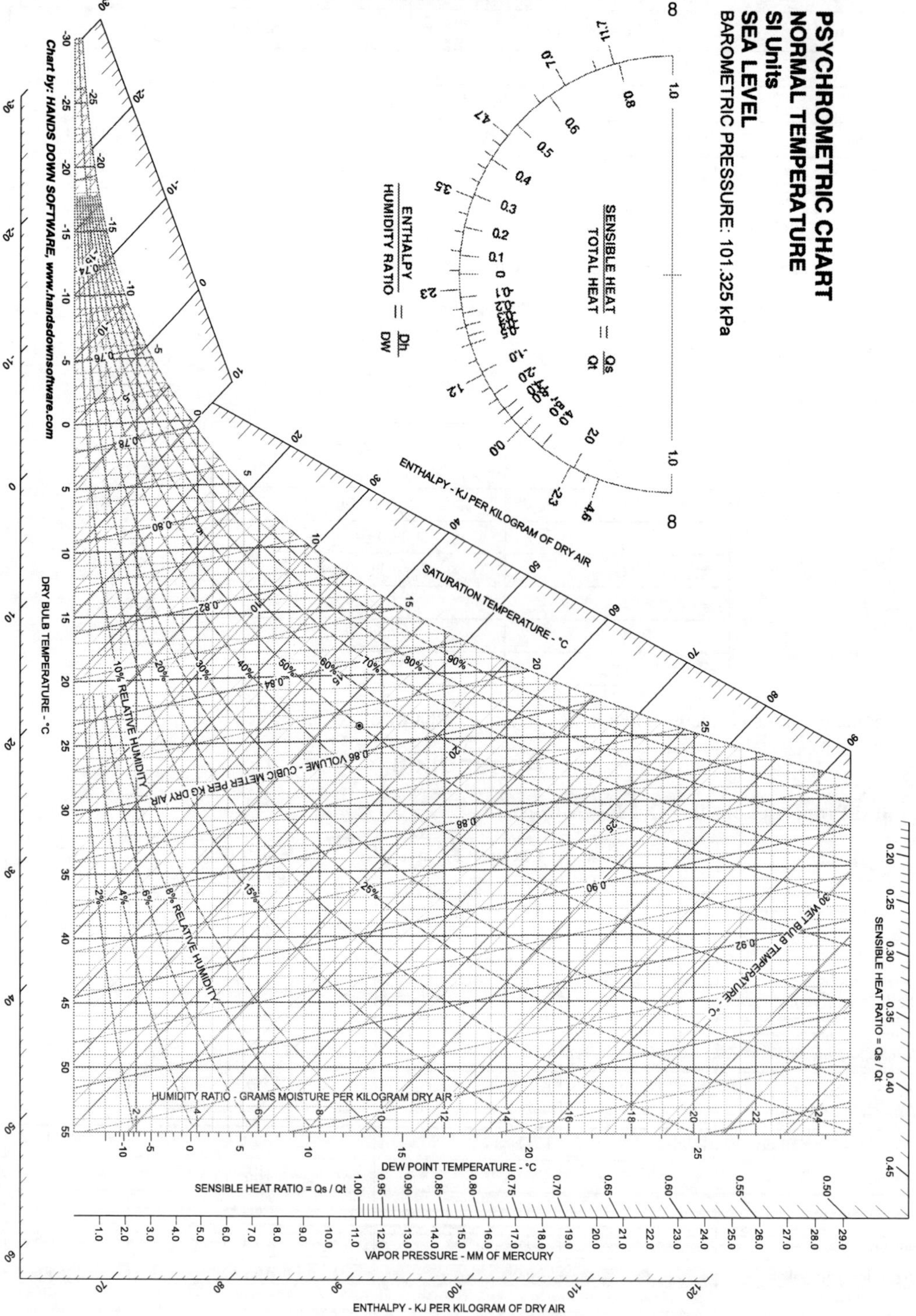
PSYCHROMETRIC CHART
NORMAL TEMPERATURE
SI Units
SEA LEVEL
BAROMETRIC PRESSURE: 101.325 kPa
SENSIBLE HEAT / TOTAL HEAT = Qs / Qt
ENTHALPY / HUMIDITY RATIO = Dh / DW
ENTHALPY - KJ PER KILOGRAM OF DRY AIR
SATURATION TEMPERATURE - °C
VOLUME - CUBIC METER PER KG DRY AIR
RELATIVE HUMIDITY
WET BULB TEMPERATURE - °C
HUMIDITY RATIO - GRAMS MOISTURE PER KILOGRAM DRY AIR
DRY BULB TEMPERATURE - °C
DEW POINT TEMPERATURE - °C
SENSIBLE HEAT RATIO = Qs / Qt
VAPOR PRESSURE - MM OF MERCURY
ENTHALPY - KJ PER KILOGRAM OF DRY AIR
Chart by: HANDS DOWN SOFTWARE, www.handsdownsoftware.com

Index